Applied Cell and
Molecular Biology
for Engineers

ABOUT THE EDITORS

GABI NINDL WAITE, PH.D., is Assistant Professor of Cellular and Integrative Physiology at Indiana University School of Medicine, Terre Haute; Assistant Professor of Life Sciences at Indiana State University, Terre Haute; and Research Professor of Applied Biology and Biomedical Engineering at Rose-Hulman Institute of Technology, Terre Haute.

LEE R. WAITE, PH.D., P.E., is Head of Applied Biology and Biomedical Engineering at Rose-Hulman Institute of Technology in Terre Haute, Indiana; President of the Rocky Mountain Bioengineering Symposium; and Director of the Guidant/Eli Lilly and Co. Applied Life Sciences Research Center.

Applied Cell and Molecular Biology for Engineers

Gabi Nindl Waite, Ph.D. Editor

Lee R. Waite, Ph.D., P.E. Editor

New York   Chicago   San Francisco   Lisbon   London   Madrid
Mexico City   Milan   New Delhi   San Juan   Seoul
Singapore   Sydney   Toronto

The McGraw·Hill Companies

Library of Congress Cataloging-in-Publication Data

Applied cell and molecular biology for engineers / Gabi Nindl Waite,
   editor, Lee R. Waite, editor.
    p.   cm.
    ISBN-13: 978-0-07-147242-5 (alk. paper)
    ISBN-10: 0-07-147242-8 (alk. paper)
    1. Cytology. 2. Molecular biology. I. Waite, Gabi Nindl.
   II. Waite, Lee.
    QH581.2.A67   2007
    571.6—dc22   2007014378

1 2 3 4 5 6 7 8 9 0   DOC/DOC   0 1 3 2 1 0 9 8 7

ISBN-13: 978-0-07-147242-5
ISBN-10: 0-07-147242-8

Sponsoring Editor: Stephen S. Chapman
Production Supervisor: Pamela A. Pelton
Editing Supervisor: Stephen M. Smith
Project Manager: Rasika Mathur
Copy Editor: Megha Roy Chowdhury
Proofreader: Anju Panthari
Indexer: Brenda Miller
Art Director, Cover: Jeff Weeks
Composition: International Typesetting and Composition

Printed and bound by RR Donnelley.

McGraw-Hill books are available at special quantity discounts to use as
premiums and sales promotions, or for use in corporate training programs. For more information, please write to the Director of Special Sales,
McGraw-Hill Professional, Two Penn Plaza, New York, NY 10121-2298.
Or contact your local bookstore.

This book is printed on acid-free paper.

Contents

Contributors   xi

Preface   xv

Acknowledgments   xix

Chapter 1. Biomolecules   1

Walter X. Balcavage

1.1   Energetics in Biology   2
     1.1.1   Thermodynamic principles   2
     1.1.2   Relationship between entropy (S), enthalpy (H),
         and free energy (E)   5
     1.1.3   Entropy as driving force in chemical reactions   7
1.2   Water   9
     1.2.1   The biologically significant molecular structure of water   9
     1.2.2   Hydrogen bonding   10
     1.2.3   Functional role of water in biology   11
1.3   Amino Acids, Peptides, and Proteins   14
     1.3.1   Peptide bonds   14
     1.3.2   Amino acids   14
     1.3.3   Polypeptides   15
     1.3.4   Proteins   18
1.4   Carbohydrates and Their Polymers   20
     1.4.1   Monosaccharides   20
     1.4.2   Oligosaccharides and polysaccharides   22
1.5   Nucleic Acids, Nucleosides, and Nucleotides   24
1.6   Fats and Phospholipids   28
     1.6.1   Fats and oils   30
     1.6.2   Phospholipids   31
Suggested Reading   35
References   35

Chapter 2. Cell Morphology   37

Michael B. Worrell

2.1   Cell Membrane   38
     2.1.1   Phospholipid bilayer   40

	2.1.2 Proteins	41
	2.1.3 Cytoplasm	42
2.2	Membrane-Bound Organelles	42
	2.2.1 Mitochondria	42
	2.2.2 Lysosomes	44
	2.2.3 Peroxisomes	46
	2.2.4 Golgi apparatus	47
	2.2.5 Endoplasmic reticulum	48
2.3	Nonmembrane-Bound Organelles	48
	2.3.1 Ribosomes	48
	2.3.2 Cytoskeleton	49
2.4	Nucleus	53
	2.4.1 Nucleolus	54
2.5	Differences in Cells	54
	2.5.1 Plant cells compared to mammalian cells	54
	2.5.2 Prokaryotes	55
	2.5.3 Tissue-specific language	55
	Suggested Reading	55
	References	55

Chapter 3. Enzyme Kinetics   57
Thomas D. Hurley

3.1	Steady-State Kinetics	58
	3.1.1 Derivation of the Michaelis-Menton equation	59
	3.1.2 Interpretation of the steady-state kinetic parameters in single substrate/product systems	63
	3.1.3 Analysis of experimental data	63
	3.1.4 Multisubstrate systems	66
3.2	Enzyme Inhibition	72
	3.2.1 Competitive inhibition	74
	3.2.2 Noncompetitive inhibition	75
	3.2.3 Uncompetitive inhibition	77
3.3	Cooperative Behavior in Enzymes	78
3.4	Covalent Regulation of Enzyme Activity	81
	Suggested Reading	83
	References	83

Chapter 4. Cellular Signal Transduction   85
James P. Hughes

4.1	Cellular Signaling	86
4.2	Receptor Binding	87
4.3	Signal Transduction via Nuclear Receptors	90
4.4	Signal Transduction via Membrane Receptors	93
	4.4.1 G-protein-coupled receptors (GPCR)	93
	4.4.2 Protein-kinase-associated receptors	99
4.5	Signaling in Apoptosis	101
	References	103

Chapter 5. Energy Conversion   105
James P. Hughes

5.1	Metabolism and ATP	106
5.2	Anaerobic Cellular Respiration	107

5.2.1   Glycolysis   107
5.2.2   Fermentation   111
5.2.3   Gluconeogenesis   111
5.2.4   Regulation of anaerobic respiration   112
5.3   Aerobic Respiration   114
5.3.1   Pyruvate oxidation   115
5.3.2   TCA cycle   116
5.3.3   Electron transport   118
5.3.4   Chemiosmosis and ATP synthesis   122
5.3.5   Usable energy   125
5.4   Photosynthesis   126
5.4.1   Conversion of light energy to chemical energy   126
5.4.2   Chloroplasts   127
5.4.3   Photosynthetic pigments   128
5.4.4   Z-scheme   131
5.4.5   Electron flow through the photosystems   133
5.4.6   Cyclic photophosphorylation   135
5.4.7   ATP synthesis   135
5.4.8   Summary of light-dependent reactions   135
5.5   Carbohydrate Synthesis   136
5.5.1 C_3 plants   136
5.5.2   Photorespiration   138
5.5.3 C_4 plants   139
5.5.4   CAM plants   142
Suggested Reading   142

Chapter 6. Cellular Communication   145
Taihung Duong

The READ Part of the Signaling Machinery   146
6.1   Membrane Receptors   147
6.1.1   Ionotropic receptors   147
6.1.2   G-protein-coupled receptors (GPCRs)   149
6.1.3   Protein kinase-associated receptors   152
6.2   Nuclear Receptors   153
6.2.1   Steroid hormone receptors   153
The WRITE Part of the Signaling Machinery   157
6.3   Signaling Molecules   158
6.3.1   Classical transmitters   164
6.3.2   Neuropeptide transmitters   168
6.4   Cell Secretion   168
6.4.1   Manufacturing   170
6.4.2   Packaging   170
6.4.3   Sorting and delivery   172
6.4.4   Regulation of secretion   173
6.4.5   Exocytosis   173
Interactions between READ and WRITE of the Signaling Machinery 174
6.5   Synaptic Interactions during Development   174
References   175

Chapter 7. Cellular Genetics   177
Michael W. King

7.1   DNA Structure   178
7.1.1   Composition of DNA in cells   178
7.1.2   Thermal properties of the DNA helix   181

7.2   Chromatin Structure   181
    7.2.1   Histones and formation of nucleosomes   182
7.3   DNA Synthesis and Repair   184
    7.3.1   Mechanics and regulation   184
    7.3.2   Postreplicative modifications   191
7.4   Transcription: DNA to RNA   193
    7.4.1   Mechanics   193
7.5   Translation: RNA to Protein   199
    7.5.1   Activation of amino acids   199
    7.5.2   Initiation   200
    7.5.3   Eukaryotic initiation factors and their functions   201
    7.5.4   Specific steps in translational initiation   202
    7.5.5   Elongation   203
    7.5.6   Termination   204
    7.5.7   Heme control of translation   204
    7.5.8   Interferon control of translation   206
Suggested Reading   207

Chapter 8. Cell Division and Growth   209

David A. Prentice

8.1   Growth of Cells: Cell Cycle   210
    8.1.1   Phases of the cell cycle   210
    8.1.2   Studying cell cycle phases   211
    8.1.3   Control of cell cycle   212
8.2   Mitosis   215
    8.2.1   Stages of mitosis   215
    8.2.2   Mechanics and control of mitosis   216
    8.2.3   Checkpoints in cell cycle control   220
8.3   Stem Cells: Maintenance and Repair of Tissues   222
    8.3.1   The problem of tissue maintenance and turnover   222
    8.3.2   Tissue stem cells (traditional view)   224
    8.3.3   Regenerative medicine with stem cells   224
    8.3.4   Sources of stem cells   224
    8.3.5   Current and potential stem cell uses and points of controversy   226
8.4   Cell Senescence: Cell Aging   229
    8.4.1   Cellular aging theories and telomerase   229
    8.4.2   Cell cycle breakdown   230
8.5   Cancer: Abnormal Growth   230
    8.5.1   Characteristics of cancer   230
    8.5.2   Mechanisms of oncogenesis   230
    8.5.3   Stem cells and cancer   231
Suggested Reading   232
References   232

Chapter 9. Cellular Development   233

Michael W. King

9.1   Primordial Germ Cells   234
    9.1.1   Eggs   234
    9.1.2   Sperm   235
9.2   Fertilization   236
9.3   Gastrulation and the Establishment of the Germ Layers   237
9.4   Specification and Axis Formation   239

9.4.1   Dorsal-ventral (DV) axis   240
9.4.2   Anterior-posterior (AP) axis   241
9.4.3   Left-right (LR) axis   243
9.5   Limb Development: A Model of Pattern Complexity   246
9.6   Apoptosis in Development   250
Suggested Reading   252
References   253

Chapter 10. From Cells to Organisms   255

Gabi Nindl Waite

10.1   From Unicellularity to Multicellularity   256
        10.1.1   Prokaryotes and eukaryotes   258
        10.1.2   Sexual reproduction and meiosis   259
10.2   Cell Features   262
        10.2.1   Common cell features   263
        10.2.2   Features that make cells different   264
10.3   Determination and Differentiation   266
        10.3.1   Cell lineage   267
        10.3.2   Size and shape of cells   268
        10.3.3   Membrane transport   270
        10.3.4   Membrane potential   272
        10.3.5   Cell polarity   275
10.4   Morphogenesis   276
        10.4.1   Cell junctions   276
        10.4.2   Extracellular matrix   278
        10.4.3   Tissues   278
        10.4.4   Organs, organ systems, and organisms   281
10.5   Systems Biology   282
        10.5.1   Homeostasis   283
Suggested Reading   285
References   285

Glossary   287
Index   311

Contributors

Walter X. Balcavage, Ph.D. (Chap. 1, Biomolecules)

Walt Balcavage is an Emeritus Professor of Biochemistry and Molecular Biology at the Indiana University School of Medicine (IUSM), Terre Haute, and Adjunct Professor of Biochemistry at Indiana State University and Rose-Hulman Institute of Technology. During his career at Indiana University, Dr. Balcavage was Associate Dean of Research and Head of the Biochemistry Section.

In his capacity as a research scientist, Dr. Balcavage has published numerous original peer-reviewed articles dealing with Intermediary and Energy Metabolism. More recently, Dr. Balcavage has studied the impact of electromagnetic fields on living organisms.

Dr. Balcavage served in the Medical Corps of the U.S. Army. He obtained his undergraduate training in the sciences at Franklin and Marshall College in Lancaster, PA, and his M.S. and Ph.D. at the University of Delaware in Newark, DE. Currently, Dr. Balcavage is President and owner of Consultants in Biotechnology, LLC, and he also holds the positions of President of Peer Medical Inc. and Director of Business Development for DesAcc Inc., two medical informatics companies.

Michael B. Worrell, Ph.D. (Chap. 2, Cell Morphology)

Mike Worrell is on the faculty of Hanover College. He is a member of the American Society of Mammalogists and the Indiana Academy of Sciences.

Worrell received his A.B. in Biology from Earlham College and Ph.D. in Cell Biology and Anatomy from Indiana University. Teaching specialties for Dr. Worrell include anatomy, physiology, introductory biology, and development.

His research interests vary from musculoskeletal assessment in typical and disabled humans, to metabolism and behavior of small mammals, and have recently focused on mutagenic effects of pollutants on amphibian development.

Thomas D. Hurley, Ph.D. (Chap. 3, Enzyme Kinetics)

Tom Hurley received his B.S. degree in biochemistry from Penn State University and his Ph.D. degree in biochemistry from the Indiana University School of Medicine. His postdoctoral work was performed at the Johns Hopkins University School of Medicine in biophysics and biophysical chemistry. He joined the faculty at the Indiana University School of Medicine in 1992, where he is Professor in the Department of Biochemistry and Molecular Biology.

Dr. Hurley has authored numerous book chapters and peer-reviewed publications and is a frequent guest speaker at national and international meetings. He is the Director of the Center for Structural Biology at Indiana University School of Medicine which utilizes synchrotron facilities for crystallographic data collection. His research is focused on understanding the mechanism of enzyme-catalyzed reactions using a variety of approaches including enzyme kinetics, x-ray crystallography, and mass spectrometry.

James P. Hughes, Ph.D. (Chap. 4, Cellular Signal Transduction, and Chap. 5, Energy Conversion)

Jim Hughes is a Professor in the Department of Life Sciences at Indiana State University, Terre Haute, IN. He received a M.S. in Biology in 1974 from University of Arizona, Tucson, AZ, and a Ph.D. in Physiology in 1979 from the University of California, Berkeley, CA. From 1979 to 1982, Dr. Hughes was awarded a Postdoctoral Fellowship in Endocrinology at the University of Manitoba, Winnipeg, Canada.

Dr. Hughes joined the faculty at Indiana State University in 1982. His research interests primarily revolve around signal transduction in endocrine systems. He teaches courses in cell biology, endocrinology, pathophysiology, reproductive physiology, anatomy and physiology, and general biology.

Taihung Duong, Ph.D. (Chap. 6, Cellular Communication)

Taihung ("Peter") Duong is Associate Professor of Anatomy and Cell Biology at the Indiana University School of Medicine, Terre Haute, and Director of the Terre Haute Center. He received a B.A. degree in Biology from Whittier College in 1977 and a Ph.D. degree in Anatomy from the University of California at Los Angeles (UCLA) in 1989. He completed 2 years in a postdoctoral fellowship in neuroanatomy at the UCLA Mental Retardation Research Center before joining the faculty at the Indiana University School of Medicine in 1991. His research interests are brain aging and Alzheimer disease.

Dr. Duong has received numerous educational honors including Outstanding Basic Science Professor Award, Trustee Teaching Award, and the Indiana University School of Medicine Faculty Teaching Award.

Michael W. King, Ph.D. (Chap. 7, Cellular Genetics, and Chap. 9, Cellular Development)

Mike King is a Professor of Biochemistry and Molecular Biology at Indiana University School of Medicine, Terre Haute, and an Executive Member of the Indiana University Center for Regenerative Biology and Medicine. Dr. King is the author/editor of 2 books, and 10 book chapters. Dr. King has expertise in both molecular and developmental biological analysis of early embryonic development and limb regeneration, having studied early *Xenopus* development for over 20 years. With colleagues at Indiana University, University of Illinois, and Eli Lily and Co., he has undertaken genomic and proteomic screens of pathways that either promote or restrict tissue regeneration in the amphibian hindlimb.

David A. Prentice, Ph.D. (Chap. 8, Cell Division and Growth)

Dave Prentice is Senior Fellow for Life Sciences at the Family Research Council. Prior to July 2004, he had spent almost 20 years as Professor of Life Sciences at Indiana State University, and Adjunct Professor of Medical and Molecular Genetics, Indiana University School of Medicine.

Dr. Prentice was selected by the President's Council on Bioethics to write the comprehensive review of adult stem cell research for the Council's 2004 publication "Monitoring Stem Cell Research." He has given frequent policy briefings, invited lectures, and media interviews regarding stem cell research, cloning, biotechnology, and bioethics.

Prentice received his B.S. in Cell Biology in 1978, and his Ph.D. in Biochemistry in 1981, both from the University of Kansas, and was a postdoctoral fellow at Los Alamos National Laboratory from 1981 to 1983. He has taught many courses including developmental biology, embryology, cell and tissue culture, history of biology, science and politics, pathophysiology, medical genetics, and medical biochemistry.

Gabi Nindl Waite, Ph.D. (Chap. 10, From Cells to Organisms, and Editor)

Gabi Waite is Assistant Professor of Cellular and Integrative Physiology at Indiana University School of Medicine, Terre Haute; Assistant Professor of Life Sciences at Indiana State University, Terre Haute; and Research Professor of Applied Biology and Biomedical Engineering at Rose-Hulman Institute of Technology, Terre Haute. She received her diploma of Biology (B.S./M.S.) in 1991 and her Ph.D. in 1995 at the University of Hohenheim, Germany. Dr. Waite teaches physiology and pathophysiology to medical students and undergraduates, and occasionally teaches a biomedical research course.

In addition to her teaching duties, Dr. Waite is Competency Director of Effective Communication at IUSM, Terre Haute. She is also a member of the Board of Directors of the Rocky Mountain Bioengineering Symposium and of the International Bioelectromagnetics Society. She is coeditor of *Clinical Science: Laboratory and Problem Solving*, and the author of five book chapters as well as numerous review articles and peer-reviewed publications. Dr. Waite's research focuses on the biophysical regulations of cell signaling, particularly cell redox signaling.

Lee R. Waite, Ph.D., P.E. (Editor)

Lee Waite is Head of Applied Biology and Biomedical Engineering, and Director of the Guidant/Eli Lilly and Co. Applied Life Sciences Research Center, at Rose-Hulman Institute of Technology in Terre Haute, IN. Dr. Waite is President of the Rocky Mountain Bioengineering Symposium, which is the longest continually operating biomedical engineering conference in North America. He is the author of *Biofluid Mechanics in Cardiovascular Systems* in McGraw-Hill's series on Biomedical Engineering.

Dr. Waite received his B.S. in Mechanical Engineering in 1980 and his M.S. and Ph.D. in Biomedical Engineering in 1985 and 1987 from Iowa State University. He has taught numerous courses such as biofluid mechanics, biomechanics, biomedical instrumentation, graphical communications, and mechanics of material. He is a registered professional engineer and an engineering consultant for a number of companies and institutions, including Axiomed Spine Corporation, and the Heart Surgery Laboratory at the University of Heidelberg in Germany. Waite has also served as visiting professor at Kanazawa Institute of Technology in Japan.

Preface

Chemistry, as a science, was the queen of science in the late nineteenth century. Likewise, physics, with the discovery and development of fission and the atom bomb, changed the world in which we lived in the early twentieth century. Biology will be the science that changes the way we live in the twenty-first century.

The areas of functional genomics and proteomics will drive discoveries in molecular medicine, gene therapy, and tissue engineering. Drug discovery will be facilitated by the clarification of new target molecules, and many pharmaceutical compounds will be produced using biological processes. Environmental management, remediation, and restoration will also benefit from advances in applied biology. The rate of growth of knowledge in the field of biology is increasing at a dizzying rate. Managing vast databases of new knowledge is almost as important as the creation of that new knowledge.

Today, it is necessary for engineers in a wide range of disciplines and for other nonbiologists by primary training to have a basic background in cell biology. Scientists increasingly work in teams comprised of engineers, scientists of other disciplines, business managers, and technicians. For each member of the team, it is necessary to understand the language of the other members. Scientists not trained in the field of cell biology need to be able to understand cell biology-associated research plans and experimental results to enable them to veto or approve proposed ideas. They must understand the societal and business impact of the research and the costs involved. The objective of this book is to present up-to-date basic cellular and molecular biology in an easily understandable way and to give examples of the manifestations of biological phenomena and of the practical results of research.

This book is the first attempt to provide a reference for cell and molecular biology that can be read by engineers and other nonbiologists, and used as a tool to become familiar with the language of biology.

Applied Cell and Molecular Biology for Engineers begins in Chap. 1 with an overview of the flow of energy that enables life and a review of biomolecules, the basic building blocks of biology. It progresses to cell morphology in Chap. 2 where we describe the anatomy and basic physiology of the cell. In Chap. 3 we address enzymes, without which cellular reactions would be too slow for life to continue. Enzymes are reusable catalysts, which speed up chemical reactions without themselves being changed.

The middle three chapters deal with cell signaling, energy conversion, and cell communication, respectively. These three chapters are related together in the sense that the cell is a transducer which converts energy from one form to another to facilitate information flow in space or in time.

The overarching theme of biology is the theme of information transfer. In Chaps. 7 and 8 on genetics and cell cycle and division, the concept of information flow across generations is presented. Chapter 9 continues the theme of information transfer and explains how it is possible for a single cell, the fertilized egg, to pass on the genetic code along with the information to build a new organism that eventually contains cells as diverse as bone, muscle, neuronal, or blood cells.

Finally, Chap. 10 tries to tie it all together to make the bridge from the cellular level to the organismal level, against a systems biology backdrop.

The Structure of the Book

Clinical and application boxes

Each chapter of this book includes at least one "Clinical Box" and one "Application Box." These boxes introduce aspects of the material that can help to motivate the reader's interest. Although the material in the box may not be critical to understanding the information presented in the chapter, it reinforces the relevance and usefulness of the material to medicine and engineering.

Glossary

This is a book about building bridges between disciplines. One of the challenges is to introduce the reader to the language of biology. Complex terms enable precise and detailed descriptions, but can be intimidating to the reader. We provide an additional tool for the reader by using a glossary at the end of the book. **Bolded terms** appear in that glossary.

Terms in italics sometimes help the reader to recognize the structure of the paragraph (e.g., when provided with a list of items or topics first, second, third, . . .). Alternatively, italics are also used to refer to terms in the text that also appear in the figures, where that reference is deemed to be helpful.

An encouragement

In spite of great effort on the part of two editors, seven authors, and many proofreaders, it is to be expected that mistakes will appear in this book. We welcome suggestions for improvement from all readers, with intent to improve subsequent printings and editions.

Acknowledgments

Writing a book requires the help and patience of many colleagues. Thanks to our colleagues at Rose-Hulman and at Indiana University School of Medicine for their ideas and intellectual contributions, most especially, the authors of each chapter. Special thanks to Ellen Hughes, who made valuable comments on the manuscript.

Thanks to Lee's mother, Charlotte Waite, and Gabi's father, Werner Hess, who played an important role in making us who we are. If it is possible for either of us to write a chapter or to edit a book, that ability began at the knees of our parents when they taught us that reading and education are important. You deserve more credit for what is written in this book than you would admit or even realize.

Finally this book is dedicated to the memory of Margarete Hess, Gabi's mother, who died of pancreatic cancer at far too young an age. We miss you very much.

Applied Cell and Molecular Biology for Engineers

Chapter

1

Biomolecules

Walter X. Balcavage, Ph.D.

OBJECTIVES

- To understand the role of physical forces for chemical reactions
- To introduce the specific biological role of water
- To present the various forms of chemical bonds
- To introduce the major categories of biomolecules

OUTLINE

1.1	Energetics in Biology	2
1.2	Water	9
1.3	Amino Acids, Peptides, and Proteins	14
1.4	Carbohydrates and Their Polymers	20
1.5	Nucleic Acids, Nucleosides, and Nucleotides	24
1.6	Fats and Phospholipids	28

Biomolecules are the fundamental building blocks of all biological matter and are necessary for the existence of all known forms of life. The knowledge of the chemical structures of biological molecules will help in understanding their biological function and the energy flow in cells. Understanding biomolecules is important for the progress of molecular biotechnology, which aims to design new drugs or autonomous nanomachines that heal wounds and perform surgery.

This book is aimed at engineering students and professionals who have only a limited background in the biological sciences but who, through need or personal desire, want a broad exposure to the principles that form the basis of the biological sciences, including medicine. With this brief disclaimer, it should be clear to the reader that this book is not intended as a comprehensive treatise on any of these disciplines but rather as a venue by which the professional nonbiologist can obtain a working knowledge of the life process.

For professionals, the game of bridging the knowledge gap between scientific disciplines is a bit like tourists trying to bridge the gap between cultures. When the tourists are successful they find that they've accomplished one of the most rewarding tasks they've ever encountered. Similarly, when technical experts bridge knowledge gaps such as those between engineering disciplines and biology, the results can lead to exceedingly rewarding personal and professional results. One of the knowledge gaps alluded to is that of the language and syntax gap that is ubiquitous between scientific disciplines. In this regard, it is fortunate that learning the language of biology is no more difficult for the engineer, or scientist from another discipline, than that encountered by a tourist making their way in a foreign country. The difference, of course, is that in the biological sciences the building blocks of our knowledge comprise a well-defined set of atoms, molecules, and chemical reactions rather than letters, words, and sentences. Additionally, the words biological scientists use often have very arcane meanings compared to their conventional usage in everyday language. For example, the term *free energy* describes a kind of energy that is anything but free.

With this introduction, it is appropriate that the first chapter of this book should focus on the very fundamentals of the language of the biological sciences. As outlined in the following, we will begin by introducing the fundamental thermodynamic principles that help us understand the way in which the flow of energy enables the life process, and then we will go on to define and illustrate the basic molecular building blocks from which all biological structures are built.

1.1   Energetics in Biology

1.1.1   Thermodynamic principles

All chemical, physical, and biological processes are ultimately enabled and regulated by the laws of thermodynamics. Thus, to understand the life processes of cells and higher life forms, we need to develop a working knowledge of thermodynamics and then use this knowledge to understand how biological processes are enabled and regulated according to classical thermodynamic principles.

Classical thermodynamics involves a consideration of the energy content of different states of systems where each system is composed of a
number of kinds of molecules or other objects and energy flows between
components of the system and between the system and its environment
with time. There are two basic kinds of thermodynamic systems: open
and closed (Fig. 1.1). Open systems are characterized by a flow of matter
(food and excreta in animals) and energy between the system (the body)
and the environment. Examples of open systems include individual
living cells and the human body, which is an aggregate of cells, and can
be considered as an open thermodynamic system. In contrast, in a closed
system, such as a bomb calorimeter, only energy is exchanged between
the system and its environment. In this discussion of thermodynamic
principles, we will review the first and second laws of thermodynamics
focusing on their relationship to energy flow in living organisms.

*The first law of thermodynamics states that the total energy of a system
plus its environment remains constant.* While not addressing the various forms in which energy can exist, this law declares that energy is neither created nor destroyed and it allows energy to be exchanged between
a system and its surroundings. In closed system, like a bomb calorimeter,
the only form of energy flow between the system and its environment
is heat. Conversely in an open system, like an animal cell, or the human
body, energy is most obviously exchanged into and out of the system in
the form of heat and energy-rich, reduced carbon-containing molecules
(e.g., sugars) and other matter (e.g., the respiratory molecules oxygen

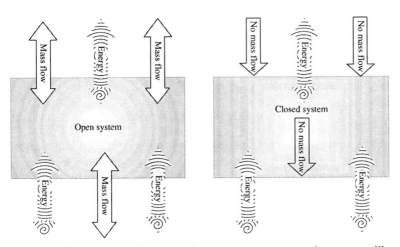

Figure 1.1   Open and closed systems. In open systems mass and energy readily
flow in and out of the system as illustrated by mass arrows and energy arrows
penetrating the boundary of the open system. In closed systems energy (heat)
moves in and out of the system but mass can neither move into the system nor
out of the system.

and carbon dioxide). In animals, it is generally the case that matter flowing into the living system contains a high energy potential and matter flowing out of the system is at a lower energy potential. The energy changes that occur between these two mass flow events are used to perform chemical and physical work processes. Some of the work processes, such as pumping molecules from compartments of low concentration to compartments of high concentration and performing biosyntheses, result in some of the energy remaining stored in the body while the remainder is used to perform mechanical work or appears in the environment as a form of heat. In summary, the ingestion of food and excretion of metabolic products represent exchanges of mass with our environment and is a hallmark of an open thermodynamic system.

The process of consuming complex substances from our environment and excreting simpler breakdown products is also a reflection of the second law of thermodynamics. *The second law of thermodynamics states that a system and its surroundings always proceed to a state of maximum disorder or maximum entropy*, a state in which all available energy has been expended and no work can be performed. **Entropy (S)** and disorder are synonymous in thermodynamics. In the absence of the transfer of mass (food) from our surroundings into the human body, we soon starve, die, and disintegrate. In the case of plants, the photon energy from the sun powers photosynthesis, providing plants (and, as a consequence, humans) with high energy-potential, reduced-carbon compounds like sugars. In these examples, the plant and the animal systems remain viable as long as a usable form of energy input is available. The systems continue to expend the available potential energy until they proceed to a state of maximum entropy with death being one waypoint on the path to maximum system entropy.

In the conversion of complex foods such as glucose [$C_6(H_2O)_6$] to simpler products such as CO_2 and H_2O, energy conversions, allowed by the first law of thermodynamics, take place.

$$C_6(H_2O)_6 + 6O_2 \rightarrow 6CO_2 + 6H_2O \qquad (1.1)$$

It is these energy changes that are available to perform the chemical and physical work that keep us alive. This energy is known as **Gibbs free energy (G)** although it might have been more profitably termed usable energy since it certainly is not free but rather is available at the cost of an aging sun. The entropy change associated with glucose oxidation, or any similar reaction, is qualitatively reflected by a change in the ordered spatial relationship of atoms as biochemical reactants are converted to products. In our example, it should be clear that the atoms of glucose [$C_6(H_2O)_6$] are much more highly structured than the product atoms in CO_2 and H_2O shown in Eq. (1.1). For any given state of a system the collective organization of the components of the system is related to its entropy. Simultaneously, as a consequence of the same chemical process, the heat content of the molecules, which is the sum of the heat associated with molecular collisions, the motion of bonding, and other electrons in the constituent atoms, also changes. This energy is known as **enthalpy (H)** and collectively, for a system in a given state, this heat energy is known as the enthalpy of the system.

1.1.2 Relationship between entropy (S), enthalpy (H), and free energy (E)

The quantitative relationship between the different forms of energy in a system, or in a reaction, going from one state to another is given by Eq. (1.2), the Gibbs equation, where Δ represents the quantitative difference in the energy forms G (Gibbs free energy), H (enthalpy), or S (entropy) between any two states of a system:

$$\Delta G = \Delta H - T\Delta S \qquad (1.2)$$

The Gibbs equation applies to all reactions and processes. A general example is the equilibrium equation (1.3):

$$A + B \rightleftarrows C + D \qquad (1.3)$$

In the Gibbs expression [Eq. (1.2)], where T is the Kelvin temperature of the system, it is clear that the magnitude of the entropic contribution to the free or usable energy is dependent on temperature (TΔS). The enthalpy, or heat content of the system, is in principal also dependent on temperature, but in our biological world, and especially in the human body where reactions take place at constant temperature, molecular motions and collisions remain the same from one state of the system to the next. As a consequence, these kinds of contributions to enthalpy change are generally considered to be negligible. At constant temperature, the remaining and principal enthalpic source of energy is that associated with the chemical bonding between atoms in systems. This energy is known as **internal energy (E)** and in organic molecules it can be recognized as the **covalent bonds**, or forces, that stabilize atoms in the molecule. At constant temperature, enthalpy can be taken to be equal to internal energy, and thus the relationship between internal energy and enthalpy changes between two states can be expressed as shown in Eq. (1.4).

$$\Delta H = \Delta E \qquad (1.4)$$

Combining Eqs. (1.2) and (1.4) yields Eq. (1.5).

$$\Delta G = \Delta E - T\Delta S \qquad (1.5)$$

Equation (1.5) states that as a consequence of a reaction or process going from one constant temperature state to another, the available useful energy (ΔG) equals the difference between the changes in internal or bonding energy (ΔE) and the changes in organization (ΔS) of the atoms involved in the reaction.

The sign, $+$ or $-$, and the magnitude of each term in the Gibbs equation is important in determining if a specified reaction or process described according to the Gibbs equation will proceed spontaneously. If $T\Delta S$ is positive and greater than ΔE, then ΔG will be negative and reactions such as the reaction in Eq. (1.3) will proceed spontaneously to the right as written. Reactions or processes having a negative ΔG are called **exergonic**. Reactions having a positive ΔG are called **endergonic** and these reactions require the input of energy in some form for the reaction to proceed in the direction written. For example, Eq. (1.1), the oxidation of glucose, can be written in the reverse direction, shown in Eq. (1.6), as a synthesis reaction in which CO_2 and H_2O are combined to form glucose.

$$6CO_2 + 6H_2O \xrightarrow{\text{Sunlight}} C_6(H_2O)_6 + 6O_2 \qquad (1.6)$$

However, notice that in this case we invoke the energy of sunlight (and implicitly all the photosynthetic machinery of a green plant) to reverse the entropic and enthalpic changes that result from oxidizing glucose. In biochemical systems, the free energy decrease of exergonic reactions is usually associated with a corresponding increase in entropy, although internal energy changes can also be important. Table 1.1 summarizes the preceding relationships.

TABLE 1.1   Relationship Between Thermodynamic Constants K_{eq}, $\Delta G°$, and the Terms Exergonic, Endergonic, and Spontaneous.

Exergonic reactions	Endergonic reactions
$\Delta G°$ is negative	$\Delta G°$ is positive
Equilibrium constant (K_{eq}) is greater than 1	Equilibrium constant (K_{eq}) is less than 1
Spontaneous as written	Spontaneous in the reverse of the direction written

The equilibrium constant (K_{eq}) is described in Sec. 1.1.3, Eq (1.10). The terms exergonic, endergonic, and spontaneous are often used to describe the thermodynamic character of reactions. For example, if the reaction A ⇨ B has a negative $\Delta G°$ or an equilibrium constant greater than 1, it is called exergonic and will proceed to equilibrium spontaneously, as written, provided a reaction pathway is available.

To illustrate these thermodynamic relationships, we can consider more carefully the oxidation of 1 mol of glucose [Eq. (1.6)] where the initial state of the system/reaction is at the so-called standard concentration state and proceeds to the equilibrium concentration state under standard conditions of temperature and pressure according to Eq. (1.7).

$$C_6(H_2O)_6 + 6O_2 \rightleftarrows 6CO_2 + 6H_2O \quad \Delta G^{\circ\prime} = -686 \frac{kcal}{mol} \quad (1.7)$$

In Eq. (1.7), the term ΔG° has an added prime mark. However, recall that in classical physical chemistry ΔG° is the symbol for the standard free energy change of a system that proceeds from the standard chemical state (as defined in physical chemistry) to the equilibrium state. However, bioscientists have defined a somewhat different set of standard state conditions that reflect the aqueous pH-neutral conditions under which the life process takes place. Thus, standard biological conditions are defined as 760 mm Hg (1 atm), a hydrogen ion concentration of 10^{-7} molar (M) (i.e., pH 7.0), 298 Kelvin (K), 55.5 M water, and 1 M concentration of all other reactants and products. The symbols for energy changes that take place going from these standard biological conditions to the equilibrium state are given a prime mark as indicated in Eq. (1.7) (i.e., $\Delta G^{\circ\prime}$) to signify that they refer to reactions taking place at 55.5 M water, pH 7.0, and 298 K. Thus, the biochemical standard free energy ($\Delta G^{\circ\prime}$) is that available as the reaction proceeds from the biological standard state (1 M glucose, 55.5 M water, 1 atm oxygen, 1 atm CO_2, pH 7.0, 298 K) to the chemical equilibrium state under otherwise standard biological conditions.

The units of free energies are either calories per mole (cal/mol) or joules per mole (J/mol). Since calories and joules are both currently in common use, it is important to recall that 1 cal is equal to 4.184 J. Thus, the change in free energy, or $\Delta G^{\circ\prime}$, for Eq. (1.7) is approximately $-686,000$ cal/mol, or $-2,870,000$ J/mol. The value of -686 kcal/mol (-2870 kJ/mol) for glucose oxidation is a large negative standard free energy change, which indicates that if a reaction mechanism or pathway is available, the reaction will proceed vigorously in the forward direction (as written). This is of course also the direction in which it proceeds in living organisms that possess abundant mechanistic pathways that enable organisms to utilize glucose as a source of energy. Standard free energy changes have been tabulated for most of the known biochemical reactions and can be found in many reference texts as well as on the Internet.

1.1.3   Entropy as driving force in chemical reactions

As a consequence of the preceding considerations, it follows that energy available from a reaction having an initial concentration state different than the standard concentration state will not be equal to that specified as $\Delta G^{\circ\prime}$. To evaluate the actual energy available from a reaction that is at other than the standard state, the free energy needs to be evaluated taking into account the prevailing reaction conditions. Thus, the free energy of a reaction when the initial reactant concentrations are other than 1 M ($\Delta G'$) is given by Eq. (1.8).

$$\Delta G' = \Delta G^{\circ\prime} + 2.303 \, RT \log \frac{[\text{products}]}{[\text{reactants}]} \tag{1.8}$$

In this expression, the brackets signify that we mean concentration, in molarity, and R is the gas constant, 1.987 cal/mol·K, or 8.134 J/mol·K. An important observation related to Eq. (1.8) is that when the reaction is in the standard state, the ratio of reactants to products is 1, the log of 1 is 0 and thus $\Delta G' = \Delta G^{\circ\prime}$!

Equation (1.8) reflects, in mathematical terms, the fact that the actual energy available from a reaction depends on its standard free energy plus (or minus) an energy contribution determined by the prevailing concentration of reactants and products. Clearly, in the human body, the concentrations of reactants and products for metabolic reactions are almost never that of the standard state; therefore, *in vivo* $\Delta G^{\circ\prime}$ is almost never representative of the actual energy available from a metabolic reaction.

Reactions that are at equilibrium under biological standard state conditions of temperature, pressure, and pH cannot proceed spontaneously to any new biologically relevant state, and thus they cannot provide any useful biological work. Consequently, it can be said that the thermodynamic cause of death is that the reactions responsible for maintaining life have come to biological equilibrium. A more precise way of expressing these ideas is to note that at equilibrium, $\Delta G' = 0$. As a consequence of this relationship, Eq. (1.8) can be algebraically modified to yield Eq. (1.9).

$$\Delta G^{\circ\prime} = -2.303 \, RT \log \frac{[\text{products}]}{[\text{reactants}]} \tag{1.9}$$

Since, at equilibrium, the ratio of [product]/[reactant] is equal to the equilibrium constant (K_{eq}), Eq. (1.9) is often written as shown in Eq. (1.10).

$$\Delta G^{\circ\prime} = -2.303 \, RT \log K_{eq} \tag{1.10}$$

From Eq. (1.10), it is apparent that at 298 K, $\Delta G^{\circ\prime}$ is equal to K_{eq} multiplied by a collection of constants. In Eq. (1.11), we explicitly identify all the constants that impact K_{eq}. and then simplify Eq. (1.11) as shown in Eq. (1.12).

$$\Delta G^{\circ\prime} = -2.303 \times 8.134 \, \frac{\text{joules}}{\text{degrees} \times \text{mole}} \times 298 \text{ degrees} \times \log K_{eq} \quad (1.11)$$

$$\Delta G^{\circ\prime} = -5582 \, \frac{\text{joules}}{\text{mole}} \times \log K_{eq} \quad (1.12)$$

Thus, at room temperature, or 25°C, the standard free energy of a biological reaction is simply -5.58 kJ/mol multiplied by the log of the equilibrium constant. The equivalent value for body temperature, 37°C or 310 K, is 5.80 kJ/mol.

1.2 Water

1.2.1 The biologically significant molecular structure of water

In our biological view of thermodynamics, we defined a set of standard biological conditions that are different from those usually encountered in classical discussions on chemistry or physics. One of the more prominently different set of conditions we noted were those applied to water, which is the universal biological solvent. Thus, it seems appropriate to turn our attention to water and consider the basis for its special treatment in biological chemistry.

First, water is the dominant molecular species in most animals, plants, and cells comprising about 65% of the mass of the human body. Water is the main biological solvent, with most of the biochemical reactions in microbes, plants, and animals taking place in the aqueous compartments that make up cells and complex organisms. The special properties of water that make it the universal biological solvent are related to the spatial distribution of electrons in the two covalent bonds that exist between the oxygen atoms and the hydrogen atoms that comprise molecular water (Fig. 1.2). The nucleus of the oxygen atom with its eight protons is strongly electrophylic compared to the hydrogen

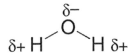

Figure 1.2 Dipolar water molecule. The figure shows the two covalent H-O bonds and the partial charges on the hydrogen and oxygen atoms which are the result of bonding electrons being displaced toward the oxygen nucleus with its relatively high positive charge.

atom with its single positive nuclear proton. As a consequence of these unequal nuclear charges, the electron pairs that constitute the covalent bonds are, on average, markedly displaced toward the oxygen atom and generate a partial negative charge ($\delta-$) on the oxygen as illustrated in Fig. 1. 2. The displacement of the bonding electrons away from the hydrogen atoms asymmetrically unshields the protons that comprise the hydrogen nuclei. It results in the appearance of a partial positive charge ($\delta+$) on each of the hydrogen atoms, and that partial charge is directed away from the oxygen atom. Because of these partial charge separations, water molecules are often described as molecular **dipoles** having partial charges on opposite sides of the molecule.

1.2.2   Hydrogen bonding

In bulk water, the charged regions of individual water molecules interact electrostatically with oppositely charged regions of neighboring molecules resulting in a highly structured matrix of molecules where partial positive charges on one molecule interact with partial negative charges on neighboring molecule, as shown in Fig. 1.3. The resulting dipole-dipole interaction is unusually strong having some of the qualities of both electrostatic and covalent bonds, and thus has acquired a special name, the **hydrogen bond** (H bond). The enthalpic contribution to the energy of a system comprised of hydrogen bonds is in the range of 1 to 10 kcal/mol of H bond. The energy of H bonds in water is generally being taken to be about 5 kcal/mol. The latter value, although about 10 times lower than the covalent O-H bonds in water molecules, imparts considerable stability to aqueous biological systems since the room temperature (25°C) energy available to thermally destabilize bonds is about 10 times lower than the water-water H bond energy.

In the crystalline form of water that we deal with in everyday life, it is generally concluded that each water molecule is hydrogen bonded to four other water molecules resulting in formation of the familiar material known as ice. As the temperature of water ice is raised above the

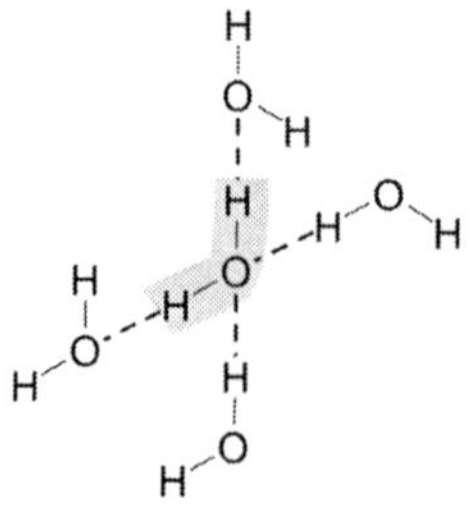

Figure 1.3 Tetrahedrally bonded crystalline ice. The central water molecule is shaded. The broken lines represent hydrogen bonds.

freezing point, hydrogen bonds break resulting in the formation of evanescent clusters of hydrogen-bonded molecules that wink in and out of existence as H bonds form and reform with a half-life of about 10 picoseconds (ps). The loss of the stable crystalline structure results in liquid water having a higher density than ice-state water and provides the basis for the fact that ice, with its lower density, always floats in water. The continuing presence of considerable intermolecular H bond structure in room temperature water and its practical significance is easily evidenced by the inexpert high divers who learn very rapidly not to have large areas of their body, like their belly, flop into the water, fast and all at the same time.

1.2.3 Functional role of water in biology

In biological systems, where water is the solvent and solutes are present in relatively low concentration, the water is generally considered to be present at a concentration that is not significantly different than that of pure water (55.5 M). This high concentration of water is very effective in dissolving and solvating ions such as sodium (Na^+) and chloride (Cl^-). In these organized complexes, water dipoles form highly oriented spherical shells around each individual ion with the countercharge of the water dipoles oriented toward the charge of the solvated ion. The spatial extent and stability of these oriented hydration spheres are directly proportional to the surface charge density of the ion. Likewise, polar organic molecules such as alcohols and sugars, as well as ionized organic molecules such as acids and bases, dissolve in water as a consequence of similar hydration effects. As we will see later in our examination of biological membranes (Sec. 1.6.2), the interactions of water dipoles is also important in helping stabilize the organization of nonpolar molecules like fats and lipids into organized lipid structures like biological membranes.

While pure water and water in biological systems are largely in the form of rapidly shifting molecular clusters, some water dissociates producing equivalent amounts of hydrogen (H^+) and hydroxyl (OH^-) ions. The extent of this dissociation is small but significant, resulting in equal H^+ and OH^- concentrations of 10^{-7}. The special importance of hydrogen ions in biology and chemistry has led to the development of the **pH** scale of H^+ measurement to facilitate the discussion and description of H^+ concentrations. The pH of a hydrogen ion containing solution is simply the negative log of the hydrogen ion concentration as shown in Eq. (1.13).

$$pH = -\log [H^+] \qquad (1.13)$$

Thus, the pH of a 10^{-7} M solution of H^+ is 7.0. Although there are striking individual exceptions, it is generally the case that the intracellular and extracellular compartments associated with living cells and organisms are tightly regulated in the range between pH 5.0 and pH 7.5, and hydrogen ion concentrations outside this range are unfavorable for the vast majority of life forms. One biologically important exception to this generalization is the pH of the stomach, in which the gastric juice has a pH of about 0.7 to 3.8 (see Clinical Box 1.1).

CLINICAL BOX 1.1
Why Doesn't Your Stomach Digest Itself?

Tight regulation of the body's pH around 7.4 is a hallmark of homeostasis, the maintenance of a stable internal environment. This is important since biomolecules including catalytic proteins (enzymes) have evolved to function best near a neutral pH, say between pH 7 and 8. For example, higher or lower pHs result in abnormal protonation or deprotonation of protein R groups, which often leads to marked changes in the protein's normal function with the result that strong acids and bases are often considered to be cell toxins.

Consequently, it is striking that parietal cells, located in the stomach wall of humans and other animals, secrete vast quantities of H^+ in response to consumption of food. The role of acid secretion by parietal cells is to hydrolytically destroy ingested microorganisms and to denature (unfold) and help hydrolyze (i.e., digest) proteins. In this process, parietal cells create an extracellular pH adjacent to the interior of the stomach wall that is in the vicinity of pH 1, with a resultant pH of about pH 2 to 3 when the secreted acid is mixed into the chyme, or homogenized contents, of the stomach. Exposure of most body tissues to such high acid conditions would result in severe acid burns. This raises the question: How does a gastric mucosa cell then protect itself from the gastric acidity? The answer lies in the fact that there are millions of goblet cells in the gastric mucosa, or stomach lining, that secrete a viscous, aqueous solution of mucus, which also contains HCO_3^- ions. The mucus with its acid neutralizing HCO_3^- buffer forms a thick gel layer that covers the surface of the gastric mucosa and prevents the epithelial cells from contacting the acid chyme.

The weakening of these mucosal defense mechanisms results in ulcerations and eventually gastric ulcer disease. A variety of factors including excessive alcohol and tobacco consumption, stress, and nonsteroidal anti-inflammatory drugs such as aspirin can lead to erosion in the lining of the stomach. Additionally, there is also a positive correlation between *Helicobacter pylori* (*H. pylori*) bacterial infection and the incidence of gastric and ulcers of the small intestine. *H. pylori* produces large quantities of the enzyme urease, which hydrolyzes urea to produce ammonia. The ammonia neutralizes the gastric acid in the bacteria's immediate environment thus protecting the bacteria from the toxic effects of its normally toxic acid environment. It is remarkable how some cells find a way to survive even in the deadliest environment.

The importance of pH to the life processes becomes even clearer when we consider the fact that there are a series of weak electrolyte chemical systems in organisms whose principal role is to maintain a viable pH. The main players in this complex system of acid/base **buffers** are phosphoric acid and carbonic acid, which help maintain a relatively constant internal chemical state, different from the equilibrium state, and which is compatible with life. This life-compatible, internal state is known as **homeostasis** and applies to all conditions such as pH, osmolarity, temperature, energy supply, and so on. Intracellular and extracellular pH homeostasis is maintained by a complex interplay of the two sets of weak acid dissociation reactions shown in Eqs. (1.14) and (1.15).

$$H_3PO_4 \rightleftarrows H_2PO_4^{-1} + H^+ \rightleftarrows HPO_4^{-2} + 2H^+ \rightleftarrows PO_4^{-3} + 3H^+ \quad (1.14)$$

$$CO_2 + H_2O \rightleftarrows H_2CO_3 \rightleftarrows HCO_3^{-1} + H^+ \rightleftarrows CO_3^{-2} + H^+ \quad (1.15)$$

For aqueous systems, we generally understand a chemical system to be a most chemically effective acid buffer when there are nearly equal amounts of H^+ donors and H^+ acceptors available, and when the tendency for H^+ to dissociate from the donor species or associate with the acceptor species is equivalent. Chemically these best buffering conditions can be related to the equilibrium constant for the buffering reaction. Phosphoric acid can dissociate 3 H^+ each having a different affinity (or equilibrium constant) for the parent molecule so that they each dissociate at a different H^+ concentration. Optimal buffering usually occurs at pH values that are numerically related to the equilibrium constant as shown in Eq. (1.16).

$$pH_{(max\ buffer\ capacity)} = pK_a \text{ where } K_a \text{ is the equilibrium constant} \quad (1.16)$$

Phosphoric acid dissociates 3 H^+, and the three equilibrium constants or pK_as for each dissociation are close to 2, 7, and 12, respectively, for the first, second, and third dissociation reactions shown in Eq. (1.14). Since the homeostatic pH is closest to 7.0, it should be clear that in biological systems the bolded reaction step in Eq. (1.14), which dissociates the second of the three H^+, is the biologically significant buffer reaction and that $H_2PO_4^{-1}$ and HPO_4^{-2} are the biologically important H^+ donor and H^+ acceptor, respectively.

The situation with the carbonic acid buffer reaction [Eq. (1.15)] is complicated by the fact that this buffer system involves gaseous CO_2, which is in abundant supply in the atmosphere and is also a main intracellular product of energy metabolism. Both of these sources of CO_2 impact the distribution of the components of the carbonic acid buffer system.

When the partial pressure of CO_2 is taken into account in physiologically relevant pH calculations, it is found that under normal atmospheric CO_2 partial pressure the carbonic acid/bicarbonate system is a potent contributor to the regulation of human pH homeostasis slightly above pH 7.0.

1.3 Amino Acids, Peptides, and Proteins

1.3.1 Peptide bonds

While hydrogen bonds are relatively weak, most biological materials including proteins, sugars, fats, and nucleic acids are composed of molecules in which the constituent atoms are linked together by covalent bonds. In covalent bonds, two electrons are shared between the bonding orbitals of the joined atoms. These bonds range in enthalpic energy from about 250 to 400 kJ/mol with the exact value depending on the atoms involved. For example, carbon-carbon bonds have an average energy of 348 kJ/mol, while carbon-nitrogen bonds and carbon-oxygen bonds have energies of 293 kJ/mol and 358 kJ/mol, respectively.

In the biological sciences, there are a number of specially named covalent bonds that have achieved this recognition as a consequence of their biological importance. However, no other covalent bond has received the attention of that which joins the amino acid monomers that comprise the polymeric structures known as peptides and proteins. The basis for the extensive study of covalence in protein structure lies in the fact that proteins are primarily responsible for catalyzing the innumerable array of chemical reactions that maintain life. Thus, it is fitting that we begin our consideration of biological molecules with the study of the amino acids and the **peptide bond**, which is the specially named bond that links amino acids into covalent polymeric structures. These structures include small peptide hormones with fewer than 10 amino acids, up to commonly encountered proteins with an amino acid content ranging from several hundred to more than one thousand. With the 20 different amino acids having an average molecular weight (M.W.) of 120 daltons (Da), the later values correspond to approximate molecular weights of 1200 Da for a 10 amino acid peptide and 60,000 Da for a 500 amino acid protein. One of the largest proteins discovered is titin, a muscle cell protein comprised of 27,000 amino acids with a corresponding M.W. of about 3.2×10^6 Da.

1.3.2 Amino acids

Although there are a large number of amino acids known in nature and in the laboratory, animal proteins are almost exclusively constructed from 20 well-known amino acids. These 20 amino acids are more appropriately referred to as stereospecific, L form, (α) amino acids. Structurally, each

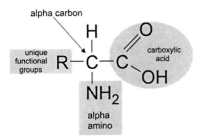

Figure 1.4 Generic structure of amino acids (AA). An H atom, an amino group (NH₂), a carboxylic acid group, and an R group, that is different for each amino acid, are covalently bound to a central α-carbon (arrow). The unique R group provides each AA its unique chemical properties.

of them can be characterized as having a hydrogen atom (H), an amino (NH₂) group, a carboxylic acid (COOH) group, and a unique functional group (generalized as R) attached to the tetrahedral α-carbon as shown in Fig. (1.4). According to standard chemical nomenclature, the atom next to the carbon atom that bears the molecule's main functional group is known as the α-carbon. In the case of the common amino acids, the main functional group is the dissociable, acidic OH group of the carboxylic acid. By extension, the next further carbon from the main functional carbon is the β-carbon, and so on.

Other than glycine, all the remaining amino acids have four different substituents attached to the α-carbon, and because of this they can exist in two structurally different chiral or optically active forms, a dextrorotatory (d or +) form, and a levorotatory (l or –) form. The defining feature of **chiral molecules** is that they are nonsuperimposable mirror images of each other. The l form, which is that found in most human proteins and peptides, rotates plane-polarized light counterclockwise while the d form rotates plane-polarized light in the clockwise direction. These two forms (d and l) are also known as optical isomers or **enantiomers**.

1.3.3   Polypeptides

Figure 1.5 depicts the structure of the 20 common amino acids grouped according to the chemical functionality of their R groups, which remain unmodified during polypeptide synthesis and provide the characteristic molecular reactivity of the final polymer. In contrast, the invariant carboxylic acid and amine group associated with the α-carbon react with corresponding groups on linked amino acids to produce the carboxy-amino linkage, which in organic chemistry is known as amide bond and in biology as a **peptide bond**. The dehydration reaction between alanine and tyrosine to produce water and the peptide bond of alanyl-tyrosine is shown in Fig. 1.6. All peptide bonds are formed via this mechanism. In Fig. 1.6, the four key atoms associated with the peptide bond, and which provide the peptide bond a set of unique structural features, are emphasized by shading. The carbon-nitrogen, or peptide bond in the shaded region has a partial double bond characteristic resulting in all of the atoms in the

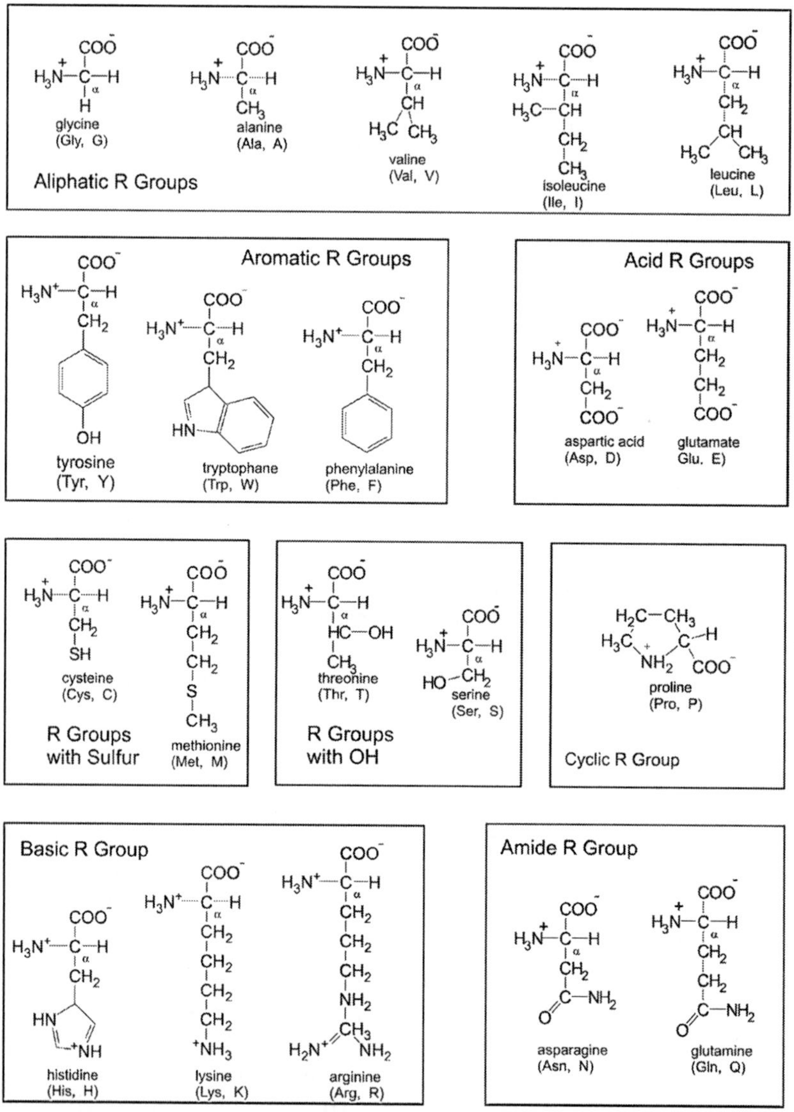

Figure 1.5 The 20 common amino acids. The amino acids are grouped according to the chemical reactivity of their R groups.

shaded region of Fig. 1.6 being rigidly coplanar. Consequently, molecular rotation about the peptide bond is severely limited. This latter steric restriction and the additional steric hindrance donated by bulky or charged R groups are key factors that define the final three-dimensional structure of polypeptides and proteins. It is that native, chemically induced structure that critically defines a polypeptide's biological function

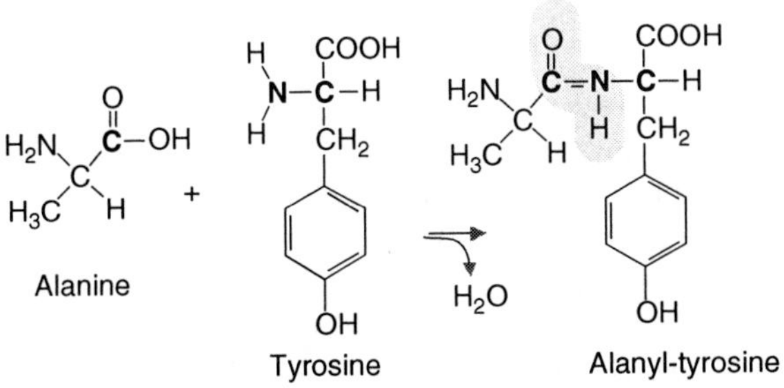

Figure 1.6 Formation of a peptide bond between alanine and tyrosine. The acid function of the amino acid alanine condenses with the α-amine of the amino acid tyrosine to split out a molecule of water resulting in amide, or peptide bond, formation. The shaded region in the dimer highlights the atoms of the peptide bond, all of which remain coplanar as a consequence of electron distribution among the shaded atoms.

as will be outlined in the following chapters. Polypeptides and proteins are presented as critical elements for cell metabolism (Chaps. 3 and 5), cell communication (Chaps. 4 and 6), and for regulation of cell division (Chaps. 7 and 8) and cell development (Chaps. 9 and 10).

Figure 1.7 illustrates the arrangement of four amino acids prior to bonding, with arrows indicating carboxyl and amino groups where dehydration and peptide bond formation can take place. In this illustration, the atoms involved in forming the backbone structure of the polymer and the relationship of the R groups to the nascent backbone become readily apparent.

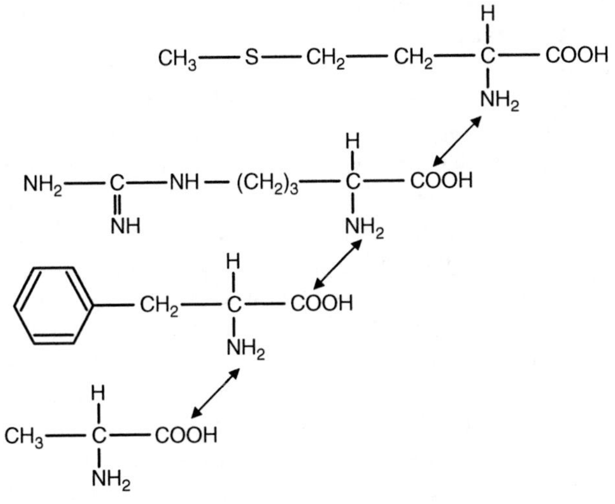

Figure 1.7 Relationship of amino acids in a nascent tetrapeptide.

1.3.4 Proteins

As can be noted from the structure of the R groups of the amino acids, many contain substituents that are either fully ionized at physiological pH (aspartate and lysine for example) or possess partial charges and are thus polar, such as the tyrosyl hydroxyl. The consequence of these facts are that fully ionized R group substituents such as $-NH_3^+$ and $-COO^-$ can form intramolecular electrostatic bonds and cause polypeptide chains to fold back on themselves. Similarly, R groups that are distant from each other in the polymer chain and that are capable of forming hydrogen bonds can interact intramolecularly, further stabilizing the folded three-dimensional conformer of a protein. As can be gleaned from the latter, distant intramolecular protein reactions are dependent on the amino acid sequence of the molecule which of course differs from protein to protein.

Thus, the combination of hydration spheres surrounding exposed ionized R groups, hydrogen bonds formed between water and exposed R groups, intramolecular hydrogen bonding, intramolecular electrostatic bonding, and the steric conformational restrictions imparted by the rigidly configured peptide bond atoms, all combine to determine a protein's final three-dimensional structure which is normally considered to be its energetically most stable state.

Didactically, proteins are said to have four levels of structure, with the first three levels leading to the three-dimensional configuration of the protein. The first level, or *primary structure*, simply refers to the linear sequence of amino acids in the structure. The *secondary structure* reflects the three-dimensional arrangement of the amino acids with respect to its peptide bonded neighbor. The most well-known secondary protein structure is the **α-helix**. Figure 1.8 shows the structure of hemoglobin, where most of the amino acids form α-helices connected by short nonhelical elements. In α-helices, the string-like peptide backbone is arranged in the form of a right-handed coil or helix, and this structure is favored by the steric constraints of the peptide bond and hydrogen bonds that bridge the gap between adjacent turns of the helix. Thus, the coil, or helix structure, is favored by the steric constraints of the peptide bond discussed earlier and stabilized by hydrogen bonds between adjacent coils. Each 360° turn of the α–helix is comprised of about 3.6 amino acid residues whose R groups are oriented perpendicularly to the axis of the helix and allow each different primary structure to present a unique reactive surface to its environment. The insertion of certain amino acids, such as proline, known as α–helix breakers, destroy the continuity of the α–helix because of steric inability of those amino acids to take part in helix structure. When this occurs, bends and folds can and do occur in the helix so that sections of α–helix, known as domains, can fold back on

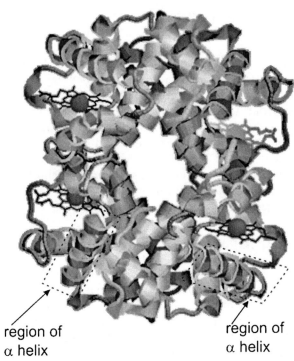

region of
α helix

region of
α helix

Figure 1.8 Hemoglobin. Four individual globularly folded polypeptide chains (two α-chains and two β-chains) that comprise the tetramer protein known as hemoglobin. The four individual monomers are linked by van der Waals forces, electrostatic bonds, and hydrogen bonds to result in the quaternary structure illustrated in the figure. (Public domain figure produced by R. Liddington; Z. Derewenda; E. Dodson; R. Hubbard; G. Dodson and displayed in H.M. Berman, J. Westbrook, Z. Feng, G. Gilliland, T.N. Bhat, H. Weissig, I.N. Shindyalov, P.E. Bourne "The Protein Data Bank" *Nucleic Acids Research* 2000, *28*, 235–242.)

each other much like sausages in the grocery store. Domains are stabilized in a variety of parallel and antiparallel configurations by interhelix bonds, including electrostatic bonds, hydrogen bonds, and **van der Waals forces**. The latter are forces between molecules, which arise from the polarization of molecules into dipoles as explained earlier. Occasionally, cysteine groups that are distant from each other in the primary structure find themselves to be neighbors upon protein folding. In many of these instances, the SH groups associated with the two cysteine residues oxidize to form an S-S or **disulfide bond**, thus adding covalent stabilization to the folded structure. The net result of the primary structure forces, the secondary structure forces, and all interhelix interactions plus solvent effects lead to a final three-dimensional configuration of a protein, the protein's *tertiary structure*.

Many enzymes have a spherical or globular tertiary structure while other kinds of proteins with fewer helix-breaking residues form long fibrous helical stretches. Collagen, which is the dominant structural protein in the human body, is one such long fibrous molecule. In the final structure of collagen, the long helical collagen molecules are arrayed parallel to each other. The bundles of collagen molecules are twisted to form a super helix in which the neighboring molecules are locked together by intermolecular bonding, leading to a discrete multimolecular structure known as *quaternary structure* (the fourth level of protein structure). The folded structure of hemoglobin, which is a molecular tetramer, is shown in Fig. 1.8. The interactions between each of the four proteins that make up hemoglobin stabilize the molecule resulting in the quaternary structure of hemoglobin.

1.4   Carbohydrates and Their Polymers

1.4.1   Monosaccharides

Like amino acids there are a large number of significant carbohydrates found in nature. Carbohydrates are known either as aldoses or as aldehydic polyols because of the aldehydic carbonyl at carbon atom 1 (C1) and extensive hydroxyl (OH) substitution. Carbohydrates also exist as ketoses having an internal carbonyl or keto group (=O) generally at C2, and again, many alcohol (OH) substituents on the carbon chain. Figure 1.9 illustrates these structures for a 6-carbon aldohexose and a 5-carbon aldoribose. Although there are many different kinds of carbohydrate monomers, three members of the family, glucose, ribose, and 2-deoxyribose dominate in biological importance. Glucose is important because it is the key product of photosynthesis, and thus ultimately the main dietary source of energy for nonphotosynthesizing forms of life. Glucose is most familiarly known as the dominant monomer of the very large molecular weight, polymeric compounds glycogen, starch, and cellulose. Ribose and 2-deoxyribose (lacking the OH on C2) dominate because of their role in the polymeric backbone structure of the genetic materials ribonucleic acid (RNA) and deoxyribonucleic acid (DNA), respectively.

A single molecule of glucose or any other similar sugar is known as a monosaccharide. Figure 1.9 shows the open-chain chemical structures of the D enantiomer of glucose, one member of a large family of 6-carbon monosaccharides (or hexoses), and that of the D enantiomer of the 5-carbon monosaccharide ribose (generically a pentose), along with their biologically most active cyclized D and L enantiomers. Chemically, glucose and ribose are aldehydes as indicated by the aldehydic oxygen on C1 of both open-chain structures, and thus these compounds and other isomers of their respective families are often known as aldohexoses and aldopentoses, respectively. In both cases, the open-chain structures

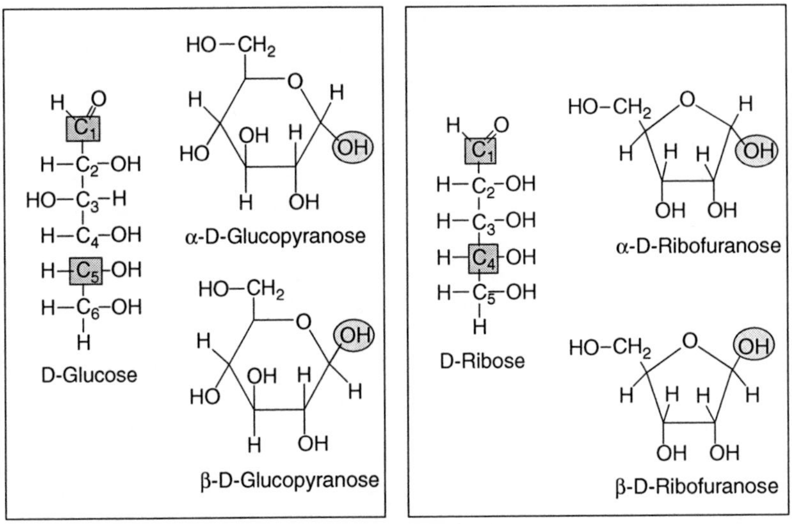

Figure 1.9 Structure of common sugars. Glucose is the most common pyranose (6 atom ring) monomer in many polymeric carbohydrates and the most important carbohydrate source of energy for living organisms. Ribose and 2-deoxyribose are the most prominent furanoses (5 atom rings) because of their prominence in DNA and RNA formation. 2-deoxyribose (not shown) lacks the OH group on the 2 position. The shaded atoms on the open chain structure indicate the atoms involved in intramolecular ring closure and the shaded OH groups on the pyranose and furanose rings indicate the alternative location of the OH residues that define the α- and the β-stereochemical forms of the molecules.

readily undergo intramolecular cyclization to produce the closed-ring structures shown, with the new intramolecular bond being chemically known as an acetal.

The closed-ring acetal forms of the 5-carbon, 6-carbon, and 7-carbon monosaccharides are the forms involved in biological reactions such as synthesis of polymeric glucose structures including starch, glycogen, and plant cell walls. Likewise, the cyclic structures shown are the forms of the compound involved in energy and nucleic acid metabolism and which are ultimately broken down to CO_2 and H_2O (mainly glucose) during the course of metabolism producing usable energy to maintain life.

Given the complexity of substitutions on the carbon chain, it should be apparent that the monosaccharides like glucose and ribose can exist as a number of structurally different isomers. The main isomer of glucose that is found in nature is the compound shown in Fig. 1.9, which is known as the D-enantiomer. The D-enantiomer form of monosaccharides is determined by the configuration of substituents around the asymmetric carbon atom farthest from the main functional group of the molecule. In glucose the main functional group is the C1 aldehyde. Consequently, in glucose C5 is the asymmetric carbon farthest from the aldehyde and with the hydroxyl written to the right as shown, we have the D form. The L form (not shown) has the OH group on C5 written to the left with all other substituents on the molecule remaining as shown for D glucose. During intramolecular cyclization or acetal bond formation, ring closure takes place between C1 and C5 and results in C1 becoming a new asymmetry center, with the newly asymmetric C1 atom being known as the anomeric carbon. Since ring closure can take place with the new OH group on either side of the plane of the ring, two isomers are possible. These isomeric forms have been designated α and β. Figure 1.9 illustrates the structure of the α and β isomers, or anomers, around C1 for the pyranose forms of glucose and the furanose forms of ribose. In the α forms, the OH groups on C1 and C2 are on the same side of the ring, generally depicted downward, while in the β forms the C1 and C2 OH groups are on opposite sides of the ring with the C1 OH generally being depicted upward. Although we will not discuss them here, there are a very large number of other structural isomers of hexoses and pentoses as well as isomeric forms of trioses (3 carbon), tetroses (4 carbon), and heptoses (7 carbon), many of which have significant albeit specialized roles in biology that will be encountered throughout this book.

1.4.2 Oligosaccharides and polysaccharides

Clearly, with the multitude of functional groups, monosaccharides have the ability to undergo a variety of chemical reactions to produce many new compounds. Among the more common of these are additions of carboxyl, acetyl, and amino groups at various locations on the parent monosaccharide. However, by quantity, the most important kinds of reactions that monosaccharides undergo are their reaction with other monosaccharides to produce a family of familiar disaccharides and the very large carbohydrate polymers known as glycogen, starch, and cellulose.

The most familiar dietary disaccharides are maltose, lactose, and sucrose whose monosaccharide reactants and disaccharide product structures are shown in Fig. 1.10. As in earlier discussions of important biological covalent bonds, the covalent link between the anomeric carbon of one monosaccharide and the OH group of a second saccharide (or another molecule containing an OH) is known as a **glycosidic bond**, and the product molecules are generically referred to as glycosides. For example, lactose, maltose, and sucrose can be referred to as glycosidic molecules with the bond being designated as "1⇨4," "1⇨4," and "1⇨2" glycosidic bonds, respectively. The latter nomenclature is used to identify the carbon atoms involved in the glycosidic bond. Many other disaccharides, trisaccharides, and tetrasaccharides, collectively known as oligosaccharides, are found in nature with some having specialized biological

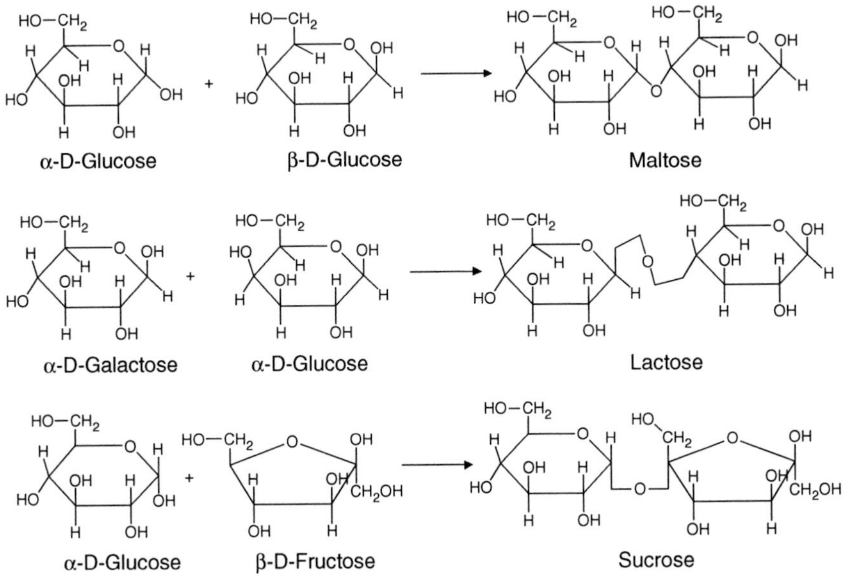

Figure 1.10 Formation of glycosidic (sugar-sugar) bonds to form the familiar disaccharides maltose, lactose, and sucrose. Notice that sucrose is formed from glucose, an aldohexose, and fructose, an example of a ketohexose. Glycogen and starch can be viewed as maltose polymers.

functions while others are merely partial degradation products of larger molecules such as glycogen.

In humans and other animals, the most important polysaccharide is glycogen, also known as animal starch because of its similarity in molecular size, structure, and composition to plant starches which were discovered earlier than glycogen. In the human body, glycogen is found principally in liver and muscle where it acts as a reservoir of glucose monomers, acting much like a glucose buffer in an animal's energy economy. Glycogen comprises about 1% of human muscle mass, and an adult human liver contains about 110 g of glycogen. This huge polymer has a molecular weight range of 10^6 to 10^7 Da, depending on the exact number of glucose residues in the molecule. Glycogen along with fat provide the main energy stores of the body. Glycogen and starch are constructed entirely of glucose monomers linked mainly by α 1⇨4 linkages like those that connect the glucose monomers in the disaccharide maltose (Fig. 1.10). Although shown more frequently in our glycogen model (Fig. 1.11), in nature 1 out of every 12 glucose monomers in a glycogen chain also is connected to a second glucose monomer in a 1⇨6 linkage, each such linkage producing a branch off the main glycogen chain as shown in Fig. 1.11. Multiple branches like those shown in the figure produce a highly compact spherical molecule.

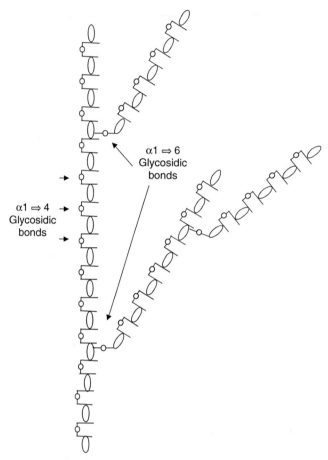

α1 ⇒ 6
Glycosidic
bonds

α1 ⇒ 4
Glycosidic
bonds

Figure 1.11   Section of a glycogen or starch molecule. The initiation point for synthesis is at the base of the chain and 1⇒6 branch points are added subsequent to main chain formation. The final glycogen molecule has a molecular weight of 1×10^{6} to 1×10^{7} Da, is spherical in three-dimensional structure, and highly hydrated.

1.5   Nucleic Acids, Nucleosides, and Nucleotides

The most important biological role of ribose and 2-deoxyribose is to act as structural components of polymeric RNA and DNA, respectively. Thus, this is an opportune time to shift our attention to the role of 5-carbon furanoses in the structure and formation of nucleosides, nucleotides, and the polymeric nucleic acids formed from the ribo and deoxyribo forms of the nucleotides. However, before we turn our attention to these genetically crucial macromolecules we need to introduce the heterocyclic compounds, or so called, nucleic acid bases, into our discussion.

It is important to identify their role in forming the reactive species known as nucleosides and nucleotides, the nucleosides being the precursors to their respective nucleotide and the nucleotides being the building blocks of the nucleic acids, DNA and RNA.

The main biologically significant heterocyclic organic compounds are known in biology as **nucleic acid** bases and are derived from the parent compounds purine and pyrimidine whose structures are shown in Fig. 1.12. Substitutions on the purine heterocyclic result in formation of the purine bases adenine and guanine, while substitutions on pyrimidine lead to the common pyrimidine bases uracil, cytosine, and thymine. In cells, the purines and pyrimidines mainly exist covalently bound to either ribose or 2-deoxyribose and phosphate to produce the biologically active form of the bases.

When coupled to ribose (as shown in Fig. 1.12) or to 2-deoxyribose (not shown) the resulting compounds are known as ribonucleosides (e.g., adenosine and uridine) and deoxyribonucleosides (e.g., deoxyadenosine and deoxyuridine). In most biosynthetic reactions, the **nucleosides** enter reactions in the form of their mono-, di-, or triphosphate adducts on the number 6 carbon of the furanose ring, which are then known as mono-, di-, or triphospho **nucleotides**. Examples of the latter structures are illustrated in Fig. 1.13 for the adenosine nucleotide series, adenosine monophosphate (AMP), adenosine diphosphate (ADP), and adenosine triphosphate (ATP). It is important to note that the phosphate moiety in AMP is bound to ribose via an ester link while the interphosphate linkages in ADP and ATP are formally **anhydride bonds**. The significance of this observation is that the standard free energy of hydrolysis of esters is about -3 kcal/mol, while under comparable conditions the standard free energies of hydrolysis of the anhydride bonds in ATP are about -7.5 kcal/mol, and thus the **phospho anhydride bonds** of all nucleotides are known as high-energy phosphate bonds.

Nucleotide triphosphates, mainly ATP, provide much of the thermodynamic driving power for metabolic reactions as shown in Chap. 5. For example, glucose is a relatively inert molecule and requires activation before it can participate in metabolic reactions. Biologically, activation is effected by reaction with ATP as shown in Eq. (1.17).

ATP + Glucose $\rightleftarrows$ ADP + Glucose-6-phosphate

$$G^{\circ\prime} = -4 \frac{\text{kcal}}{\text{mol}} \quad (1.17)$$

Since $\Delta G^{\circ\prime}$ for hydrolysis of ATP into ADP and phosphate is about -7 kcal/mol and since $\Delta G^{\circ\prime}$ for the coupled reaction shown in Eq. (1.17) is about -4 kcal/mol, it follows that about 3 kcal/mol of ATP energy is

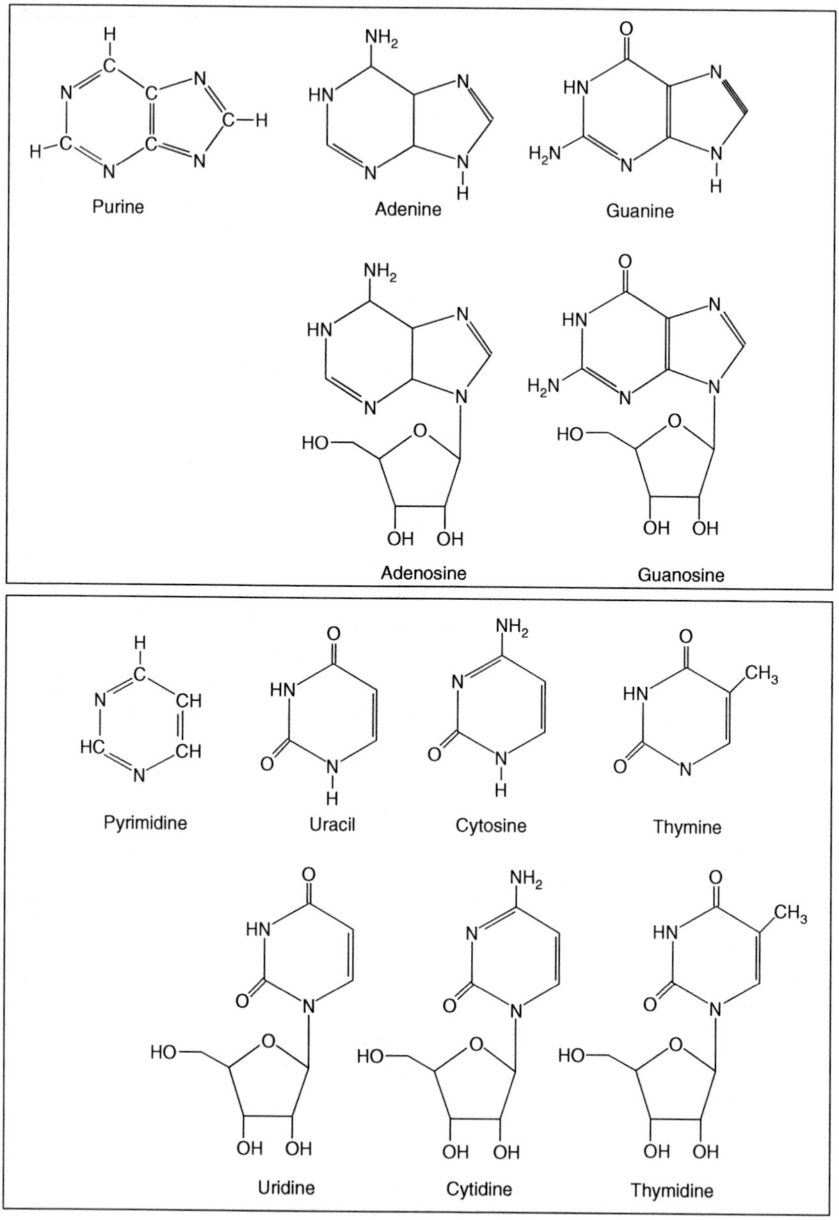

Figure 1.12 Nucleic acids and nucleosides. The nucleic acid bases adenine and guanine are derived from purine, the nucleic acid bases uracil, cytosine, and thymine are derived from pyrimidine. When covalently bonded to a furanose, either ribose or 2-deoxyribose, they are called nucleosides.

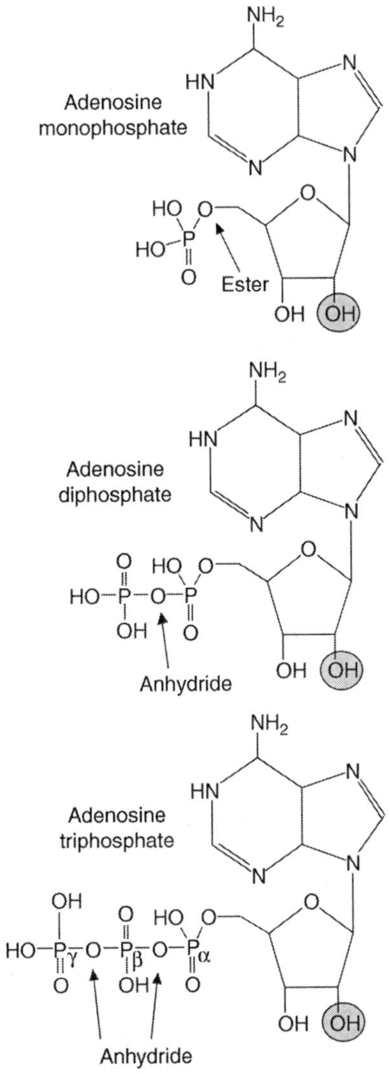

Figure 1.13 Examples of mono-, di-, and triphosphonucleotides. Adenosine triphosphonucleotide, or ATP, has a very high (negative) free energy of hydrolysis and thus is used to provide usable free energy to drive many biosynthetic reactions. The deoxy forms of the nucleotides (which lack the shaded OH groups) are involved in the formation of DNA while the hydroxy forms are used to construct RNA. The "high energy" phosphoanhydride bonds of the nucleotides are used to drive the synthesis of the polymers.

used to drive the synthesis of glucose-6-phosphate and thus allows glucose to be "activated" and enter metabolism. Reactions of this nature, where energy resident in a nucleotide triphosphate bond is used to drive an otherwise unlikely, biological reaction, are ubiquitous in nature.

From a reproductive biology perspective, the high-energy bonds of the triphosphonucleotides are likely to be viewed as most important in driving the synthesis of genes, that is, DNA, as well as the various species of RNAs that are critical in expressing the biological messages of inheritance that are coded in DNA (see Chap. 7). The general reaction involved in forming the nucleic acids is shown in Eq. (1.18) where "n" high energy triphosphonucleotides are shown to combine to yield a polymer linked by ester bonds that contain the same number of bases plus twice the number of free phosphate residues as in the original triphosphonucleotide reactant pool.

$$\text{n (triphosphonucleotides)} \rightleftarrows \text{(phosphodiester polymer)}_{n\ bases}$$
$$+ \text{ 2n (phosphate)} \tag{1.18}$$

A 4-nucleotide section of a single strand of a DNA molecule is diagramed in Fig. 1.14 illustrating individual phosphoester bonds and the phosphodiester linkages involved in producing nucleic acid polymers. Notice that the polymer backbone is composed of alternating $2'$-deoxyribose furan rings and phosphodiester linkage, and that the acid OH group on the phosphates are largely in the dissociated form (O^-) resulting in a highly negatively charged nucleic acid backbone structure. Unlike glycogen, there are no regular covalent branch points in nucleic acid polymers and thus the nucleic acid molecules, ranging in base content from several thousand to many millions, are exceedingly long molecular strands which in cells are very efficiently packaged into a variety of compact structures that are described in Chap. 7. In cells, nucleic acids exist in many molecular forms and are highly modified by chemical reactions on the purine and pyrimidine rings. Chapter 7 outlines the structure and role of the nucleic acids including DNA, messenger RNA (mRNA), transfer RNA (tRNA), and ribosomal RNA (rRNA), as well as other more recently recognized species that are critical players in regulating the expression of genes.

1.6 Fats and Phospholipids

While there are a multitude of kinds of biomolecules, the final discrete class that we will deal with in this chapter includes the fats and lipids. These molecules are classed together mainly as a consequence of the fact that they contain moderately long stretches of mostly saturated, covalently bonded carbon atoms, are relatively insoluble in water, and are generally soluble in organic solvents. There are two main biological functions of this class of molecules. First, saturated hydrocarbon chains, mainly the fats, represent the principal and most chemically efficient storage form of biological energy. Second, a special class of lipids, known as phospholipids, are principal players in forming the hydrophobic barrier of cell membranes which separate the aqueous extracellular milieu from the mainly water-soluble material that comprises the cytosol or intracellular compartment of cells.

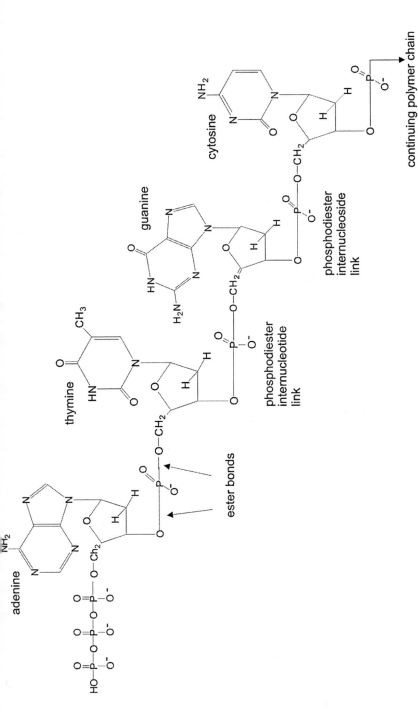

Figure 1.14 Four nucleotide section of a DNA molecule. Note the 2-deoxy form of the ribofuranose rings used in DNA formation and the residual triphosphate residue on the upper terminal adenine nucleotide residue.

adenine

thymine

guanine

cytosine

ester bonds

phosphodiester
internucleotide
link

phosphodiester
internucleoside
link

continuing polymer chain

1.6.1   Fats and oils

Fats and oils are fatty acid triesters of the 3-carbon molecule glycerol and are often known as triglycerides or just glycerides. The formation and structure of a triglyceride is outlined in Fig. 1.15. Hexanoate, octanoate, and decanoate (6, 8, and 10 carbon atoms, respectively) are all saturated fatty acids (no intrachain double bonds) and the resultant triglyceride is known as a saturated triglyceride. Figure 1.16 illustrates the structure of a triglyceride constructed from unsaturated forms of hexanoate, octanoate, and decanoate, each containing one double bond. Fats found in animals often contain such unsaturated carbon-carbon bonds, and in these cases the hydrocarbon chains on either side of the double bond are in the cis arrangement as shown in Fig. 1.16. As a result a bend or kink is introduced into the otherwise linear hydrocarbon chain. This reduces the ability of the molecule to pack tightly with contiguous triglycerides. In general, the greater the amount of unsaturation and the shorter the hydrocarbon chains, the less tightly the triglycerides can pack. The result is that at room temperature bulk quantities of short and/or unsaturated triglycerides exist in the liquid state and are known as oils, while the longer chain, more saturated triglycerides, known as fats, are solid at room temperature.

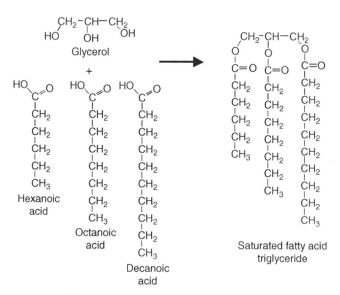

Figure 1.15   Formation of a saturated fatty-acid triglyceride. The 6, 8, and 10 carbon saturated fatty acids shown form ester bonds with the three alcohol functions on glycerol to form the compact, highly saturated, or chemically reduced, triglyceride. In nature, the carbon chains of the fatty acids are often found in the range of 15 to 25 carbon atoms long.

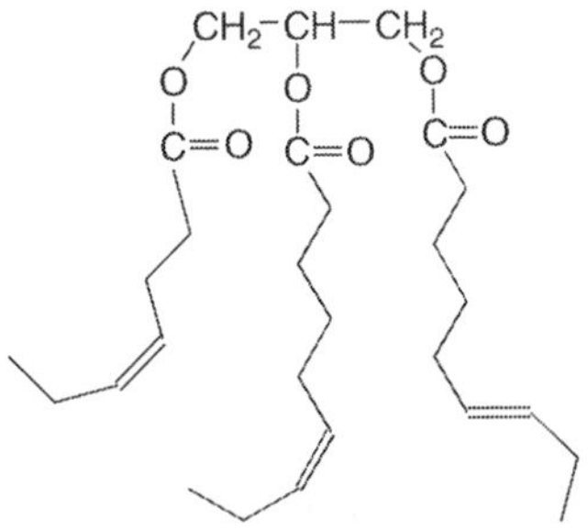

Figure 1.16 Unsaturated fatty acid triglyceride. The same triglyceride as in Fig. 1.15 is shown except containing one unsaturated carbon-carbon bond per fatty acid. Here the hydrocarbon chain is represented by lines indicating the correct bond angles between carbon atoms in the hydrocarbon chains and thus better approximating the volume of space occupied by the molecule. Notice that at the location of double bonds there are "kinks" in the fatty acid chains that prevent tight intermolecular packing of such triglycerides.

Unsaturated fatty acid triglyceride

1.6.2   Phospholipids

There are a number of other fatty acid ester-containing molecules that are important in biology but in the remainder of this chapter we will focus on the most common of those complex molecules that are involved in forming the hydrophobic, or water-repelling interface that comprises biological membranes and which are known as phospholipids. The common precursors to the glycerol-based phospholipids are shown in Fig. 1.17. Glycerol is shown to react with phosphoric acid to form phosphatidyl-glycerol. The latter can esterify two moles of fatty acid, as shown in Fig. 1.17, to yield the simplest phospholipid, which is known as phosphatidic acid.

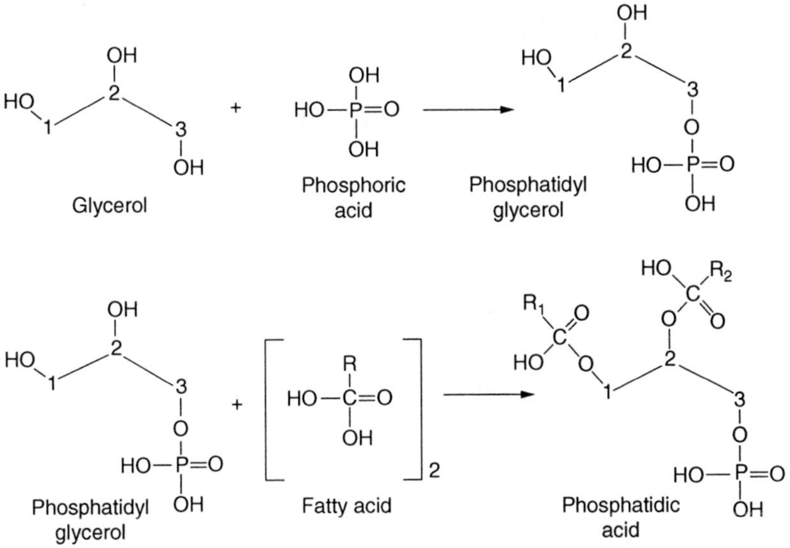

Figure 1.17   Formation of phosphatidyl glycerol and phosphatidic acid. The compounds are precursors to the formation of the common phospholipids. Note that all of the reactions shown here result in ester bond links between the reactants. The water produced as a consequence of the reactions is not shown. The glycerol carbon atoms are represented by the numerals 1, 2, and 3.

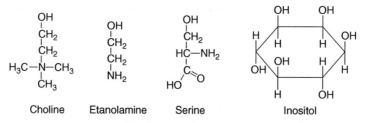

Choline   Etanolamine   Serine   Inositol

Figure 1.18 Structure of the four common alcohols that form esters with phosphatidic acid. The results are common phospholipids which are major components of all biological membranes as shown in Fig. 1.20. The structure of one such ester, phosphatidylserine, is illustrated in Fig. 1.19.

The dominant phospholipids of biological membranes are composed of phosphatidic acid esterified to the alcohol function of choline, serine, ethanolamine, or inositol. The structures of these molecules are shown in Fig. 1.18. Figure 1.19 illustrates the structure of one of the most common phospholipids, phosphatidylserine. To understand the basis for biological membrane formation, it is important to note that one aspect of the molecule (the upper face in Fig. 1.19) contains the R groups that represent the hydrophobic fatty acid residues of the phospholipid. The opposing aspect of the molecule (the lower face in Fig. 1.19) contains the negatively charged phosphoric acid and the zwitterionic (balanced plus and minus charge) form of serine. This aspect of the molecule is relatively highly polar and thus highly water soluble. The combination of a polar/charged aspect and a highly hydrophobic fatty acid containing aspect is typical of most phospholipids and is the basis for their incorporation into biological membranes (Fig. 1.20). The remaining glycerol phospholipids are known as phosphatidylcholine, phosphatidylethanolamine, phosphatidylserine, and phosphatidylinositol.

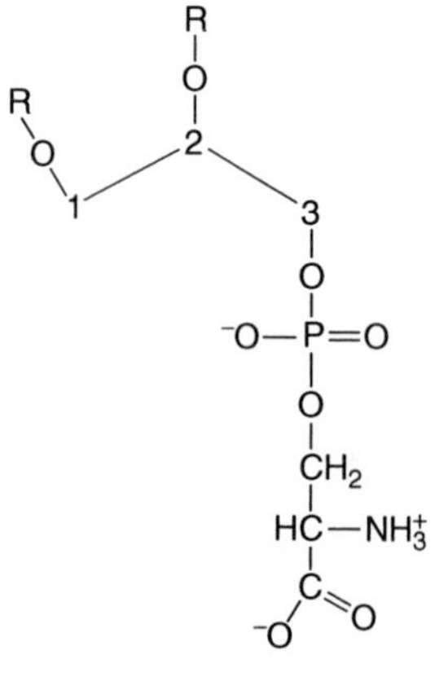

Figure 1.19 Phosphatidylserine. The structure is shown as the charged species that exists at physiological pHs. The acid functions of the phosphate and the carboxylate of serine are shown dissociated and thus have a negative charge. The α-amino group of serine is shown protonated and positively charged. The glycerol atoms are represented by numbers.

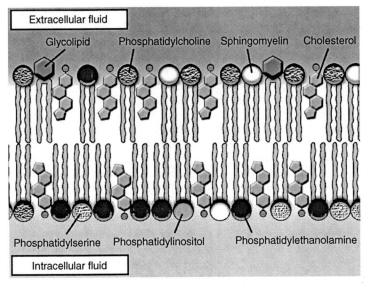

Figure 1.20   Phospholipids within a biological membrane. The charged phosphate group is represented by a circle, the two hydrocarbon chains are shown as tails. The tails are highly hydrophobic and oriented towards each other while the charged phosphate groups face the hydrophilic environment. This lipid bilayer is about 5nm thick, permeable to lipid-soluble low-molecular weight molecules, but impermeable to large and charged molecules.

Depending on the exact mixture of lipids, phospholipids can be dispersed in water where they can form complex structures ranging from simple spherical aggregates, known as **micelles**, to complex double membrane aggregates, known as vesicles, that are often used as models of biological membranes and as pharmacological carriers (see Application Box 1.1). In all cases, including in biological membranes, the hydrophobic

---

APPLICATION BOX 1.1
Phospholipid Nanoparticles and Drug Delivery

Nanoparticles formed from materials as diverse as titanium dioxide and phospholipids are highly attractive to biomedical engineers as vehicles to deliver a variety of cosmetics, drugs, and diagnostic materials to targeted sites in the human body. Nanoparticles employed in the pharmaceutical industry are typically formed from elongated biomolecules, such as phospholipids, having one hydrophilic end and one hydrophobic end. Molecules having the latter properties are known as **amphiphilic** (also called amphipathic) molecules or amphiphiles. Phospholipids are typical amphiphiles with phosphatidylcholine being the favored phospholipid used to form biological nanoparticles. Phospholipid-based nanoparticles are particularly attractive as *in vivo* drug/diagnostic delivery devices since they are formed from nontoxic substances normally found in cells and cell membranes.

Nanoparticles bioengineered from customized combinations of phospholipids with one or two fatty acid substituents, cholesterol, and other polyols are known as micelles, vesicles, or liposomes depending upon the geometry of the structure formed. **Micelles** are small spherical aggregates typically formed from phospholipids having one fatty acid substituent. In micelles the interior of the sphere is packed with the hydrophobic hydrocarbon chain of the fatty acid with the charged or polar end of the phospholipid exposed to the polar, aqueous environment. The hydrophobic interior allows micelles to be stably loaded with water insoluble, and/or toxic drugs, or diagnostics and injected into the blood stream where they circulate and potentially deliver the carried compounds to target sites in the body. However, due to their small size, micelles have a limited loading capacity and most lipid-based nanoparticulate delivery devices have focused on the use of vesicles or liposomes.

Liposomes and vesicles are typically formed from phospholipids plus cholesterol with the phospholipid, usually phosphatidylcholine, containing two fatty acid substituents. Like micelles, liposomes and vesicles form spontaneously when the mixed lipid formulations are dispersed in a polar solvent, such as water, with the lipid formulation and method of dispersion determining the geometry of the resultant particles. Although the nomenclature is not consistent, spherical particles with the phospholipids oriented into a single well-defined bilayer membrane, much like the biological membrane illustrated in Fig. 1.20, are generally referred to as liposomes while particles of ambiguous shape and form are generally referred to as vesicles. The key attributes of liposomes or vesicles are that they are formed of spherically closed bilayer membranes, with the hydrophobic acyl chains of phospholipids from the two sides of the membrane oriented to the inside of the bilayer and the polar portion of the phospholipids oriented toward the aqueous/polar environments on the exterior and interior of the nanoparticles. The figure below illustrates one hemisphere of a spherical, hollow, unilamellar-membrane liposome.

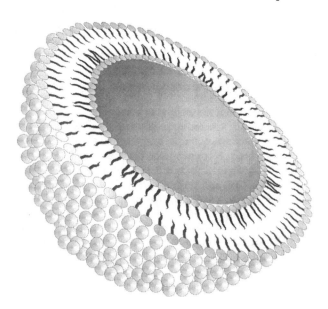

In pharmaceutical applications, customized lipid formulations are dispersed in aqueous solutions containing a drug or diagnostic compound and liposome formation proceeds with the interior aqueous compartment trapping a few nanoliters of drug solution ready to be injected into the blood stream or other body site.

Early researchers using liposome-encapsulated anticancer drugs found that higher drug levels could be delivered to solid tumors but that most of the liposomes were removed from the circulatory system by reticuloendothelial system (RES) cells (e.g., phagocytic macrophages and liver Kupffer cells). Subsequently, "stealth liposomes" cloaked or coated with compounds like polyethylene glycol have been shown to be capable of largely avoiding destruction by the RES. At the same time, animal studies have shown that by bonding targeting molecules, like immunoproteins against cell surface markers, to the liposome surface, tissue-specific targeting can be markedly improved. In the near future many new, targeted, liposome/drug formulations can be expected to appear in clinical medicine and other industries as diverse as cosmetics and nutraceuticals.

fatty acid chains are sequestered in one plane, away from the water phase, and the polar head group of the phospholipid is oriented into the water phase. Although clearly highly structured, these complex structures are stable mainly as a consequence of the fact that the thermodynamic system comprising the water phase and the oriented hydrophobic lipid phase actually represents the most entropic state of the system.

Suggested Reading

Berg, J.M., Tymoczko, J.L. and Stryer, L. (2006). *Biochemistry*, 6th ed. W.H. Freeman.
Devlin, T.M. (2005). *Textbook of Biochemistry with Clinical Correlations*, 6th ed. Wiley-Liss.
King, M.W. (2006). The Medical Biochemistry Page. (*http://isu.indstate.edu/mwking/*).

References

Walde, P. and Ichikawa, S. (2001). Enzymes inside lipid vesicles: preparation, reactivity and applications. *Biomol. Eng.* **18**: 143–177.

Cell Morphology

Michael B. Worrell, Ph.D.

OBJECTIVES

- To introduce the cell concept
- To recognize the basic functions of individual organelles
- To understand the complexity and organization of cells
- To relate the cell to real-life issues

OUTLINE

2.1 Cell Membrane   38
2.2 Membrane-Bound Organelles   42
2.3 Nonmembrane-Bound Organelles   48
2.4 Nucleus   53
2.5 Differences in Cells   54

This chapter presents the structural components of cells and forms the basis for understanding cell behavior as presented in the following chapters. Cell morphology is the foundation for cellular and tissue engineering, which are concerned with the influence of physical factors on cell structure and function. Understanding cell morphology is indispensable for cell modeling and development of biological substitutes such as artificial organs.

The cell is a fantastic representation of biology. It literally encapsulates life in its elemental form. By understanding the components of a cell, we are able to discern and decipher many of the complexities of living organisms. Each component of a cell has larger-scale counterparts which can be examined in multicellular organisms such as a human. For instance, the endosome in a cell filters, excretes, or saves different components of a cell just as the digestive system filters, excretes, and saves components of humans. Another example is cellular respiration. While it is a complex process at the biochemical level, it is, in the end, taking in oxygen and releasing carbon dioxide similar to the actions of the lungs in a human. The circulatory system in the human consists of a series of interconnected tubes powered by a muscular pump, the heart. While not emulated with interconnected tubular structures, in the cell, a comparable variety of molecular pumps and channels are used to regulate fluids and electrolytes in the cell.

The largest human cell is the oocyte, or ovum, or the female sex cell. Its functioning for development of an organism is presented in Chap. 9. The dimension of an oocyte borders on the mesoscopic (visible with minimal magnification such as a hand lens). At roughly 100 μm in diameter, or $^1/_{10}$ of a millimeter, it is huge in comparison to the average cell size of 10 to 12 μm, or $^1/_{100}$ of a millimeter. Thinking about the dimensions of cells, organelle, and molecules can be elusive and mind-boggling. Light microscopy can resolve structures of a few hundred nanometers. A nanometer is $^1/_{1000}$ of a micrometer. The cell membrane about to be discussed is roughly 60 nm thick. At the time of this writing, electron microscopes can reliably resolve images of about 2 nm. This is about the size of the double-stranded DNA molecule. Some electron microscope images with resolutions as low as 0.2 nm are possible. This is nearly the size of individual atoms. Nanotechnology is the novel technology that operates at dimensions below 100 nm, most often at the single-digit nanometer scale. It is used for fabrication of materials and devices on the scale of atoms and molecules.

2.1 Cell Membrane

From the early days of the microscope, the cell has been differentiated as having an outer boundary membrane (the cell or **plasma membrane**) containing a heterogeneous soup (cytoplasm) and a nucleus. As microscopy has improved, more details have emerged from the cellular morass to be seen as distinct objects. Through study of pathologic cells and experimentation, we have come to learn about the function of these objects called **organelles**. For instance, a pathologic cell deficient in the organelle called rough endoplasmic reticulum (RER) will demonstrate an inability to produce many proteins. Therefore, RER is involved with protein production.

Figure 2.1 shows a generic animal cell. The outer boundary, or cell membrane, forms a compartment that is biochemically distinct from

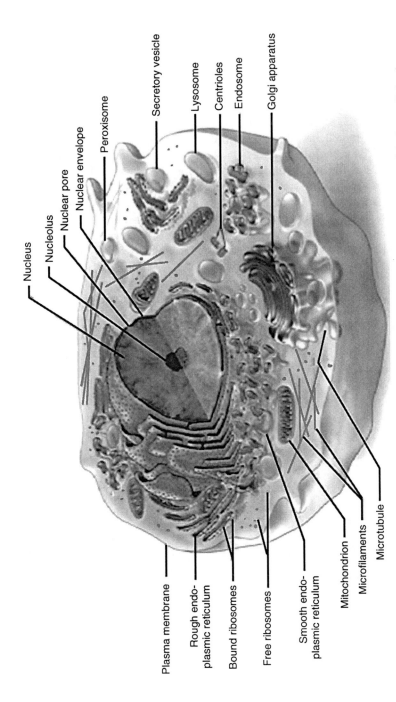

Figure 2.1 Typical animal cell showing characteristic organelles and cellular inclusions. The arrangement of the intracellular features and the shape of the cell vary from cell to cell. (Reproduced with permission from McKinley and O'Loughlin, *Human Anatomy*, 1st ed. McGraw-Hill, New York, 2006.)

Nucleus

Nucleolus

Nuclear pore

Nuclear envelope

Peroxisome

Secretory vesicle

Lysosome

Centrioles

Endosome

Golgi apparatus

Plasma membrane

Rough endo-plasmic reticulum

Bound ribosomes

Free ribosomes

Smooth endo-plasmic reticulum

Mitochondrion

Microfilaments

Microtubule

the outside environment. The chemical soup enclosed by the cell membrane is regulated by the membrane and its components allowing some substances into and out of the cell. This **differential permeability** is a hallmark of a cell membrane.

2.1.1 Phospholipid bilayer

When examining the cell membrane via ultramicroscopy (magnification greater than 1000×), details of the membrane structure can be seen (Fig. 2.2). The cell membrane is actually a bilayer composed of phospholipids. The structure and formation of phospholipids were introduced in Sec. 1.6.2. Different types of phospholipids are shown in Fig. 1.20. The **amphipathic** (having two natures) character of phospholipids helps to maintain a hydrophilic (polar or water soluble) and hydrophobic (nonpolar or non-water-soluble) orientation to the cell membrane. This feature of the membrane causes water and other polar (water-soluble) molecules and compounds to stay on one side of the membrane. The amphipathic nature of phospholipids also allows for self-assembly as described in Application Box 2.1. Some hydrophobic molecules (lipid-soluble) are able to pass through the membrane with ease. This character of the cell

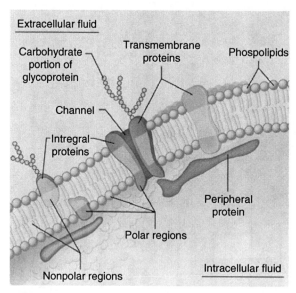

Figure 2.2 Cell membrane. Membranes are composed of a phospholipid bilayer and associated proteins. Proteins include embedded, or integral proteins, as well as peripheral proteins on a surface of the membrane. Membrane proteins serve a variety of purposes including cell communication and identification. (Reproduced with permission from McKinley and O'Loughlin, *Human Anatomy*, 1st ed. McGraw-Hill, New York, 2006.)

APPLICATION BOX 2.1
Self-Assembly in Biology and Technology

The amphipathic character of the phospholipid molecules (see Figs. 1.19 and 1.20) allow for the property of *self-assembly*. If individual phospholipid molecules are dropped into a polar solution such as water, they will merge together to form either a sphere or a membrane with polar heads oriented toward the water and the nonpolar lipid ends oriented inward or away from the water. (Think of drops of grease in water.) The characteristic is termed self-assembly and can be used within the cell to segregate lipid or hydrophobic substances from the watery interior of the cell. This is often seen in cells which manufacture lipid secretions (such as the hormone testosterone). The lipid-based hormone is segregated from the watery cytoplasm in a membrane-bound vesicle until it is released from the cell.

Self-assembly is an important technique in nanotechnology to form systems out of objects and devices without the necessity of external actions. Each individual component contains enough information to form a multiple unit structure. A challenge of nanotechnology is the development of bridges between nanoscale entities and the micro- and macroworld. One approach is to use lipid/carbon mixtures that self-assemble into *carbon nanotubes*. These tubes have remarkable electronical and mechanical properties which would allow them to serve as conducting fibers that connect subcellular structures to larger scaled structures.

membrane is important in regulating the internal environment of the cell and in creating and maintaining concentration gradients between the internal cell environment and the extracellular environment.

2.1.2 Proteins

The phospholipid bilayer of the cell also includes a variety of proteins added to the membrane (Fig. 2.2). The membrane is literally studded with proteins serving a myriad of functions. Membrane proteins fall into two broad categories: *integral* (incorporated into the membrane) proteins and *peripheral* (not embedded in the membrane) proteins. Integral proteins are also amphipathic like phospholipids. Integral proteins are not easily extracted from the membrane; however, they can move within the membrane. The so-called **fluid-mosaic model** of the cell speaks to the mobility of proteins within the cell's membrane. Integral proteins that span the thickness of the membrane are called transmembrane proteins and are often engaged in communication between the intracellular and extracellular compartments (channel proteins). Nontransmembrane integral proteins may be oriented to either the intracellular or extracellular side of the membrane alone. These proteins may be involved with cell recognition or act as anchors for attachment inside or outside of the cell.

Some transmembrane proteins act as anchors between adjacent cells. These anchors or junctions not only prevent the cells from wandering off, as some cancer cells do, but they may also allow for intercellular communication (see Sec. 10.4.1 and Table 10.1).

Peripheral proteins as the name implies are applied to a surface of the membrane though not embedded in the membrane. These proteins are polar like the surfaces of the membrane and found mostly on the intracellular side. These proteins are involved with the cellular skeleton (cytoskeleton) to be discussed below. For more details on the chemical nature of different membrane lipids and proteins, see Chap. 1 on Biomolecules.

2.1.3 Cytoplasm

The **cytoplasm** or cytosol is contained by the cell membrane. Cytoplasm is an archaic term based on primitive understanding of the cellular plasm or fluid. Current understanding shows the cytoplasm to be mostly water containing a variety of solutes. Many ions such as calcium, sodium, and potassium ions are found in the cytoplasm and engage in initiating and terminating cellular functions. In fact, the cytoplasm is a **semifluid** because of the volume and characteristics of its components. In some portions of the cell, the cytoplasm is gelatinous, in other portions, watery. Additionally, numerous compounds including proteins, carbohydrates, and lipids are distributed in the cytoplasm. Unlike the soup analogies used earlier, the ingredients of the cytoplasm are often distributed and arranged in specific portions of the cell, and while they may move within the cell, this movement is frequently directed by cellular machinations and chemical gradients. In an elegant video animation called *The Inner Life of a Cell* by Alain Veil (Viel and Lue, 2006), the cytoplasm is shown as a constantly changing and complex matrix. Proteins aggregate in one portion of the cell, thereby thickening the cytoplasm while making another, less protein-laden, portion more fluid.

2.2   Membrane-Bound Organelles

Membrane-bound organelles are characterized by having at least one layer of phospholipid separating the contents of the organelle from the rest of the cytoplasm. This membrane allows for the maintenance of a different biochemical environment for the contents of the organelle, thereby allowing for specialized processes to take place.

2.2.1   Mitochondria

Mitochondria are colloquially called the "powerhouse" of the cell (Fig. 2.1). While metaphorically correct, it belittles the myriad of biochemical events occurring in this tiny organelle. For instance, it is now well known that these events are not only producing energy but also playing intriguing roles for immunity and cell death (Schmidt, 2006). A large mitochondrion is usually less than a micron in size in any orientation. The mitochondrion is composed of two layers of phospholipids making an outer smooth membrane that forms a characteristic ball- or bean-shaped structure and a highly folded inner membrane. The inner membrane marks the location for the biochemical events (the electron transport system), which will produce energy for the cell (see Sec. 5.3 and Fig. 5.6). By folding the inner membrane, the surface area is increased two to three times compared to

---

CLINICAL BOX 2.1
Graft Versus Host Disease

A main reason for the shortage of transplant organs is the fact that the tissues of the donor and the recipient need to "match" for successful transplantation. Tissues are matched when they have a similar pattern of cell surface proteins. Cell surface proteins are in fact glycoproteins due to the carbohydrates (sugars) attached to their surface. The carbohydrate acts as a flag designating the cell as belonging to the individual. The cellular glycoprotein pattern may specify an individual within a species or specify a species. If a particular sugar is missing from the surface of a cell, the immune system may recognize this cell as foreign and try to kill it. An attack on self tissue may lead to autoimmune diseases, where the autoimmune reaction can be directed against a specific tissue such as the brain in multiple sclerosis or the digestive tube in Crohn disease. In other cases, the overly active immune system may attack many cells and tissues so that various organs are affected such as in systemic lupus erythematosus.

Histocompatibility, the condition in which the cells of one tissue can survive in the presence of cells of another tissue, is also the most important criterion for successful organ transplantation. The reason for organ rejection is that different markers on the cells of the graft (tissue from the donor) act as antigens, which can trigger an immune response in the host (the organ recipient). These markers are called human leukocyte antigens. They are present on the surface of most body cells, but were first discovered on white blood cells (leukocytes). Clinically, a much more common problem is that functional immune cells in the graft recognize host tissue as foreign and cause an immune response resulting in clinical symptoms that are called graft versus host disease.

Blood transfusions can also cause transfusion reactions (agglutination) if the donor red blood cells (erythrocytes) have glycoproteins (antigens) that are attacked by the recipient's antibodies. The most well-known antigens are the ones that define the ABO blood type system. But there are more than 600 additional red blood cell antigens known defining other blood type systems, which are important to consider for some blood transfusions.

the flat surface area, and there by more energy can be produced. The folds of the inner membrane form compartments called **cristae**. Cristae are ultramicroscopic features of mitochondria. The energy produced is in the form of a molecule called adenosine triphosphate (ATP), which effectively stores energy in its chemical bonds between phosphate molecules. Making and breaking those bonds stores and releases energy, respectively (see Sec. 1.5). The ATP formed by mitochondria can be distributed throughout the cell to facilitate other chemical processes. A cell rich in mitochondria is likely to be a very energetic cell such as a muscle cell. A cell poor in mitochondria is less energetic such as a white fat (adipose) cell. An exception is a brown adipose cell, which is a unique cell in human infants and certain hibernating creatures utilizing nonshivering thermogenesis to produce much energy. As the name implies, nonshivering thermogenesis produces heat through the unusual mechanism of cell metabolism. Shivering is the rapid and repeated contraction of skeletal muscle (somewhat like revving an engine) resulting in the production of heat. Brown adipocytes are packed with mitochondria. These mitochondria, rather than producing ATP as energy for use elsewhere in the cell, disconnect the ATP production pathway, literally spilling energy from the system and releasing heat in the process. Imagine leaving a door open into a steam engine. The energy that leaves the system in the form of heat is wasted energy, which means that it does not produce useful work.

2.2.2 Lysosomes

The term "some" means "body" and "lyse" means "destroy," so, a lysosome is a body which destroys. A lysosome is a membrane-bound vesicle (fluid-filled sack) containing enzymes (lysozymes) which can disrupt chemical bonds. The lysosomal membrane keeps enzymes segregated from the rest of the cell contents so that the degradative process is regulated. To access the lysozyme contents of the lysosome proteins on the surface of a lysosome are contacted by proteins on the membranous vesicle of an endocytosed vesicle (see below). This recognition allows the two vesicles (lysosome and endocytosed vesicle) to combine to form an **endosome**. The lytic enzymes of the lysosome then act on the contents of the endosome often resulting in digestion or destruction of the matter. The regulation of lysosomal activity is essential to the survival of the cell. Cells can die from exposure to the contents of a lysosome. The process of apoptosis (programmed cell death; see Secs. 4.5 and 9.6) regulates how long some cells live and is sometimes brought about by rupture of lysosomes inside of the parent cell.

Endocytosis is the process of bringing into a cell something from the outside (Fig. 2.3). In a multistep process, a pocket is formed in the cell membrane allowing some extracellular material to be engulfed. The pocket of cell membrane containing extracellular material is pinched off

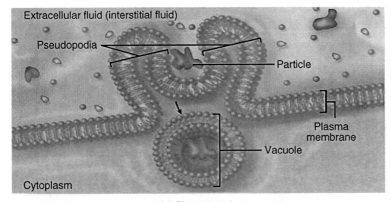

(a) Phagocytosis

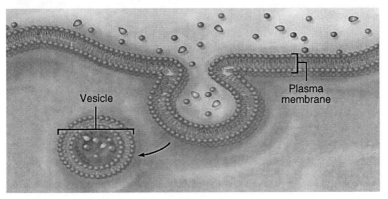

(b) Pinocytosis

Figure 2.3 Endocytosis. It occurs at the cell membrane to bring extracellular material into the cell. (a) Phagocytosis, or cellular "eating," of an extracellular particle; (b) Pinocytosis, or cellular "sipping," of extracellular fluid with small solutes included. (Reproduced with permission from McKinley and O'Loughlin, *Human Anatomy*, 1st ed. McGraw-Hill, New York, 2006.)

and allowed to enter the cytoplasm. A lysosome may fuse with the endocytosed vesicle to form an endosome. The lytic enzymes then can break down the contents of the endocytosed vesicle, and the resulting components are made available for use in the cell or for excretion. Forming small endocytotic pockets is referred to as **pinocytosis** (cellular sipping), and forming larger pockets is called **phagocytosis** (cellular eating).

Cells also excrete numerous products and waste materials from the cell. This process is referred to as **exocytosis** (Fig. 2.4). Exocytosis is especially important for polar or water-soluble materials which cannot freely pass through the cell membrane. These polar molecules are packaged in a layer of cell membrane by a structure called the Golgi apparatus (see Sec. 2.2.4). The membrane around the material to be excreted can fuse with the

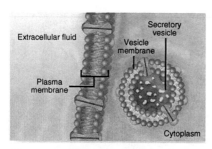

① Vesicle nears plasma membrane

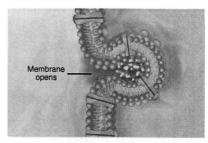

③ Exocytosis as membrane opens externally

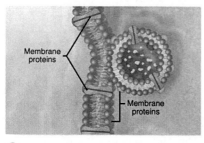

② Fusion of vesicle with membrane

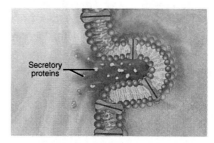

④ Release of vesicle components into the extracellular fluid and integration of vesicle membrane components into the plasma membrane

Figure 2.4   Exocytosis. It can be described as a four-step process. A membrane-enclosed intracellular vesicle moves to the cell membrane (step 1). The two membranes fuse (step 2), allowing the contents of the vesicle to be expelled from the cell (step 3). The vesicle membrane is incorporated into the cell's membrane (step 4). Exocytosis is used to dispose of waste and to excrete cell products such as hormones and extracellular proteins. (Reproduced with permission from McKinley and O'Loughlin, *Human Anatomy*, 1st ed. McGraw-Hill, New York, 2006.)

intact cell membrane. In the fusing process, the contents are released into the extracellular environment.

2.2.3 Peroxisomes

Like lysosomes, peroxisomes are enzyme-containing vesicles in the cytoplasm of the cell (Figs. 2.1 and 2.5). Peroxides (reactive molecules which can disrupt chemical bonds) contained in the vesicle are compartmentalized from the rest of the cytoplasmic contents in order to protect the cell. Peroxides are by-products of routine, as well as aberrant, cell metabolism. They have been associated with a variety of diseases including cancer. Accordingly, a dysfunction of peroxisomes may be associated with some forms of cancer. Peroxisomes, like lysosomes, can combine with endocytosed vesicles to form endosomes. Endosomes are the product of an endocytosed vesicle combining with an intracellular vesicle to make a new, single, membrane-bound package containing the ingredients of the two.

Rough ER   Peroxisome   Free ribosomes

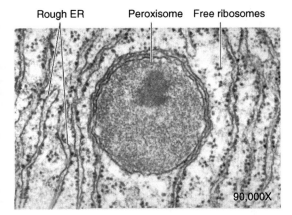

90,000X

Figure 2.5 Peroxisome. These organelles contain oxidative enzymes and resemble lysosomes. Peroxisomes are responsible for detoxification of harmful substances from the cell such as peroxides and hydrogen peroxide, which is converted to water. (Electron micrograph, reproduced with permission from McKinley and O'Loughlin, *Human Anatomy*, 1st ed. McGraw-Hill, New York, 2006.)

2.2.4   Golgi apparatus

Named after a highly productive histologist of the nineteenth century, the Golgi apparatus, or Golgi body, is responsible for packaging cellular products for exocytosis (Figs. 2.1 and 2.6). The stacks of membranes that make up the Golgi apparatus are located between the

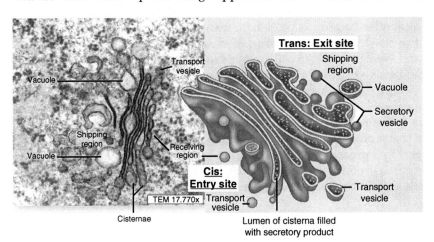

Figure 2.6 Golgi apparatus. Composed of a series of stacked, folded membranes, the Golgi apparatus serves as shipping and receiving center for the cell. The receiving region (*Cis*) allows membrane-encased vesicles to deliver their contents into the lumen of the Golgi. As the contents move through the Golgi apparatus, they are modified, for instance through the addition of sugars. The contents are then released inside vesicles from the shipping region (*Trans*). (Reproduced with permission from McKinley and O'Loughlin, *Human Anatomy*, 1st ed. McGraw-Hill, New York, 2006.)

nucleus of the cell and the cell membrane in an orderly fashion. This organization facilitates the delivery of certain protein and lipid products to the Golgi apparatus. Transport vesicles fuse with the **cis** face of the Golgi apparatus and empty their content into the Golgi lumen. The proteins and lipids are then transported through the Golgi and modified by the addition of sugar, carbohydrate, or phosphate groups, which can change their chemical natures. The glycoproteins (sugar or carbohydrate added to proteins) or phosphoproteins are packed in a membrane vesicle at the **trans** site of the Golgi and exocytosed from the cell. Once released, the membrane-encased secretory product has one of three fates: modification of the plasma membrane, lysosome formation, or excretion at the plasma membrane.

2.2.5 Endoplasmic reticulum

The last type of membrane-bound organelle is the endoplasmic reticulum of which there are two types: smooth and rough (Fig. 2.1). **Smooth endoplasmic reticulum (SER)** represents a series of folded membranes found in abundance in cells that produce many lipids for exocytosis. These cells include those making steroid-type hormones (chemical messengers). The processes that occur in the SER are myriad and beyond the scope of this discussion. It is sufficient to say that complex lipids are assembled from component parts while in the SER. Most of the lipid products can then be released from the SER and allowed, due to their nonpolar nature, to simply pass through the cell membrane and into the extracellular milieu. In the example of steroid hormones, the lipid is most often bound to a protein for transport in the aqueous environment of the blood. Because of its own lipid nature, the SER appears as a void in the typically stained cell's cytoplasm. With standard methods of fixation, lipids are largely lost from tissues during processing.

    Rough endoplasmic reticulum (RER) is so-called due to its bumpy appearance as seen using ultramicroscopy. The rough appearance is due to the presence of ribosomes (Fig. 2.5). Ribosomes are responsible for the production of proteins from a messenger RNA (mRNA) transcript in a process called translation (see Sec. 7.5). In cells responsible for much protein synthesis, the RER is abundant and can be seen as a dark area in the cell's cytoplasm using light microscopy.

2.3 Nonmembrane-Bound Organelles

2.3.1 Ribosomes

As mentioned above, ribosomes are organelles used to help form proteins. Instead of being composed of, or contained in, a membrane, ribosomes are proteins combined with RNA molecules. The ribosome is made of two subunit pieces which come together to facilitate formation of a chain of amino acids into a peptide and then often into a larger protein. Ribosomes have a large subunit and a small subunit. A ribosome that is attached to the endoplasmic reticulum (ER) is called bound ribosome, while those ribosomes scattered in the cytoplasm or joined together in small clusters in the cytoplasm are called free ribosomes. Messenger RNAs from the cell's nucleus travel into the cytoplasm and encounter ribosomes. The ribosomes then bring together amino acids to form an amino acid chain, or peptide, according to the code of the mRNAs (see Sec. 7.5). If the peptide is formed on free ribosomes it is usually used for intracellular processes. Peptides formed by ribosomes bound to endoplasmic reticulum are often for use outside of the cell or in the cell membrane. The peptides, once formed in the RER, may be transferred to the Golgi apparatus for packaging into a vesicle for extracellular transport.

2.3.2 Cytoskeleton

As the name implies, most cells have a matrix of tubules and filaments, which can maintain or change the cells' shape as well as move substances around within the cytoplasm. There are three main types: microfilaments, intermediate filaments, and microtubules (Fig. 2.7).

Microfilament as the name implies, are small linear fibers which form much of the framework of a cell and are used to maintain structure. They are sometimes called actin filaments since they are composed

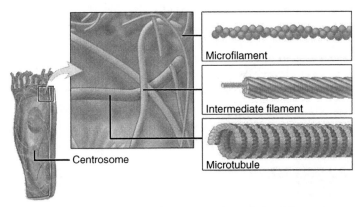

Figure 2.7 Cytoskeleton. The three types of cytoskeletal filaments are microfilaments with a diameter of around 7 nm, microtubules which may exceed 25 nm, and intermediate filaments with an intermediary size. The filaments have in common that they are all composed of repeating protein subunits. The cytoskeleton provides for the form of the cell, for transport within the cell, and, in some cells, for the motility of the cell. (Reproduced with permission from McKinley and O'Loughlin, *Human Anatomy*, 1st ed. McGraw-Hill, New York, 2006.)

APPLICATION BOX 2.2
The Muscle Machine

Hugh E. Huxley elegantly described in 1969 what has been come to be known as the *Sliding Filament Theory* of muscle contraction. The theory describes the movement of a series of proteins in conjunction with calcium and energy from ATP.

The sliding between the linearly arranged proteins *actin* and *myosin* creates a subcellular machine. This machine changes the shape of the muscle cell, and when combined with other muscle cells, the shape of the whole muscle. Changing the shape of a muscle ultimately creates, in conjunction with bones, movement of a limb. It is fascinating to understand how the actions of the molecular muscle machine composed of actin, myosin, and other proteins translate into movement of an organism. This information of cell and molecular biology is useful in robotics and bionics, the combination of biology and technology, and in biomechanics in general. Muscle, generally working with a third-class lever within the skeletal system, is roughly 25% efficient when performing work. By comparison, an internal combustion engine operates at roughly 20% efficiency.

---

of the protein actin. Cell extensions filled with a network of cross-linked actin filaments are called **microvilli** (Fig. 2.8). They serve to increase surface area for absorption and secretion across the cell membrane.

A special type of microvilli are **stereocilia**. They are specialized cell protrusions with a rigid internal framework composed of actin on the apical membrane of **hair cells**. These hair cells are part of the inner ear (Fig. 2.9) and are responsive to sounds transmitted to the ear through the air. The hair cells are arranged on a platform called the *basilar membrane*. The membrane vibrates in different regions of the inner ear depending on the frequency of the sound wave. When a segment of the membrane vibrates, the hair cells resting on that membrane also vibrate. The stereocilia (hairs)

Microvilli

Figure 2.8 Microvilli. These are protrusions on a cell's apical surface that serve to increase surface area for absorption or secretion. For instance, intestinal lining cells have rich microvilli layers for the vast amounts of absorption occurring in the small intestine. (Reproduced with permission from McKinley and O'Loughlin, *Human Anatomy*, 1st ed. McGraw-Hill, New York, 2006.)

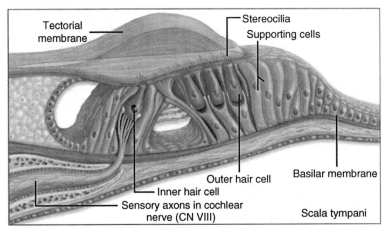

Figure 2.9 Stereocilia on hair cells within the spiral organ of the inner ear. Sound waves travel through the air to the chambers of the ear, move fluid of the ear, and cause a wave motion of the basilar membrane. The movement of the basilar membrane causes movement of the hair cells relative to their rigid stereocilia. The bending of the stereocilia causes activation or inhibition of signals in the cochlear nerve leading to the brain. This neuronal activity is detected as sound. (Reproduced with permission from McKinley and O'Loughlin, *Human Anatomy*, 1st ed. McGraw-Hill, New York, 2006.)

of these now vibrating hair cells are anchored into yet another membrane called the *tectorial membrane*. The other end of the stereocilia is tightly anchored to the inner surface of the hair cell membrane. Tension is applied to the stereocilia as the rest of the cell moves with the basilar membrane. When this movement occurs relative to the stereocilia, a signal is sent to the central nervous system, which then is interpreted as a specific sound.

Intermediate filaments (Fig. 2.7), intuitively, are in between the size of microfilaments and microtubules. These fibers give further shape to the cell. They are stable and durable and hence very prominent in cells that withstand mechanical stress. There are five different types of intermediate filaments. One of them, called keratin, gives strength to the cells of the skin (Freeman, 2005).

Microtubules (Fig. 2.7), are composed of repeating protein subunits called tubulin. Microtubules can grow by adding subunits at either end or shorten by the removal of subunits. In this way, a substance that is attached to the microtubule can be moved through the cytoplasm. Additionally, motor protein such as kinesin can transport vesicles, granules, and organelles such as mitochondria along a microtubule by forming and breaking bonds with the energy supplied by ATP. The energized kinesin molecule then "walks" along the length of the microtubule (Freeman, 2005; Viel and Lue, 2006).

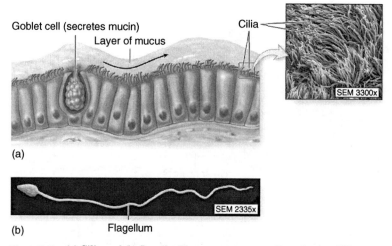

(a)

(b)   Flagellum

Figure 2.10 (a) Cilia and (b) flagella. These are common cell surface modifications composed of microtubules. Cilia often move extracellular material along the surface of cells, such as particles on top of airway cells. The sperm's flagellum (tail) is used to help drive through the surface of an oocyte to accomplish fertilization. SEM, scanning electron micrographs. (Reproduced with permission from McKinley and O'Loughlin, *Human Anatomy*, 1st ed. McGraw-Hill, New York, 2006.)

Two frequent cellular appendages associated with microtubules are cilia and flagella. **Cilia** (Fig. 2.10*a*) are fine cellular foldings of the cell membrane typically associated with the apical (top) surface of a cell responsible for absorption and motility. Microtubules create movement of the cilia moving extracellular material along the surface of the cells. An example of this is the airways leading to and from the lungs. Mucus accumulates particulate matter inhaled from the air, and the cilia move the contaminated mucus toward the throat for swallowing and disposal by the digestive system. The cilia in the apical surface membrane appear as "fuzz" on a cell at the light microscopic level giving rise to the identification of "hair" cells in certain cases, not to be mixed up with the hairs (stereocilia) of the inner ear hair cells described above.

Microtubules are also the functional component of **flagella** (Fig. 2.10*b*). Flagella are long cell processes or cell surface modifications. Flagella are longer than cilia and are used for motility by some cells. The most recognizable flagellated cell is the sperm. The sperm flagellum is used to help motivate the cell through the female reproductive tract. Inside the flagellum is an arrangement of microtubules in a figure-of-8 which represents the fused pairs of microtubule doublets. Projecting from these figure-8 microtubules are protein called dynein. The dynein arms make and break bonds with the adjacent microtubules in a spiraling

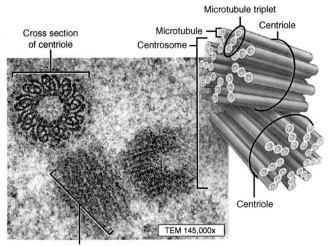

Figure 2.11 Centrosome. The centrosome includes the two perpendicularly oriented centrioles and the space immediately around them. Centrosomes and centrioles serve as microtubule-organizing centers in nondividing cells and are involved in facilitating mitosis in dividing cells. (Reproduced with permission from McKinley and O'Loughlin, *Human Anatomy*, 1st ed. McGraw-Hill, New York, 2006.)

fashion, which produces a swishing of the sperm tail. There are nine figure-8 microtubule structures surrounding a central pair of microtubules in the typical eukaryotic flagella in a 9 + 2 arrangement.

Centrosomes (Fig. 2.11) are microtubule structures involved in movement of chromosomes (DNA) during cell division. **Centrioles** are paired structures in the centrosome. Each centriole is comprised of nine triplet microtubules. The centrioles are organizing centers for microtubule formation since microtubules originate there and are grown and shortened from there as needed in the cell. When cells divide, genetic material is packaged together to form chromosomes. These chromosomes are then divided between daughter cells when centrioles attached to the chromosomes pull the duplicated genetic material apart through microtubule shortening (see Sec. 8.2).

2.4 Nucleus

The nucleus houses most of the genetic material of the cell (Fig. 2.1). [Mitochondria have their own genetic material, which is inherited from the mother in humans and remains distinct from the so-called nuclear DNA (Gilbert, et al. 2003)]. The DNA, or genetic material, guides the cellular machinery. Through the processes of transcription and translation described in Secs. 7.4 and 7.5, genetic material guides the formation of RNA and subsequently proteins, which turn on and off a multitude of functions in the cell. The appearance of the nucleus varies. In metabolically active cells producing many proteins, the chromatin is dispersed for easy transcription and gives the nucleus a pale appearance. The dispersed chromatin is called **euchromatin**. Tightly packed chromatin that is more visible with typical stains is called **heterochromatin**. Cells with heterochromatic nuclei are less actively making proteins.

2.4.1 Nucleolus

Inside the nucleus is found the nucleolus, or little nucleus. This spherical mass is composed of RNA and proteins. The nucleolus is the source of the ribosomes (see Sec. 2.3.1). Ribosomal subunits are made and released from the nucleolus and then transported into the cytoplasm from the nucleoplasm (fluid matrix of the nucleus). Passageways, nuclear pores, in the nuclear envelope allow ribosomal subunits and mRNA to leave the nucleus and enter the cytoplasm or lead directly into the endoplasmic reticulum.

Also entering and leaving the nucleus are nucleic acids used to make DNA and RNA. A central dogma of biology states that DNA leads to RNA, which leads to the production of protein (Raven, et al., 2005).

2.5   Differences in Cells

2.5.1   Plant cells compared to mammalian cells

Plant cells differ from human/mammalian cells on several fronts. Plants, in addition to the cell membrane, have an outer coating called a **cell wall**, which helps to maintain the structure of the cell and to hold on to water within the cell. Some single-celled organisms also have a cell wall (see Fig. 10.1).

Plants contain **chloroplasts** where photosynthesis occurs. Chloroplasts are organelles which are not present in animal cells. Photosynthesis is the conversion of CO_2 and water in the presence of sunlight into sugar as a food source for the plant and O_2 as a by-product (see Sec. 5.4 and Fig. 5.12). The product of cellular and organismal respiration in humans and many multicelled animals is CO_2. The CO_2 expired by humans and other organisms is then used by plants to make sugar used as metabolic fuel. In return, the plant releases O_2 as an end product of its metabolism. The dependence of humans on plants for the production of O_2 is inescapable.

2.5.2 Prokaryotes

Prokaryotes (more primitive, "pro" = early, organisms) are single-celled organisms which have many cellular commonalities with the **eukaryotes**, which include humans. These prokaryotes lack a defined nucleus. The genetic material is suspended in the cell's cytoplasm rather than enclosed by a nuclear membrane. Some prokaryotes have many fewer organelles than eukaryotes, too. It is the common view that prokaryotes were first to arise in evolution and that eukaryotes developed from prokaryotes (see Sec. 10.1.1).

2.5.3 Tissue-specific language

Throughout the biological sciences, the use of different vocabulary in different cells for similar structures can be daunting. For instance, in the case of muscle and muscle cells, the label "**sarco-**" is used. Endoplasmic reticulum becomes sarcoplasmic reticulum, and the sarcoplasm is the cytoplasm of a muscle cell. Another example is the RER in nerve cells (neurons), which is called **Nissl substance** after the histologist who recognized the mass of organelle in the early years of microscopy before RER was characterized. Microfilaments in a neuron are sometimes called neurofilaments. While the scientific vocabulary is cumbersome, it allows for precision of communication. It is important therefore to master the "jargon" of a scientific discipline in order to assure meaningful and accurate communication.

Suggested Reading

Freeman, S. (2005). *Biological Science,* 2nd ed. Pearson, Prentice Hall, New Jersey.

McKinley, M. and O'Loughlin, V. (2006). *Human Anatomy,* 1st ed. McGraw-Hill, New York.

Nowak, T.J. and Handford, A.G. (2004). *Pathophysiology,* 3rd ed. McGraw-Hill, New York.

Raven, P.H., Johnson, G.B., Losos, J.B. and Singer, S.R. (2005). *Biology,* 7th ed. McGraw-Hill, New York.

Widmaier, E., Raff, H. and Strang, K. (2006). *Vander's Human Physiology,* 10th ed. McGraw-Hill, New York.

References

Gilbert, M.T.P., Willerslev, E., Hansen, A.J., Barnes, I., Rudbeck, L., and Lynn, N. (2003). Distribution patterns of postmortem damage in human mitochondrial DNA. *Am. J. Hum. Genet.* 72(1): 32–47.

Schmidt, K.F. (2006). The powerhouse and sentinel of the cell. *HHMI Bulletin.* 19(2): 28–32.

Viel, A. and Lue, R.A. (2006). Cellular Visions: *The Inner Life of a Cell* (video). Studiodaily (*www.studiodaily.com*), *http://www.studiodaily.com/main/technique/tprojects/6850.html.*

Enzyme Kinetics

Thomas D. Hurley, Ph.D.

OBJECTIVES

- To introduce the basic concepts of steady-state kinetics
- To understand the significance of the Michaelis-Menton equation
- To introduce the concept of multisubstrate systems
- To develop the concept of enzyme inhibition
- To introduce cooperative interactions and regulation of enzymes
- To illustrate control mechanisms of enzyme activity

OUTLINE

3.1 Steady-State Kinetics   58
3.2 Enzyme Inhibition   72
3.3 Cooperative Behavior in Enzymes   78
3.4 Covalent Regulation of Enzyme Activity   81

Enzymes are biological catalysts that enable thousands of biological reactions that occur every second within cells. The kinetics of enzymatic reactions is important for the spatial and temporal regulation of cell processes and its measurement is used for diagnosis of diseases. Biochemical engineers manipulate biological processes with the use of enzymes in order to produce the desired products.

By and large, enzymes are the engines that drive the metabolic functions that underlie cellular functions. Enzymes are extraordinary **catalysts**, often accelerating chemical reactions by more than 10^8 times over the rates of the uncatalyzed reactions and they do so at relatively low temperature (37°C) and low pressure (~1 atm). They achieve their remarkable rate enhancements without sacrificing their selectivity for particular substrates, and these rates can be regulated through numerous mechanisms, providing the cell with precise control over what reactions to catalyze and the rate at which those reactions will occur. In this chapter, we will explore the manner in which enzymes interact with their substrates and how these reactions may be controlled through both noncovalent and covalent modifying agents.

3.1   Steady-State Kinetics

Much of our understanding of **enzymes, E,** and the reactions that they catalyze, has been obtained from a study of their behavior under conditions in which the disappearance of **substrate, S,** and the appearance of **products, P,** is constant per unit time; in other words, the reaction has reached a steady state (Fig. 3.1). The importance and significance of the **steady-state condition** is that these analyses allow us to investigate the behavior of enzymes in solution under reaction conditions that can be similar to those in the cell; namely that the enzyme is generally in a steady-state condition in the cell where there is a relatively constant supply of the substrate and the cell is generally constantly removing the product of the reaction. The parameters obtained from these analyses permit us to model the effect that changes in substrate concentration or the addition of inhibitors of the reaction will have on catalysis. They also help us understand how to design approaches to study the function of the enzyme under actual cellular conditions so that we can better design selective inhibitors for a particular reaction of interest.

Additional and more detailed information can be obtained through experiments that probe the characteristics of the reaction prior to

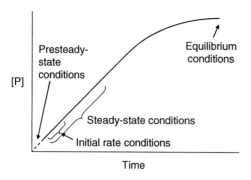

Figure 3.1 Conversion of substrate to product. The hypothetical graph shows the production of product during an enzyme-catalyzed reaction in which a substrate, S, is converted to a product, P. The presteady state condition can last anywhere from 10^{-6} to 10 seconds depending on the enzyme in question. The steady-state condition is said to have been obtained when the slope of product formation is a constant for a significant fraction of the time interval of the experiment.

reaching the steady-state condition—so-called *presteady state kinetics*. These experiments can isolate particular steps in the reaction, such as substrate binding and product release, or closely examine the chemical step. However, the design and analysis of these experiments are outside the scope of this chapter and the reader is referred to texts that address these aspects of enzyme catalysis referred to at the end of this chapter for further information.

The steady-state treatment of enzymatic catalysis and the analysis of its properties rely on a series of assumptions that define and limit this condition. The *first assumption* is that the concentration of substrate, [S], does not change significantly over the course of our observation. This assumption permits us to utilize the rate obtained at a particular substrate concentration as the rate obtained at the starting concentration. That the amount of substrate must change in order to observe the enzymatic rate is a given, but if we only concern ourselves with the initial rate—say within the first minute or so—the substrate concentration will not have changed significantly from the starting concentration. The *second assumption* is that the concentration of [E] is much lower than [S]. As we will see below, the enzyme forms a complex with S, called the **ES complex**. This assumption permits us to neglect the loss of S to the ES complex and thereby treat the starting S concentration as the true solution concentration. The *third assumption* is that the concentration of [P] is essentially zero. This assumption is related to the first assumption. The rates measured are thus those that occur during the initial period of catalysis—termed **initial rate conditions** (Fig. 3.1)—much before equilibrium conditions are approached. This assumption allows us to neglect any tendency for the enzyme to catalyze the conversion of P back to S. The violation of any of these conditions does not preclude the analysis of the enzymatic reaction, but the equations necessary to account for the loss of S to catalysis, or to the ES complex or competition with the reverse reaction, are more complex and are generally avoided unless features of the reaction in question require their use. The reader is referred to more dedicated texts for the treatment of these enzymatic phenomena referred to at the end of this chapter.

3.1.1   Derivation of the Michaelis-Menton equation

For the following discussion of the steady-state approach to the analysis of an enzyme-catalyzed reaction, we will use the conversion of a single substrate, S, to a single product, P, by an enzyme, E. In 1913, Michaelis and Menton derived an equation that describes the concentration dependence of the reaction rate for this simple system. Their work was based on the work of Henri (1903) 10 years earlier and provided algebraic solutions to the equations he derived. It was understood by this time that a binding site on the surface of the enzyme existed and that this site was saturable. It was also known that the formation of the ES complex from their free components in solution exhibited an inherent **equilibrium constant** and that catalysis occurred from this ES complex. In addition to the steady-state conditions above, other assumptions are imposed on the enzymatic system to simplify the analysis. One assumption is that the concentration of the complex between enzyme and substrate, ES, is in thermodynamic equilibrium with the free forms of both and that the ES complex is held together by noncovalent forces. In other words, the rate constants for the formation of this complex (k_1, units of $M^{-1}s^{-1}$) and its breakdown to the free components (k_{-1}, units of s^{-1}) are considerably faster than the catalytic step (k_{cat}, units of s^{-1}) that leads to product formation. In addition, we assume that the breakdown of the EP complex following catalytic conversion is so fast that we may neglect its formation. Thus, the slow step is the chemical step and we denote that rate constant as k_{cat}.

$$E + S \underset{k_{-1}}{\overset{k_1}{\rightleftharpoons}} ES \xrightarrow{k_{cat}} E + P \tag{3.1}$$

In the absence of P, this process is unidirectional from left to right. From simple chemical equilibria, we can write the equilibrium dissociation constant for the formation of the ES complex.

$$K_M = \frac{k_{-1}}{k_1} = \frac{[E][S]}{[ES]} \tag{3.2}$$

From simple chemical rate laws, we can write that the velocity of the forward reaction will be equal to the concentration of ES times the rate constant for its breakdown in that direction.

$$v = k_{cat}[ES] \tag{3.3}$$

From these relations, and a little bookkeeping with respect to the enzyme species present in solution during initial rate conditions, we can derive the **Michaelis-Menton equation**. Equation (3.8) is its final form, and Eqs. (3.4) to (3.7) demonstrate how to obtain it.

$[E]_0 = [E] + [ES]$ is the relation that expresses the total concentration of enzyme in solution, $[E]_0$, as related to the free enzyme concentration, $[E]$, and the concentration of the ES complex. Rearranging this relationship to $[E] = [E]_0 - [ES]$, and substituting the right-hand side of this equation for $[E]$ in Eq. (3.2), we get the following equation.

$$K_M = \frac{([E]_0 - [ES])[S]}{[ES]} \tag{3.4}$$

If we multiply through on the numerator and then multiply both sides by [ES], we get the following equation.

$$K_M[ES] = [S][E]_0 - [ES][S] \qquad (3.5)$$

Bringing [ES][S] to the left-hand side and grouping terms gives us the following equation.

$$[ES](K_M + [S]) = [S][E]_0 \qquad (3.6)$$

Dividing both sides by $(K_M + [S])$ gives us the following relationship that expresses the concentration of the ES complex in terms of the concentration of enzyme, substrate, and the dissociation constant for the formation of the ES complex.

$$[ES] = \frac{[S][E]_0}{(K_M + [S])} \qquad (3.7)$$

By substituting the right-hand side of this relationship for ES in Eq. (3.3), we get the final form of the Michaelis-Menton equation.

$$v = \frac{k_{cat}[S][E]_0}{(K_M + [S])} \qquad (3.8)$$

If we plot the velocity dependence on substrate concentration that is given by this relationship, we get a rectangular hyperbola in which the reaction velocity reaches a maximal value, V_{max}, that is equal to $k_{cat}[E]_0$ (Fig. 3.2). The value of V_{max} is always an observable value in an enzymatic experiment, whether or not the enzyme is of a pure form. However, we cannot know the true value of k_{cat}, the inherent catalytic rate of a single enzyme, unless the enzyme concentration is known and the amount of enzyme that gives rise to a particular V_{max} value can be determined.

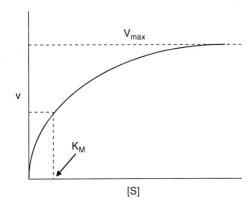

Figure 3.2 Substrate saturation curve. The plot shows the saturation curve given by the Michaelis-Menton equation, where the observed velocity, v, is plotted against the substrate concentration, [S], that gave rise to the observed velocity. The position of the V_{max} value given by extrapolation of the hyperbola is shown along with the value of [S] that gives rise to one-half the maximal velocity, termed the K_M value.

The substrate/velocity curve produced by the Michaelis-Menton equation can be seen to have two nearly linear portions. The first is at very low substrate concentrations ($[S] <<< K_M$), where the velocity is essentially equal to:

$$v = \frac{k_{cat}[E]_0[S]}{K_M} = \frac{V_{max}}{K_M}[S] \tag{3.9}$$

This is a **first-order velocity expression** ($[S]^1$), where the observed velocity of the reaction is linearly dependent on the substrate concentration. The constant, V_{max}/K_M, or k_{cat}/K_M for single enzymes, is often used to describe the *efficiency* with which an enzyme catalyzes its reaction, since enzymes with low K_M values and high maximal velocities will yield higher velocities at a fixed substrate concentration. The second linear portion of the curve is at very high substrate concentrations ($[S] >>> K_M$), where the velocity is essentially equal to $v = k_{cat}[E]_0 = V_{max}$. This is a **zero-order reaction** ($[S]^0$), where the observed velocity is completely independent of the substrate concentration. Thus, the Michaelis-Menton equation, and the substrate/velocity curve it produces, essentially describes the t*ransition from a first-order reaction to a zero-order reaction.*

The other property of the Michaelis-Menton curve is that K_M is a true **dissociation constant.** Its value is equal to the substrate concentration that produces one-half of the maximal velocity. This can be shown in the following way.

If $K_M = [S]$ then,

$$v = \frac{k_{cat}[S][E]_0}{([S] + [S])} = \frac{k_{cat}[S][E]_0}{(2[S])} = \frac{k_{cat}[E]_0}{2} \tag{3.10}$$

$$v = \frac{1}{2}V_{max}$$

This is the treatment of steady-state enzyme catalysis in its most simple form. However, rarely do the features of an enzyme-catalyzed reaction actually obey those imposed on the reaction in this treatment. Briggs and Haldane (1925) derived a more general approach, one that can be adapted to describe most enzyme-catalyzed reactions. Their approach requires no assumption as to what step in the reaction sequence is rate-limiting. Consequently, no particular rate constant is associated with the observed catalytic rate. In this treatment, the enzymatic expression is modified as shown below and then called the **Briggs-Haldane relationship.**

$$E + S \underset{k_{-1}}{\overset{k_1}{\rightleftharpoons}} ES \overset{k_2}{\longrightarrow} E + P \tag{3.11}$$

For this treatment, our steady-state assumptions must now also include the provision that [ES] remains constant over time. Therefore, $d[ES]/dt = 0 = k_1[E][S] - k_2[ES] - k_{-1}[ES]$. Since, $[E] = [E]_0 - [ES]$ and $v = k_2[ES]$ still apply, we can substitute into K_M equation. After substitution and rearrangement, the following equation is obtained.

$$v = \frac{k_2[S][E]_0}{\left\{ \dfrac{(k_2 + k_{-1})}{k_1} + [S] \right\}} \quad \text{where } K_M = \frac{(k_2 + k_1)}{k_1} \quad (3.12)$$

The Michaelis-Menton simplification is restored if, as we originally stipulated, $k_2 \ll k_{-1}$ so that k_2 disappears from the numerator of the K_M term. When $k_2 \gg k_{-1}$, then we can see that k_{cat}/K_M reduces to k_1, the association rate constant for free enzyme and free substrate.

3.1.2 Interpretation of the steady-state kinetic parameters in single substrate/product systems

K_M—Only if $K_M = K_S$ does this **Michaelis constant** represent the true dissociation constant for the ES complex. However, a more practical description of K_M can be provided when K_M is thought of as the apparent dissociation constant for all enzyme-bound species along the reaction pathway (e.g., $K_M = [E][S]/\Sigma[ES]$).

k_{cat}—If only one ES complex is formed along the reaction pathway and all the binding and release steps are fast, k_{cat} becomes the first-order rate constant for the conversion of ES to EP. If the enzyme system is more complex, k_{cat} is a function of all first-order rate constants along the reaction sequence and is not associated directly with any particular rate constant. The term k_{cat}, or the **turnover number** for the enzyme, is the maximum number of product molecules produced by each enzyme molecule per unit of time.

k_{cat}/K_M—As mentioned above, this term is often used to define the **specificity constant** for the enzyme with respect to any particular substrate. As such, k_{cat}/K_M is a second-order rate constant (with units of $M^{-1}s^{-1}$) that describes how favorable the association between free E and free S is in solution.

3.1.3 Analysis of experimental data

Before we attempt to analyze experimental data, we should outline some important practical considerations in performing the experiment, since the setup of the experiment can dramatically affect the quality of the data obtained from the analysis. It is important to the experiment that *the substrate concentration ranges ~tenfold* and in such a way that the K_M for the substrate occurs within the concentration range used. From a practical perspective, it is often useful to perform a series of test assays using different substrate concentrations to establish roughly where half of the maximal activity is obtained and then set that concentration as the midpoint of a range of substrate concentrations. As long as [E] << [S] and there is no significant product inhibition, this approach should provide a reasonable first experiment. Subsequent experiments can use this initial one as a guide for how to best modify the experimental approach, if needed. The drawback to deviation from these design parameters is that if the initial concentration range of the substrate is too low, then extrapolation to V_{max} becomes problematic as one is essentially fitting a hyperbola to the initial linear portion of the substrate saturation curve (Fig. 3.2). If the initial concentration range is too high, then extrapolation to K_M becomes problematic as one is again essentially fitting the hyperbola to the essentially linear part of the curve at high concentrations (Fig. 3.2). This becomes clearer as we examine the methods for plotting the data.

Prior to the advent of the modern personal computer, most investigators used rearrangements of the steady-state kinetic equation that produced straight line relationships, so that the kinetic constants could be obtained through linear regression and extrapolations to x- and y-axis intercepts. The two most commonly used linear forms of the Michaelis-Menton equation are those produced by Lineweaver and Burk (1934) and Hanes-Woolf/ Eadie-Hofstee (Hanes, 1932). Even though today the analysis of data from enzyme kinetics experimets, such as determining K_M and V_{max}, is usually done by fitting the data to nonlinear kinetic expressions using the computer, fitting the data to linear plots is still useful. The linear plots make the clearest representation of kinetic data so that it is the best way to recognize potential deviations of enzyme behavior from standard behaviors.

Lineweaver-Burk equation (double reciprocal plot). For the analysis of kinetic data, this is the simplest method to obtain estimates of the kinetic constants that describe the experiment. For this analysis, the Michaelis-Menton equation is inverted such that $1/v = (K_M + [S])/(V_{max}[S])$, rearranging this we can obtain the following relationship:

$$\frac{1}{v} = \frac{K_M}{V_{max}}\left(\frac{1}{[S]}\right) + \left(\frac{1}{V_{max}}\right) \tag{3.13}$$

This gives a linear relationship between $1/v$ and $1/[S]$ with a slope of K_M/V_{max}. Thus, one plots the reciprocal of the measured velocity in the experiment versus the reciprocal of the substrate concentration that

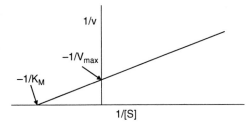

Figure 3.3 Lineweaver-Burk plot. In this plot, the reciprocal of the observed velocity is plotted against the reciprocal of the substrate concentration that gave rise to the observed velocity. Values for V_{max} and K_M can be determined from the graph as indicated.

gave rise to that velocity (Fig. 3.3). We often refer to this plot as a *double reciprocal plot*.

- The y-intercept is the reciprocal of the maximal velocity V_{max}.

- Extrapolation of the line to the x-intercept gives the negative reciprocal of the K_M value.

Thus, the experimenter can easily extract the salient features of the kinetic experiment from this plotting procedure. In addition, deviations from standard Michaelis-Menton behavior can easily be spotted, since those deviations generally lead to either downward or upward curvature of the experimental points from this straight line relationship. In fact, this latter feature is likely the most useful aspect of this plot, since quantitative determination of the kinetic constants from this plot is not robust. This is because the plot places too much weight on those data points with large experimental error (low [S] and low v) since the points at high v and high [S] are compressed near the 1/v axis and have less influence on the slope calculations.

Eadie-Hofstee or Hanes-Woolf plot.   This plotting procedure (Fig. 3.4) uses the following rearrangement of the Michaelis-Menton equation in which the experimenter would plot v versus v/[S].

$$v = V_{max} - K_M \left(\frac{v}{[S]} \right) \tag{3.14}$$

- The slope of the resulting line is equal to the negative of the K_M value.
- The y-intercept is equal to the V_{max} value.
- The x-intercept is equal to V_{max}/K_M.

The main advantage of this plot is that it does not compress or overemphasize experimental data points. Consequently, the kinetic constants obtained from this analysis are more accurate than those from the

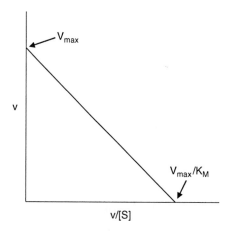

v

v/[S]

Figure 3.4 Eadie-Hofstee/Hanes-Woolf plot. In this plot, the observed velocity, v, is plotted against the value of the observed velocity divided by the substrate concentration that gave rise to the observed velocity. Values for V_{max} and K_M can be determined from the graph as indicated.

Lineweaver-Burk plotting procedure. The disadvantage of this procedure is that deviations from nonstandard kinetic behavior are not as easy to spot, or at least interpret, and the plotting procedure is slightly more cumbersome.

Example problem. For the following data, produce Lineweaver-Burk and Eadie-Hofstee plots and calculate approximate K_M and V_{max} values.

Velocity (μmol/min)	[Substrate] (mM)
1.063	0.075
1.068	0.075
1.248	0.111
1.269	0.111
1.461	0.150
1.534	0.150
2.075	0.300
1.960	0.300
2.481	0.600
2.548	0.600

Lineweaver-Burk answer: K_M = 0.138 mM; V_{max} = 2.94 μmol/min
Eadie-Hofstee answer: K_M = 0.148 mM; V_{max} = 3.05 μmol/min
Non-linear fit answer: K_M = 0.162 mM; V_{max} = 3.16 μmol/min

3.1.4 Multisubstrate systems

The kinetic expressions described above are useful only for those enzymatic systems that utilize single substrates and produce single products. However, many of the enzymes in the cell bring together more than one substrate and produce more than one product. Consequently, the kinetic expressions become more complex since more than one ES complex is formed during the reaction cycle and the reaction velocity is dependent on the concentration of both substrates. There are a myriad of substrate/product combinations; for simplicity we will look at three simple systems and go over the basic aspects of two of these systems. These systems, while generally more simplified than many actual enzymatic mechanisms, provided the basis for most of the common multisubstrate enzyme systems. Enzymes that use two substrates and produce two products are referred to as **Bi Bi systems** (bireactant, biproduct).

$$E + A + B \leftrightharpoons E + P + Q \tag{3.15}$$

As a matter of convention, the substrates are now referenced as A and B and the products are referenced as P and Q. There are three basic ways that two substrates and two products can interact with the enzyme's **active site:** *ordered-sequential, ping-pong sequential,* and *random.* These terms describe the way in which the substrates bind to and are released from the enzyme surface, as shown in Eqs. (3.16) to (3.19).

Ordered-sequential Bi Bi reaction mechanism. The ordered-sequential Bi Bi mechanism is shown in Eq. (3.16), where the substrate binding to, or releasing from, a particular enzyme species is indicated below the bidirectional arrows.

$$E \underset{[A]}{\rightleftharpoons} EA \underset{[B]}{\rightleftharpoons} EAB \rightleftharpoons EPQ \underset{[P]}{\rightleftharpoons} EQ \underset{[Q]}{\rightleftharpoons} E \tag{3.16}$$

An alternative means of presenting this enzymatic process is shown in Eq. (3.17), where the substrates are bound to and released from a surface which represents the various complexes during the reaction cycle.

$$
\begin{array}{ccccccc}
A & B & & & P & Q & \\
\updownarrow & \updownarrow & & & \updownarrow & \updownarrow & \\
E & EA & EAB & \rightleftharpoons & EPQ & EQ & E
\end{array} \tag{3.17}
$$

In an ordered-sequential mechanism, the binding of the A substrate always precedes the binding of the B substrate and product P is always released before product Q. In addition, a **ternary complex** forms in which both substrates are simultaneously bound prior to the catalytic event and both products are bound prior to the release of P and then Q. The explanation for this ordered mechanism is commonly that the binding of A influences the structure of the enzyme in some manner that makes the binding of B much more likely. This is the case in most alcohol dehydrogenases (see Clinical Box 1), where the first substrate to bind, the **coenzyme** molecule NAD^+, induces a conformational change in the enzyme that actually creates the binding site for B. In many enzymes, the binding of B to the free enzyme can happen at very high concentrations, but from a practical standpoint, these concentrations are seldom seen, so the fraction of the reaction that might go through the reverse order of substrate binding is so small that it can often be neglected.

CLINICAL BOX 3.1
Alcohol Sensitivity in East Asian Populations

The majority of beverage-derived ethanol is metabolized by two enzyme systems in the liver: alcohol dehydrogenase and aldehyde dehydrogenase.

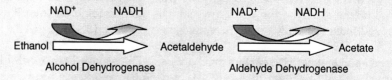

Both enzymes utilize a coenzyme, NAD^+, as the electron acceptor in order to catalyze the oxidation of their substrates. Humans possess multiple forms of both enzymes; these are termed isoenzymes. **Isoenzymes** are related gene products that catalyze similar reactions, but do so with differing affinities for substrates and/or with differing rates of catalysis. A common occurrence among individuals with East Asian ancestry (China, Japan, North and South Korea) is intolerance toward the consumption of alcoholic beverages. This ethanol intolerance is characterized by a rapid "flushing reaction" in which vasodilation of the blood vessels in the face gives rise to a reddened appearance within a few minutes of ethanol consumption. In addition to this external symptom, susceptible individuals also experience nausea, dizziness (syncope), and rapid heart rates (tachycardia) following the consumption of alcoholic beverages (Hurley, Edenberg and Li, 2002).

The primary basis for these reactions is the presence of a different form of aldehyde dehydrogenase, termed a *polymorphic variant*, in this ethnic population. The polymorphic variant of this enzyme is the result of a single nucleotide exchange in the gene that gives rise to a single amino acid change in the enzyme structure. This amino acid substitution decreases the ability of aldehyde dehydrogenase to use its coenzyme NAD^+ by 200-fold. We measure this change by reporting the concentration of NAD^+ that gives half-maximal activity—the K_M value—which, in this case, requires that we use 200 times the concentration we would use for the enzyme without this mutation.

At the same time, this mutation also decreases the rate at which the enzyme can catalyze its reaction by tenfold. As a consequence, this less active form of the enzyme has less than 10% of the activity of the more active form under the conditions that exist in the cell. The human liver cell maintains a relatively constant concentration of NAD^+ near 0.5×10^{-3} M. However, due to this amino acid change, the less active form of aldehyde dehydrogenase would require more than 100 times this concentration to reach maximal activity, which would still be tenfold lower than the more active form.

The basis for the physiological reaction to ethanol consumption is that the intracellular levels of acetaldehyde begin to rise as a consequence of the poor catalytic properties of this enzyme and spill out into the serum, where the acetaldehyde causes vasodilation, dizziness, and nausea. Ultimately, another enzyme takes over the metabolism of the acetaldehyde produced by alcohol dehydrogenase. However, the levels of acetaldehyde must accumulate before this enzyme can effectively utilize acetaldehyde as a substrate (it has a high K_M for acetaldehyde), and this accumulation of acetaldehyde is what leads to the aversive reaction to beverage ethanol consumption.

Ping-pong sequential Bi Bi reaction mechanism. In a ping-pong sequential mechanism, the substrate A is chemically converted to P, and the enzyme, E, is modified in some manner by this reaction to become F. Then substrate B binds to this modified form of the

$$E \underset{[A]}{\rightleftarrows} EA \rightleftarrows FP \underset{[P]}{\rightleftarrows} F \underset{[B]}{\rightleftarrows} FB \rightleftarrows EQ \underset{[Q]}{\rightleftarrows} E \qquad (3.18)$$

enzyme, regenerates the original form of the enzyme, E, and is converted to the product Q in this process. Classically, the enzymes that catalyze the transfer of amino groups from one amino acid to a different alpha-ketoacid (termed *transaminases*) obey ping-pong mechanisms. A transaminase utilizes the coenzyme pyridoxal phosphate to pick up the amine group from one amino acid, for instance aspartate, to generate the alpha-ketoacid oxaloacetate. **Coenzymes** are small molecules that are essential for the normal activity of the enzyme. During the transamination process, the coenzyme pyridoxal phosphate is chemically modified to pyridoxamine phosphate, which remains tightly bound to the enzyme active site. In the next step, a second alpha-ketoacid, for instance alpha-ketoglutarate, binds to this modified form of the coenzyme and picks up the amino group from pyridoxamine phosphate to regenerate the pyridoxal phosphate in the active site and produce the second product, glutamate.

Random Bi Bi reaction mechanism.

$$(3.19)$$

In a random mechanism, the enzyme still goes through a "ternary complex" as observed in the ordered-sequential situation, but the substrates can bind in either order and the products can dissociate in either order. In this case, there is no structural or chemical pressure to select for preferential binding and release of the ligands to the enzyme surface as occurs in the ordered-sequential and ping-pong sequential mechanisms.

Kinetic expressions for Bi-reactant systems

Ordered Bi Bi mechanism:

$$E \underset{k_{-1}}{\overset{k_i}{\rightleftarrows}} EA \underset{k_{-2}}{\overset{k_2}{\rightleftarrows}} EAB \underset{k_{-p}}{\overset{k_p}{\rightleftarrows}} EPQ \underset{k_{-3}}{\overset{k_3}{\rightleftarrows}} EQ \underset{k_{-4}}{\overset{k_4}{\rightleftarrows}} E \qquad (3.20)$$

For this enzymatic mechanism, the complete velocity expression, in the absence of P and Q (initial rate, steady-state assumptions), is as follows:

$$v = \frac{[A][B]V_{max}}{K_{iA}K_{MB} + K_{MB}[A] + K_{MA}[B] + [A][B]} \qquad (3.21)$$

These kinetic constants are defined in Eqs. (3.22) to (3.24) in terms of their individual rate constants; k_p and k_{-p} are the forward and reverse "product forming" rate constants.

$$K_{iA} = \frac{k_{-1}}{k_1} \text{ (true dissociation constant for substate A)} \qquad (3.22)$$

$$K_{MA} = \frac{k_3 k_4 k_p}{k_1(k_3 k_4 + k_3 k_p + k_4 k_p + k_4 k_{-p})} \qquad (3.23)$$

$$K_{MB} = \frac{k_4(k_{-2}k_3 + k_{-2}k_{-p} + k_3 k_p)}{k_2(k_3 k_4 + k_3 k_p + k_4 k_p + k_4 k_{-p})} \qquad (3.24)$$

$$V_{max} = \frac{k_3 k_4 k_p}{k_3 k_4 + k_3 k_p + k_4 k_p + k_4 k_{-p}} \qquad (3.25)$$

As you can see, the complexity of the kinetic expression increases tremendously as does the interpretation of the physical meaning of the individual kinetic constants. However, the practical descriptions for K_M, k_{cat}, and k_{cat}/K_M outlined in Sec. 3.1.2 still apply here and serve as a basis for understanding what these more complex Michaelis constants actually describe in these enzymatic systems. However, due to their increased complexity, the kinetic experiment that properly determines the kinetic constants of multisubstrate enzymes becomes two-dimensional. This means that the concentration of both substrates must be varied simultaneously. Practically, this is carried out by varying the concentration of one substrate at various constant concentrations of the second. The kinetic plots are thus a series of saturation curves, in which the apparent saturation for one substrate is plotted at various concentrations of the second substrate (Fig. 3.5). Enzymatic mechanisms that proceed through a ternary complex show a set of intersecting lines when plotted using the method of Lineweaver and Burk. This is because the apparent saturation level of one substrate influences the apparent saturation level of the other substrate and vice versa.

Ping-pong sequential mechanism:

$$E \rightleftarrows_{[A]} EA \rightleftarrows FP \rightleftarrows_{[P]} F \rightleftarrows_{[B]} FB \rightleftarrows EQ \rightleftarrows_{[Q]} E \qquad (3.26)$$

The complete velocity expression for the formation of P and Q from A and B for this enzymatic mechanism is as follows; for simplicity we have omitted the rate constants from these relationships.

$$v = \frac{[A][B]V_{max}}{K_{MB}[A] + K_{MA}[B] + [A][B]} \qquad (3.27)$$

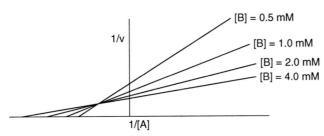

Figure 3.5 Ordered Bi Bi Lineweaver-Burk plot. The double reciprocal (Lineweaver-Burk) plot shows the results of a covary kinetic experiment for a ordered Bi Bi kinetic system. In this plot, each line represents a single fixed concentration of the second substrate, B. The reciprocal of the velocities obtained at this concentration of B and at varying concentrations of A are plotted against the reciprocal of the concentration of A that gave rise to the observed velocity.

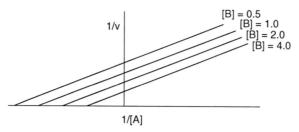

Figure 3.6   Ping pong Lineweaver-Burk plot. The double
reciprocal (Lineweaver-Burk) plot for a ping-pong reaction
mechanism results from a series of experiments with fixed
concentrations for substrate B at varying concentrations
of A.

As in the ordered mechanism, the proper way to evaluate the kinetic constants for enzymes that obey this mechanism is to vary both substrates simultaneously and to thereby obtain a set of activity versus substrate concentration curves. However, unlike enzyme mechanisms that proceed through a ternary complex, ping-pong mechanisms do not show interdependence of substrate binding. Consequently, plotting the data using the Lineweaver-Burk equation yields a set of parallel lines with constant slope values and differing x- and y-intercepts (Fig. 3.6). This is because substrate binding to the different enzyme forms (E or F) influences only the maximal velocity that can be obtained, but does not affect the apparent affinity of the different enzyme forms for the second substrate.

The reader is referred to other textbooks and references for information on how to obtain the relevant kinetic constants from these covariation experiments.

3.2   Enzyme Inhibition

From a formal perspective, an **enzyme inhibitor** is *any* substance that reduces the velocity of an enzyme-catalyzed reaction, which could include changes in temperature or pH. However, since all enzymes will respond at some level to these nonspecific changes in the environment, we are most interested in mechanisms that confer some sort of specificity to the observed change in enzyme activity. Inhibition is the *major regulatory mechanism* for the control of enzyme activity within the cell. Often the end product of a metabolic pathway regulates its own production by acting as an inhibitor on an enzyme near the beginning of a pathway. This type of metabolic regulation is often referred to as **end-product inhibition**. Inhibitors are used to investigate specificity and kinetic mechanism of enzymes (order of substrate addition, i.e., which substrate is A or B). Many of the currently prescribed pharmaceuticals are inhibitors of enzymes. For example, aspirin, acetaminophen, and ibuprofen are all inhibitors of enzymes in the inflammatory pathway.

The statin class of drugs inhibits HMG-CoA synthase in the pathway of cholesterol synthesis (see Clinical Box 3.2), and various antibiotics are inhibitors of cell wall synthesis or of bacterial protein synthesis. There are *three* basic types of inhibitors: *competitive, noncompetitive,* and *uncompetitive.* These classifications are based on which enzyme species the inhibitor binds to, relative to the substrate molecule.

CLINICAL BOX 3.2
Inhibitors of Cholesterol Biosynthesis

It is estimated that approximately 30% of the population in the United States has circulating cholesterol levels that exceed desirable levels (>200 mg/dL) (American Heart Association). Elevated cholesterol levels are a major indicator for the development of coronary artery disease and a number of factors contribute to the high prevalence of excess circulating cholesterol, including genetic disposition and lifestyle issues. While familial hypercholesterolemia is one of the most common genetic diseases in the United States (prevalence between 1/500 and 1/700), it accounts for only a small fraction of patients with elevated cholesterol. Changes in diet and exercise among patients diagnosed with elevated cholesterol levels can lead to a reduction in circulating cholesterol levels for about 10% of the population. These simple changes work because circulating cholesterol levels arise from two distinct sources—from our diet and direct synthesis by the body. If we take in less cholesterol in our diet, we can reduce the total cholesterol to which we are exposed and reduce our serum cholesterol values. In addition, regular exercise has been shown to decrease the fraction of cholesterol associated with serum proteins that are associated with the development of arteriosclerosis.

If diet and exercise are insufficient to reduce cholesterol levels to less than 200 mg/dL, additional interventions are required in order to lower the risk for coronary complications. Cholesterol is synthesized in the liver in a complex and multistep process. However, the committed step in this process is the production of mevalonate from 3-hydroxy-3-methylglutaryl-CoA by 3-hydroxy-3-methylglutaryl-CoA reductase (HMG-CoA reductase). Consequently, inhibition of this committed step would directly reduce the production of cholesterol by our liver and dramatically lower circulating cholesterol levels in conjunction with changes in diet.

A whole class of drugs has been developed over the past 20 years that targets HMG-CoA reductase as important and central enzyme in cholesterol synthesis. These drugs are generically termed *statins* (lovastatin, simvastatin, pravastatin, and so forth) and function as competitive inhibitors of HMG-CoA reductase. These are competitive inhibitors because each of the statin drug molecules has a portion that closely resembles the structure of 3-hydroxy-3-methylglutaryl-CoA and competes with substrate binding, but cannot participate in the reaction. Consequently, the inhibitors reduce the amount of mevalonate, and ultimately cholesterol, produced by the liver.

3.2.1 Competitive inhibition

A competitive inhibitor is a compound that combines with the free enzyme to form an EI complex, and in doing so directly prevents substrate binding to form the ES complex (the binding of S and I are mutually exclusive, Fig. 3.7).

For a single substrate enzyme, the kinetic mechanism is as follows:

$$
\begin{array}{c}
\mathrm{E + S \underset{K_M}{\rightleftarrows} ES \underset{k_p}{\rightleftarrows} E + P} \\
+ \\
\mathrm{I} \\
K_i \updownarrow \\
\mathrm{EI}
\end{array}
\tag{3.28}
$$

where $K_M = \dfrac{[E][S]}{[ES]}$ and $K_i = \dfrac{[E][I]}{[EI]}$

However, the initial velocity of the reaction will still be proportional to $k_p[ES]$. An infinite concentration of S will drive the equilibrium completely in favor of the productive reaction. Thus, it follows that V_{max} should be unchanged, but a higher concentration of S will be required to attain a particular enzymatic rate than in the absence of the inhibitor (Fig. 3.8).

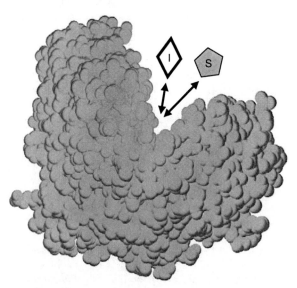

Figure 3.7 Competition between substrate and inhibitor. Shown is a graphical representation of the competition for the substrate-binding site between the inhibitor molecule (I) and the substrate molecule (S). The protein structure shown is that of human glucokinase (PDB code 1V4S *www.rcsb.org*).

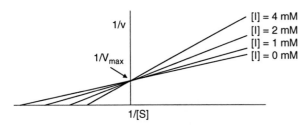

Figure 3.8 Competitive inhibition plot. The double reciprocal plot shows the effect that different fixed concentrations of a competitive inhibitor, I, have on the observed velocity of an enzymatic reaction at varying concentration of substrate, S.

The velocity expression for the interaction of a competitive inhibitor with an enzyme is as follows:

$$v = \frac{[S]V_{max}}{K_M(1 + \frac{[I]}{K_i}) + [S]} \tag{3.29}$$

Examination of this equation shows that the apparent K_M is increased by $(1 + [I]/K_i)$, without an effect on V_{max}.

It is an important principle that the K_i for competitive inhibition is equivalent to the [I] that doubles the slope of the $1/v$ versus $1/[S]$ plot, but that it is *not* equivalent to the [I] that yields 50% inhibition.

3.2.2 Noncompetitive inhibition

A *classical* noncompetitive inhibitor has no effect on substrate binding. It binds reversibly to two different enzyme species, E and ES, in the reaction pathway (Fig. 3.9). However, the ESI complex is inactive, such that I must dissociate prior to proceeding with the catalytic step.

The kinetic mechanism for the simplest system, where the dissociation constant of I from EI and ESI is identical, is shown below.

$$
\begin{array}{ccccc}
E + S & \rightleftarrows & ES & \rightleftarrows & E + P \\
 & K_M & & k_p & \\
+ & & + & & \\
I & & I & & \\
K_i \updownarrow & & \updownarrow K_i & & \\
EI + S & \rightleftarrows & ESI & & \\
 & K_M & & &
\end{array} \tag{3.30}
$$

where $K_M = \dfrac{[E][S]}{[ES]}$ and $K_i = \dfrac{[E][I]}{[EI]} = \dfrac{[ES][I]}{[ESI]}$

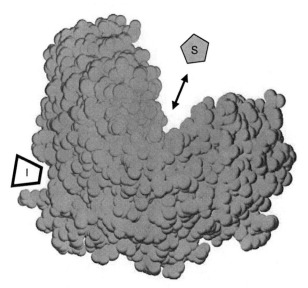

Figure 3.9 Noncompetitive inhibitor binding. Shown is a graphical representation of one possible mode of inhibitor binding in a noncompetitive system. Here the inhibitor, I, binds to a site remote from that of the substrate-binding site and by some mechanism slows the velocity for the conversion of substrate (S) to product (P). The binding site for I here is hypothetical, as there is no known regulator of glucokinase that binds at the indicated site. The figure is used here only to illustrate how noncompetitive or uncompetitive inhibitors might bind to an enzyme.

An examination of the species in this diagram shows that an infinite concentration of S will never completely drive the reaction toward product formation. Thus, the V_{max} of the reaction will be affected in the presence of I, but the K_M value will be unaffected. The complete velocity equation for this type of inhibition is as follows:

$$v = \frac{[S]V_{max}}{K_M\left(1 + \frac{[I]}{K_i}\right) + [S]\left(1 + \frac{[I]}{K_i}\right)} \tag{3.31}$$

Since both terms in the denominator are multiplied by $(1 + [I]/K_i)$, it is the same thing as expressing the numerator divided by $(1 + [I]/K_i)$; thus, V_{max} is reduced by $(1 + [I]/K_i)$ (Fig. 3.10).

Quite often, however, the dissociation constants from the EI and ESI complexes are not identical. This is termed *mixed-type noncompetitive inhibition*, because both the V_{max} and K_M are affected by the presence

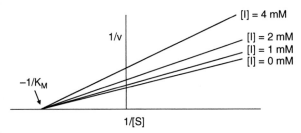

Figure 3.10 Noncompetitive inhibition plot. The double reciprocal plot shows the effects of different fixed concentrations of a noncompetitive inhibitor, I, on the observed velocity of an enzymatic reaction at varying concentrations of substrate, S.

of the inhibitor, but to different degrees. Now the form of the equation is modified as shown below.

$$v = \frac{[S]V_{max}}{K_M\left(1 + \dfrac{[I]}{K_{is}}\right) + [S]\left(1 + \dfrac{[I]}{K_{ii}}\right)} \tag{3.32}$$

where K_{is} defines the dissociation constant from the EI complex and K_{ii} is the dissociation constant from the ESI complex.

You might ask: "Why is it called 'mixed-type' noncompetitive inhibition?" The answer is that if I binds very much more tightly to E than to ES, then the equation essentially reduces to that for competitive inhibition (intercept at 1/v axis). If the binding of I to E and ES are equivalent, then it is the form for *classical* noncompetitive inhibition. Lastly, if I binds very much more tightly to ES than to E, then it is termed *uncompetitive inhibition*.

3.2.3 Uncompetitive inhibition

$$E + S \underset{K_M}{\rightleftharpoons} ES \underset{k_p}{\rightleftharpoons} E + P$$
$$+$$
$$I \tag{3.33}$$
$$K_i \updownarrow$$
$$ESI$$

where $K_M = \dfrac{[E][S]}{[ES]}$ and $K_i = \dfrac{[ES][I]}{[ESI]}$

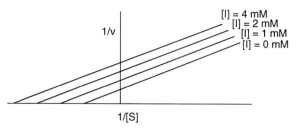

Figure 3.11   Uncompetitive inhibition plot. The double reciprocal plot shows the effects of different fixed concentrations of an uncompetitive inhibitor, I, on the observed velocity of an enzyme catalyzed reaction at varying concentrations of substrate, S.

The velocity equation for uncompetitive inhibition is found below.

$$v = \frac{[S]V_{max}}{K_M + [S]\left(1 + \dfrac{[I]}{K_i}\right)} \tag{3.34}$$

Both the K_M and V_{max} are decreased by the same amount, $(1 + [I]/K_i)$. Thus, since the slope of a line on a Lineweaver-Burk plot is equivalent to K_M/V_{max}, the slope of the lines at various [I] do not change, but their intercepts on the 1/v and 1/S axes do change, with the consequence that a series of parallel lines result (Fig. 3.11).

3.3   Cooperative Behavior in Enzymes

Many enzymes are classified as **cooperative enzymes**. A common feature of all cooperative enzymes is that they contain multiple copies of the same polypeptide chain to form the functional enzyme, or a set of different polypeptide chains combine to form the functional enzyme. A consequence of these multiple polypeptide chains is that there are *multiple active sites* that all catalyze the same reaction. Not all enzymes with multiple active sites show **cooperativity**, but all cooperative enzymes have multiple active sites. In contrast to the enzymes in the preceding section, cooperative enzymes communicate information between their active sites. This cooperativity could result in either the activation of the other active site(s) (*positive cooperativity*) or the inhibition of the other active site(s) (*negative cooperativity*). Many cooperative enzymes also contain binding sites for other ligands (*regulatory molecules*) that influence the nature or degree of cooperativity between the active sites in the enzyme complex.

The enzyme *aspartate transcarbamoylase* is a cooperative enzyme that shows allosteric regulation by components of the metabolic pathway in which it participates. This enzyme catalyzes the first step in the biosynthetic

Asp ⤨ Carbamoyl-Asp →→→→→ CTP

Figure 3.12   Feedback inhibition. The simplified diagram of the synthetic pathway for CTP shows the feedback inhibition of CTP on the first step in the pathway which is catalyzed by aspartate transcarbamoylase.

pathway for production of cytidine triphosphate (CTP) (Fig. 3.12). CTP is an essential nucleotide triphosphate for a number of metabolic reactions.

One of the more important uses of CTP is during the duplication of the cell's genetic material prior to mitosis. Consequently, prior to cell division there is a large requirement for CTP for DNA synthesis, but that requirement diminishes when DNA synthesis is not proceeding. To address this variable metabolic demand, the cell needs to sense when sufficient resources are available for CTP synthesis and when enough CTP has been synthesized for its current metabolic needs. In this regard, the enzyme is **allosterically** activated by adenosine triphosphate (ATP). Since ATP is required for the reaction catalyzed by this enzyme, a threshold level of ATP must be present in order to activate the enzyme. In contrast, the eventual product of the metabolic pathway, CTP, is an *allosteric inhibitor* of aspartate transcarbamoylase such that when sufficient levels of CTP are available the reaction is inhibited (Fig. 3.13).

These allosteric regulatory molecules bind to sites on the enzyme surface that are distinct from the active site (there is a separate ATP binding site in the active site for catalyzing the actual reaction). When molecules bind to these sites, conformational changes occur that influence the ability of the active sites to catalyze the reaction. The regulation of enzyme activity within a metabolic pathway by the pathway's ultimate product is termed **feedback inhibition**. The end product of the pathway controls the flux through the pathway, preventing too much product synthesis, which would be energetically wasteful.

Notice that the substrate saturation profiles in Fig. 3.13 are not hyperbolic. Cooperative enzymes show sigmoidal (S-shaped) saturation curves that reflect the fact that initial binding of a substrate to one active site is difficult, but its binding activates the other active sites for binding the substrates. The saturation kinetics for cooperative enzymes is governed by the following equation.

$$v = \frac{V_{max}\,[S]^n}{K_S + [S]^n} \tag{3.35}$$

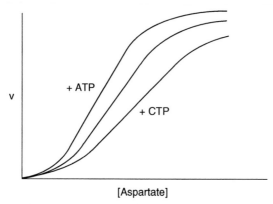

Figure 3.13 Sigmoidal substrate binding curves. The center curve is the substrate saturation curve for the enzyme aspartate transcarbamoylase. This saturation curve is shifted to the left (making the enzyme more active at any given concentration of aspartate) in the presence of the allosteric activator ATP. It is shifted to the right (making the enzyme less active at any given concentration of aspartate) in the presence of the allosteric inhibitor CTP.

The exponent, or **Hill coefficient, n**, is indicative of the type and degree of cooperativity. If $n = 1$, then standard Michaelis-Menton kinetics are observed; if $n > 1$, then the enzyme exhibits positive cooperativity; if $n < 1$, then the enzyme exhibits negative cooperativity. In order to determine the value of n in this equation, we generally perform a separate analysis of the data and fit the data obtained to the *Hill equation* [Eq. (3.35)]. A **Hill plot** is generated by plotting the logarithm of the fraction of enzyme with bound substrate divided by the fraction of enzyme free in solution versus the logarithm of the substrate concentration. This relationship produces a linear relationship in which the slope of the line in the middle of the plot is equal to n. In this equation, the quantity y is the fraction of enzyme to which substrate is bound.

$$\log \left[\frac{y}{(1 - y)}\right] = \log K_S + (n) \log [S] \qquad (3.36)$$

In the case of positive cooperativity, the maximal value for n is equal to the number of substrate binding sites in the enzyme complex, though this value does not always reach this limit. For instance, hemoglobin binds oxygen molecules cooperatively to the four binding sites present in this tetrameric protein molecule (tetramer = four copies), but its Hill coefficient is only 2.9.

3.4   Covalent Regulation of Enzyme Activity

There are two basic types of covalent control of enzyme activity: *reversible and irreversible*. An example of irreversible activation is the proteolytic cleavage of proenzymes in the digestive tract that lead to the active forms of trypsin and chymotrypsin. The activation of these proteolytic enzymes outside the cell is critical, as the presence of active enzymes within the cell could lead to the unwanted degradation of cellular components, or to the complete digestion of the contents of the secretory vesicles in which the proenzymes reside prior to secretion from the cell. Another example of irreversible inhibition of enzyme activity is the action of serpins on extracellular proteases. Proteases, in addition to their action during digestion, are essential contributors to the remodeling of the extracellular matrix and in the action of the clotting cascades. Following activation of these proteases, the duration of their activity is controlled by the presence of specific inhibitors, termed serpins (*seri*ne *p*rotease *in*hibitors), that form tight and irreversible complexes with the proteases. This complex is then rapidly cleared from the extracellular circulation and degraded.

The key reversible covalent modification of activity is phosphorylation. In eukaryotic systems, **protein kinases** utilize ATP to transfer a phosphate to the protein on the side chains of either serine, threonine, or tyrasine residues in the target protein's sequence (Fig. 3.14). The specificity of kinases determines both which amino acid is phosphorylated and which protein is phosphorylated. **Protein phosphatases** hydrolyze the phosphate moiety off of the phosphorylated protein using water to catalyze the reaction and release free phosphate groups (Fig. 3.14). Notice that if this process is not regulated, it becomes a "futile" cycle of ATP hydrolysis. Thus, the phosphorylation and dephosphorylation reactions must be highly specific and tightly controlled, and opposing activities must be coordinated by the cell. A principal way by which changes

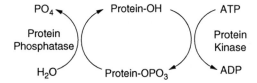

Figure 3.14   The phosphorylation cycle of proteins. Phosphorylation is a key reversible covalent modification of enzyme activity. Protein kinases utilize ATP to transfer a phosphate to the protein, while protein phosphatases are enzymes that remove phosphate groups. Addition or removal of phosphate changes protein activity, but does not necessarily correspond with enzyme activation or inhibition.

in the phosphorylation state of proteins within the cell arise is as a response to some extracellular signal, such as hormone action (see Chap. 4 on Cellular Signaling).

A good example of coordinate regulation of a pathway is the control of glycogen synthesis and its breakdown. *Glycogen synthase*, as the name implies, synthesizes glycogen from glucose, and *glycogen phosphorylase* breaks down glycogen to glucose. Phosphorylation of glycogen synthase *inactivates* the enzyme, while phosphorylation of glycogen phosphorylase *activates* the enzyme. The cell responds to the hormone epinephrine by activating protein kinases that phosphorylate both enzymes, while the hormone insulin causes the opposite effects on the phosphorylation states of these two enzymes by activating phosphatases (Fig. 3.15).

The action of these hormones is directed to different kinases and phosphatases that are specific for their target enzymes, glycogen synthase and glycogen phosphorylase, in this example. This is a common theme in the regulation of cellular processes in response to changes in the external environment. In the above example, this type of coordinate regulation of activity is crucial, otherwise the activity of phosphorylase and glycogen synthase will oppose each other and no net synthesis of glycogen or net breakdown of glycogen will occur during periods of glucose excess or glucose deficiency, respectively.

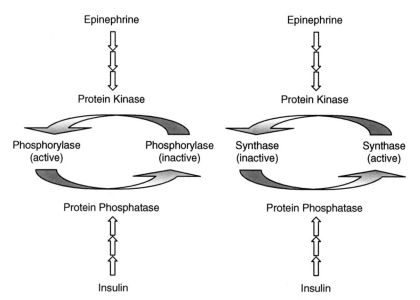

Figure 3.15 Glycogen regulation. Glycogen synthesis and glycogen breakdown are regulated by the hormones epinephrine and insulin. When glucose levels are high, the organism responds by releasing insulin which signals the activation of enzymes responsible for glucose storage. The action of epinephirine opposes that of insulin and leads to the activation of phosphorylase and inhibition of glycogen synthase.

Suggested Reading

Schulz, A.R. (1994). *Enzyme Kinetics: From diastase to multi-enzyme systems.* Cambridge University Press, NY.
Segel, I.H. (1975). *Enzyme Kinetics: Behavior and analysis of rapid equilibrium and steady-state enzyme systems.* John Wiley and Sons, Inc., NY.

References

Briggs, G.E. and Haldane, J.B.S. (1925). A note on the kinetics of enzyme action. *Biochem J.* 19:338–339.
Hanes, C.S. (1932). The effect of starch concentration upon the velocity of hydrolysis by the amylase of germinated barley. *Biochem J.* 26:1406–1421.
Henri, V. (1903). *Lois generales de l'action des diastases,* Paris, Hermann.
Hurley, T.D., Edenberg, H.J. and Li, T.-K. (2002). The pharmacogenomics of alcoholism. In: *Pharmacogenomics: The Search for Individualized Therapeutics,* (J. Licinio, M.L. Wong, eds.), Wiley-VCH, Weinheim, pp. 417–442.
Lineweaver, H. and Burk, J. (1934). The determination of enzyme dissociation constants. *J Am Chem Soc.* 56:658–666.
Michaelis, L. and Menton, M.L. (1913). Die kinetik der invertin wirkung. *Biochem Z.* 49:333–369.
The American Heart Association Fact Sheet on High Blood Cholesterol and Other Lipids-Statistics. (*www.americanheart.org*), *http://www.americanheart.org/download-able/heart/1136819539968CholesterolO6.pdf*

Cellular Signal Transduction

James P. Hughes, Ph.D.

OBJECTIVES

- To introduce receptor structure, specificity, and regulation
- To describe how ligand-receptor interaction initiates signaling
- To present representative signaling pathways used by cells

OUTLINE

4.1 Cellular Signaling   86
4.2 Receptor Binding   87
4.3 Signal Transduction via Nuclear Receptors   90
4.4 Signal Transduction via Membrane Receptors   93
4.5 Signaling in Apoptosis   101

Transducers convert energy from one form to another and signal transduction is the method by which signals are converted. The cell's molecular circuits sense, amplify, and integrate certain input signals received from its environment to generate responses such as changes in enzyme activity, gene expression, or ion-channel activity. Defects in these pathways are associated with cell dysfunction and disease. Cellular engineering aims at elucidating specific disease-related signal transduction cascades to generate novel drug targets and markers allowing for diagnosis of disease and tracking of disease progression.

Maintenance of **homeostasis** requires constant communication among cells. Messages must be sent in a timely manner, and they must be specific. Sometimes, specificity is maintained by cellular architecture. This is, for instance, the case when one neuron synapses on the next (see Chap. 6). But more often than not, specificity is determined by cellular **receptors** that recognize individual messages and translate (transduce) them into language that is meaningful to the cell. The structure of cellular receptors is introduced in Chap. 6 with emphasis on neuronal receptors. Receptor-associated recognition and translation are the complex processes that will be addressed in this chapter.

4.1   Cellular Signaling

In most cases, intercellular communication involves binding of chemical messengers to appropriate receptors. Messengers are often classified on the basis of how far they travel before reaching a target tissue (Fig. 4.1). **Autocrine** messengers act on the producing cell. An example would be *insulin-like growth factor 1 (IGF1)*, which is produced by cartilage cells and stimulates proliferation of the same chondrocytes. A **paracrine** messenger acts on neighboring cells. Neurotransmitters are classified as

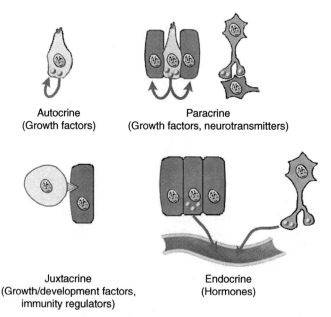

Autocrine
(Growth factors)

Paracrine
(Growth factors, neurotransmitters)

Juxtacrine
(Growth/development factors,
immunity regulators)

Endocrine
(Hormones)

Figure 4.1   Classification of chemical messengers. Chemicals can act as autocrine, paracrine, juxtacrine, or endocrine messengers, depending on how far they travel and whether they are free or cell associated. For details, see Sec. 4.1.

paracrine messengers because chemicals secreted by one neuron diffuse across a synaptic cleft to regulate the activity of a second neuron. Of course, neurotransmitters may also act in an autocrine manner by acting back on the secreting cell. A **juxtacrine** messenger acts on neighboring cells but does not diffuse from the cell producing it. Juxtacrine messengers include (1) a protein in the membrane of one cell that binds to a receptor on an adjacent cell, (2) a ligand (a molecule that binds to another substance) in the extracellular matrix secreted by one cell binding to a receptor on an adjacent cell, and (3) a signal that is transmitted directly from the cytoplasm of one cell into the cytoplasm of an adjacent cell. One example of juxtacrine signaling would be the triggering of apoptosis (see Sec. 4.5, Signaling in Apoptosis) when the *Fas ligand* on the surface of a cell, for example, a cytotoxic T lymphocyte (CTL), binds to a *Fas receptor* on an adjacent cell. Finally, an **endocrine** messenger, or hormone, is a chemical that acts at a greater distance from the site of secretion. *Estrogens*, hormones that are produced by ovarian cells, diffuse into the blood and are carried to receptors in the breast, uterus, brain, and in other distant targets. However, estrogens also act back on the secreting cells and on neighboring cells, so they could also be classified as autocrine or paracrine messengers. It is clear that classifications based on distance to the target tissue are somewhat arbitrary. Hence, classifying a messenger is far less important than understanding its interaction with a receptor.

4.2 Receptor Binding

Highly specific receptors allow cells to differentiate among closely related signaling molecules. Each receptor contains a receptor binding site, or three-dimensional binding pocket, that accommodates a **ligand** having the appropriate shape and functional groups. Upon entering the binding pocket, the ligand will form noncovalent bonds with amino-acid side chains surrounding the pocket. In many cases, the pocket is not a fixed structure. Rather, the pocket is somewhat "plastic" and tolerates some variation in ligand structure. However, critical changes in ligand structure may greatly decrease the **affinity** of the receptor for the ligand or the ability of the ligand-receptor complex to trigger a biological response (see Secs. 4.3 and 4.4 on Signal Transduction).

The equation below is a simplistic representation of the interaction between a ligand, L, and its receptor, R. The probability of achieving a critical concentration of a ligand-receptor complex [LR] depends on the concentration of the ligand [L] and the concentration of the receptor [R]. Once a critical complex concentration has been achieved, a biological response will be observed.

$$[L] + [R] \Leftrightarrow [LR] \rightarrow\rightarrow\rightarrow \text{Biological Response}$$

Receptor affinity is a measure of the avidity with which a receptor binds a ligand, just as enzyme affinity indicates the avidity of an enzyme for the substrate (see Chap. 3). Receptor affinity most often is expressed in terms of the **dissociation constant (K_d)** or the concentration of free ligand at which half the receptors are occupied. (The dissociation constant for enzymes is named K_M.) K_d values generally range from 10^{-8} to 10^{-12} Molar. It can be appreciated from the K_d values that high-affinity receptors are 50% occupied even when the ligand is present at a very low concentration. At best, however, the K_d for a receptor provides only a ballpark estimate of the concentration of ligand needed to trigger a biological response.

In some systems, activation of only a small percentage (e.g., <1%) of the total number of cellular receptors is necessary to generate a maximal biological response. When occupation of only a small fraction of the receptors is required, the K_d overestimates the concentration of ligand necessary to produce a biological response. The discovery of systems that could be maximally activated by occupancy of only a small percentage of the total receptors gave rise to the concept of *spare receptors*. This is a misleading term, because it suggests that the majority of the receptors have no real function. In truth, expression of "spare receptors" serves to increase the **sensitivity** of a cell to a particular ligand. A cell that expresses a thousand receptors is much more likely to generate the required number (e.g., 10) of ligand-receptor complexes than a cell that expresses only 20 receptors. Hence, expression of more receptors (receptor up-regulation) increases cell sensitivity, whereas receptor down-regulation decreases cell sensitivity.

Regulation of receptor expression is critical for normal cell signaling. One example is the down-regulation of *G-protein-coupled receptors* (*GPCRs*; see Sec. 4.4.1) that occurs in response to prolonged stimulation. GPCR signaling involves activation of a membrane-associated *G protein* composed of α-, β-, and γ-subunits. Upon activation, the *$\beta\gamma$ subunits of the G protein* are released and stimulate a *GPCR kinase (GRK)* that phosphorylates the receptor. GRK-*phosphorylated receptor* binds regulatory proteins called *β-arrestins* which induce down-regulation. This desensitizing mechanism is responsible in part for the tolerance that develops in response to drugs such as morphine. GPCR down-regulation in response to receptor occupancy is an example of *homologous* (i.e., by its own ligand) down-regulation. Receptor expression also can be regulated by *heterologous* ligands that act through other receptors. For example, estrogens acting through estrogen receptors (ERs) up-regulate receptors for progesterone in the endometrium of the uterus and in the breast. Conversely, progesterone acting through its own receptor, down-regulates ERs in the endometrium, preventing excessive cell proliferation. The important take-home message is that *cell signaling depends on receptor concentration as well as on ligand concentration.*

CLINICAL BOX 4.1
Partial Estrogen Agonists (Raloxifene)

Partial estrogen agonists have been used for many years in hormone replacement therapy in postmenopausal women. Raloxifene (Evista) is a **selective estrogen receptor modulator (SERM)** that is widely used in the United States because the drug shares estrogen's bone-protecting actions. Unlike estrogen, raloxifene does not increase the risk of breast or endometrial (uterine) cancer. However, raloxifene is not the perfect substitute for estrogen. Both estrogen and raloxifene increase the risk of blood clotting, and raloxifene exacerbates hot flashes.

Ligands bound by a receptor may function as agonists, antagonists, or partial agonists. **Agonists** are capable of activating the receptor and triggering a biological response. An **antagonist** is bound by the receptor, often with high affinity, but binding does not activate the receptor and may block the activity of another ligand. Clearly, K_d and biological response are completely disconnected in the case of antagonists. A **partial agonist** generates a biological response under some conditions but blocks receptor activity under different conditions. The effect exerted by a partial agonist may depend on the form of the receptor expressed, competition with other ligands, the presence of specific accessory molecules (e.g., *coactivators*), or other factors present in specific target tissues. For example, the drug tamoxifen acts as an estrogen antagonist in the breast but as an estrogen agonist in the uterus. Hence, tamoxifen can be used to treat some breast cancers, but the drug may increase the incidence of cancers of the uterine lining.

Receptor activation often involves subtle changes in receptor shape or conformation. Figure 4.2 shows changes in the binding domain of the estrogen receptor alpha (ER-α) in response to an estrogen agonist (diethylstilbestrol) and an estrogen antagonist (hydroxytamoxifen). Agonist binding shifts the position of helix 12 and exposes a hydrophobic cleft formed by helices 3, 4, and 5. The cleft recognizes an amino-acid motif (LXXLL, where X is any amino acid and L is leucine) found in several steroid-receptor coactivators. Therefore, the agonist-induced shift in helix 12 can trigger a biological response. The antagonist, though bound with high affinity by the receptor, does not induce the appropriate shift of helix 12, and thus does not trigger a biological response. The

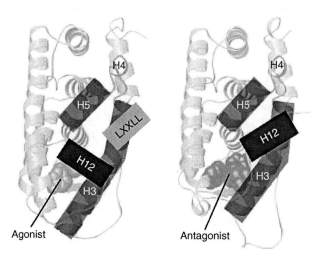

Figure 4.2   Ligand binding domain of estrogen receptor alpha (ER-α). The receptor associates with the receptor agonist diethyl-stilbestrol (left) and the receptor antagonist hydroxytamoxifen (right). H1, H2, H3, H4, H12, helices 1,2,4, and 12. LXXLL, motif for receptor coactivator. For details, see Sec. 4.2. (Modified from Fig. 2 in the publication by Shiau et. al., 1998.)

mechanism described above applies most directly to steroid receptors, but changes in conformation also underlie the signaling cascades initiated by membrane-bound receptors.

4.3   Signal Transduction via Nuclear Receptors

Nuclear receptors are transcription factors that regulate the frequency and accuracy of transcription initiation. Therefore, nuclear receptors interact with the **general transcription factors (GTFs)** and **RNA polymerase II** that assemble at the **promoter**, which is a specific DNA sequence most often found just upstream of a gene (also see Chap. 7.4). Generally, interactions between transcription factors and the GTF-polymerase complex are mediated, at least in part, by **coactivator** or **corepressor** proteins (Fig. 4.3). Coactivator proteins activate gene transcription by increasing the efficiency with which the GTF-polymerase complex assembles at promoter region of the gene. Conversely, corepressor proteins inhibit transcription.

Coactivators increase transcription in several different ways. Some coactivators alter **nucleosome** structure. The promoter region of a gene often is inaccessible to the GTF-polymerase complex because the DNA is wrapped around a core of **histone** proteins or is associated with other proteins.

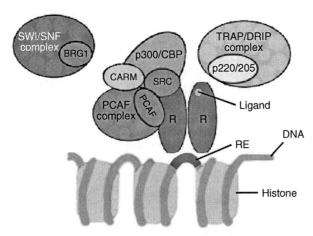

Figure 4.3   Regulation of gene transcription by ligand-activated receptors (R). Ligand-activated receptors (R) bind to specific DNA sequences (i.e., response elements, RE) and recruit various coactivator or corepressor proteins. For details, see Sec. 4.3.

One group of coactivators forms *histone-modification complexes* that can facilitate transcription by modifying histones and other chromatin proteins. For example, coactivators such as SRC (steroid-receptor coactivator), p300/CBP (CREB binding protein), and PCAF (p300/CBP-associated factor) have histone acetyltransferase (HAT) activity and add acetyl groups to lysine residues in histones (Fig. 4.3). Acetylation negates positive charges on histones and loosens their binding to DNA. Coactivators such as the coactivator-associated arginine methyltransferase (CARM) express arginine methyltransferase activity, and increase transcription by adding methyl groups to arginines in histones and in other proteins.

A second group of coactivators forms *chromatin-remodeling complexes* that use ATP to make DNA available for transcription. One example of a chromatin-remodeling complex is the switching mating types/sucrose nonfermenting (SWI/SNF) family of proteins. The actions exerted by this complex are unclear, but they may involve histone sliding or simply a loosening of the association between DNA and histones.

Finally, a third group of coactivators forms *mediator complexes that link transcription factors to the GTF-polymerase complex.* An example of a mediator complex is the TRAP/DRIP (thyroid hormone receptor-associated protein/vitamin D receptor interacting protein) complex. This mediator complex acts as an adaptor that allows transcription factors to communicate with the GTF-polymerase complex. Categorization of coactivators as histone-modifiers, chromatin-remodelers, or mediators is somewhat arbitrary because coactivators can exert multiple functions.

APPLICATION BOX 4.1
Development of the "Perfect Estrogen"

Estrogen binding and consequent signaling is complex. The nuclear forms of the estrogen receptor (ER) shown in the figure associate with DNA-binding domains (DBD) with specific regions (**response elements**, RE) of the DNA. The bound receptors then regulate the transcription of multiple genes. Regulation of transcription requires **transactivation functions** (AF) on the receptor that bind specific coactivator proteins. Estrogen also binds to cell-membrane (i.e., nonnuclear) receptors, but the signaling actions of these receptors are unclear.

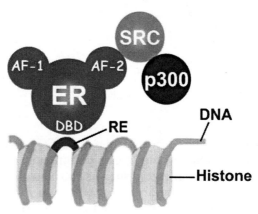

Hence, the development of the "perfect" estrogen has to take into account the existence of (1) multiple estrogen receptors (ER-α, ER-β, membrane forms), (2) multiple transactivation functions (AF-1 and AF-2) on each receptor, and (3) cell-specific coactivator proteins (i.e., SRC, p300). The ligand-binding domains of ER-α and ER-β have been well described and numerous ligands have been synthesized. Not surprisingly, receptor affinity has not always correlated well with receptor activation. Moreover, it has been difficult to develop ligands selectively recognized by ER-β. Identification of the perfect selective estrogen receptor modulators (SERMs) undoubtedly will require development of rapid screening systems. For one example of a SERM see Clinical Box 4.1. Modified two-hybrid systems can be used as screens for the effects of ligands on receptor AFs. However, cellular systems engineered to express relevant coactivator proteins and even more cumbersome bioassays will still be required to identify promising SERMs.

One such example is p300/CBP, because this histone-modifier also interacts with GTFs.

The coactivators and coactivator complexes do not assemble simultaneously on the activated receptor (transcription factor). Rather, a subset of coactivators binds transiently to the activated receptor and exerts its actions. The initial coactivator subset is then displaced by a new group of coactivators that exerts a complementary action. Sequential binding of coactivators is dictated in part by the fact that several of the coactivators bind via the same LXXLL motif (Fig. 4.2), precluding simultaneous assembly on a single receptor.

It also should be remembered that the coactivators are shared by a large number of transcription factors. Sharing allows a complex communication or cross talk between transcription factors. For instance, different transcription factors may compete for a limited pool of coactivator proteins, and thus exert mutual inhibitory actions. There are innumerable families of coactivator proteins, and expression of specific coactivators varies greatly among cell types. Hence, two cells exposed to the same nuclear receptors/transcription factors can respond very differently.

Corepressor proteins mediate inhibitory actions on transcription. For example, in the absence of thyroid hormone, the thyroid-hormone receptor binds to the corepressor molecules nuclear receptor corepressor (N-CoR) and silencing mediator of retinoid and thyroid receptors (SMRT). These corepressor proteins attract additional complexes containing histone deacetylase (HDAC) activity. The subsequent removal of acetyl groups from lysines in histones represses transcription by inducing a more compact chromatin conformation.

4.4   Signal Transduction via Membrane Receptors

Signal transduction across a cell membrane and activation of **signaling pathways** within the cell are diverse and complex processes. Indeed, even a superficial description of the many pathways is beyond the scope of this chapter. Hence, discussion primarily will be limited to receptors whose transduction mechanisms involve *guanine-nucleotide binding proteins (G proteins)*, *"direct" activation of protein kinases*, or *"direct" activation of proteases*. It will become apparent that there is considerable overlap among all signaling pathways.

4.4.1   G-protein-coupled receptors (GPCR)

The prototypical GPCR comprises seven α-helices, each of which crosses the cell membrane (i.e., a *7-transmembrane receptor*). The α-helices are connected by peptide sequences (loops) that contain binding sites for ligands and signaling molecules (Fig. 4.4). The extracellular portion of the receptor contains a ligand-binding site while the intracellular loops associate with *heterotrimeric* G proteins composed of α-, β-, and γ-subunits. G-protein subunits activated by ligand binding commonly couple the receptor to

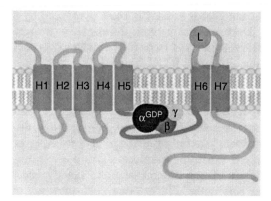

Figure 4.4   The figure shows a prototypical G-protein-coupled receptor (GPCR). GPCRs are characterized by seven transmembrane helices (H1-H7) connected by intervening loops. Extracellular loops contain a ligand-binding site (L) and intracellular loops bind a G protein composed of α, β, and γ subunits. For details, see Sec. 4.4.1.

enzymes, ion channels, or other effector molecules in the cell. Signaling through GPCRs can be quite complex, because there are many different G proteins that interact to regulate multiple pathways within a cell.

Activation of heterotrimeric G proteins primarily involves an exchange by Gα of guanosine diphosphate (GDP) for guanosine triphosphate (GTP) and dissociation of Gα from the $\beta\gamma$ subunits. This GDP-GTP "off-on" switch mechanism is encountered in many regulatory pathways that incorporate small GTP-binding proteins (proteins related to Gα) that belong to five subfamilies (i.e., Ras, Rho, Ran, Rab, and Arf). Gα remains in an activated state until **GTPase** activity intrinsic to Gα hydrolyzes the GTP to GDP. For some time, it was believed that the primary function of the $\beta\gamma$ subunits was inhibition of Gα. However, it has been shown that the $\beta\gamma$ subunits also exert important intracellular actions. One example discussed in Sec. 4.2 on Receptor Binding was that released $\beta\gamma$ subunits stimulate a GPCR kinase (GRK) that phosphorylates the receptor. The phosphorylated receptor then binds regulatory proteins called β-arrestins which induce down-regulation.

The adenylyl cyclase (AC)-cAMP pathway.   One of the first G-protein-coupled pathways to be identified was the **adenylyl cyclase (AC)-cAMP pathway** (Fig. 4.5). This pathway is activated by many G-protein-coupled receptors (GPCRs) including β-adrenergic receptors that bind epinephrine (adrenalin) and norepinephrine (noradrenalin). Binding of epinephrine to the extracellular portion of its receptor transmits a signal to the associated **G-protein** (G$_s$ or "stimulatory" G protein) that causes

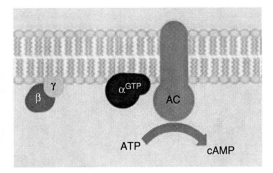

Figure 4.5   cAMP-adenylyl cyclase (AC) system. Figure shows an activated Gα (i.e., associated with GTP) bound to AC, a transmembrane enzyme. AC activated by Gα converts ATP to the "second messenger" cAMP. Released βγ subunits are free to activate additional pathways.

$G_s\alpha$ to exchange GDP for GTP. The activated G_s ($G_s\alpha$-GTP) dissociates from the βγ subunits and binds to **adenylyl cyclase (AC)**, a transmembrane enzyme. AC bound to $G_s\alpha$-GTP actively converts adenosine triphosphate (ATP) to **cyclic adenosine monophosphate (cAMP)**. Because cAMP mediates some of the actions of the receptor, it often is referred to as a *second messenger*. Although ingrained in the literature, the term second messenger is of limited value because receptors often generate multiple messengers that cannot be placed in any specific order.

The pathway leading from activation of a GPCR to generation of cAMP may seem more complex than is necessary just to create an intracellular signaling molecule. However, multistep pathways allow for *amplification* and *integration*. For example, activation of AC by $G_s\alpha$-GTP can amplify the original signal by producing numerous molecules of cAMP. Integration is achieved when other pathways modulate (1) $G_s\alpha$ activation, (2) adenylyl cyclase (AC) activation, or (3) accumulation of cAMP.

Activation of $G_s\alpha$ is limited by intrinsic GTPase activity that hydrolyzes bound GTP to GDP. However, proteins known as *regulators of G-protein signaling (RGS)* can promote GTP hydrolysis and inactivation of heterotrimeric G proteins. RGS proteins are similar to the *GTPase-activating proteins (GAPs)* (Fig. 4.6) that play important roles in regulating the activities of the closely related small G-proteins. Heterotrimeric G proteins and small G proteins also are regulated by *guanine nucleotide exchange factors (GEFs)* (Fig. 4.6) that promote exchange of GDP for GTP, and thus promote G-protein activation. The receptor generally serves as the GEF for heterotrimeric G proteins but in some cases, *activators of G-protein signaling (AGS)* proteins may stimulate GDP-GTP exchange independent of a receptor.

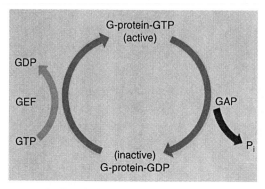

Figure 4.6 Regulation of G-protein activity. G-protein activity can be inhibited by GTPase activating proteins (GAPs) that increase hydrolysis of GTP to GDP and stimulated by guanine nucleotide exchange factors (GEFs) that promote exchange of GDP for GTP (GEFs).

Adenylyl cyclase (AC) and cAMP are also points of integration. For example, a competing pathway may activate an "inhibitory" G protein (G_i) that decreases the activity of AC or activates phosphodiesterases that degrade cAMP to 5'-adenosine monophosphate (5'-AMP). Hence, every step within a single signaling pathway represents a highly controlled intersection in a complex cellular road map.

Cellular signaling by cAMP occurs in large part through activation of **protein kinase A (PKA)**, a serine-threonine kinase. More recently, it has been shown that cAMP also activates PKA-independent pathways through the exchange protein directly activated by cAMP (Epac). Epac is a GEF for the small GTPases Rap1 and Rap2 (Ras-proximate-1, 2), small GTP-binding switch proteins that belong to the Ras family. This latter pathway is mentioned to illustrate the complexity of cellular signaling, but not discussed in detail here.

Cyclic AMP activates PKA by binding to the regulatory subunit (R) of the enzyme and releasing the catalytic subunit (C) (Fig. 4.7). The catalytic subunit is then capable of transferring a phosphate group from ATP to serine and threonine residues in specific proteins. Binding of a negatively charged phosphate can alter the conformation and activity of the protein.

Many proteins serve as targets for activated PKA (Fig. 4.8). One example is the *cAMP response-element binding protein (CREB)*. Upon phosphorylation, CREB translocates to the nucleus and binds to DNA at a specific sequence of bases known as the *cAMP response element (CRE)*. CREB then regulates gene transcription by interacting with various coactivators and/or corepressors in a manner similar to the scenario described earlier for the estrogen receptor. Many other proteins also

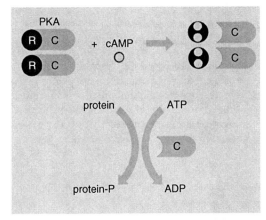

Figure 4.7 Activation of PKA by cAMP. Two molecules of cAMP bind to each of the regulatory subunits (R) of PKA and release the active catalytic subunits (C). The catalytic subunits convert ATP to cAMP.

are phosphorylated by PKA, including enzymes, G-proteins, ion channels, cytoskeletal proteins, and receptors.

Phosphorylation often alters protein function. This was also explained as important covalent modification of enzyme activity (see Sec. 3.4). One of the first PKA-regulated pathways identified was the pathway leading to glycogen breakdown. In this pathway, PKA phosphorylates the enzyme phosphorylase kinase, which in turn phosphorylates the enzyme phosphorylase. Activated (phosphorylated) phosphorylase promotes breakdown of glycogen to glucose. This enzyme cascade allows for signal amplification and provides points for signal integration. Hence, triggering the cAMP-PKA pathway can alter many cytoplasmic and nuclear events within a cell.

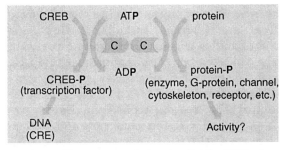

Figure 4.8 The active catalytic subunit (C) of PKA phosphorylates CREB and many other proteins in the cell. Phosphorylation can alter the activity of the proteins, and thus affect cell function.

CLINICAL BOX 4.2
Marijuana Antagonist Blocks "Munchies"

Agonist, antagonists, and partial agonists developed for G-protein-coupled receptors (GPCRs) have numerous clinical applications. Common examples associated with just the sympathetic nervous system include β_1-adrenergic receptor antagonists (e.g., atenolol) used to slow heart rate and lower blood pressure, β_2-adrenergic receptor agonists (e.g., albuterol) used as anti-asthmatics, and α_1-adrenergic antagonists (e.g., prazosin) used to lower blood pressure.

Recently, researchers have described cannabinoid (CB) receptors, which are the GPCRs that recognize tetrahydrocannabinol found in marijuana. The CB1 receptor is found in areas of the brain that control appetite and addiction to smoking (tobacco), and this receptor may be responsible for the "munchies" experienced by marijuana smokers. The munchie connection led researchers to develop CB1-receptor antagonists in hopes that they would be useful as aids for weight loss. One antagonist (rimonabant from Sanofi-Aventis) may soon be available as a diet pill and as an antismoking aid.

The phospholipase C (PLC) pathway.   G-protein-coupled receptors (GPCRs) are coupled to G-proteins from several families, including G_s, G_i, G_q, and $G_{12/13}$. Each of these G-protein families triggers different pathways. For example, binding to Gq can lead to activation of **phospholipase C (PLC)**, an enzyme that increases cytosolic calcium through cleavage of the membrane phospholipid 4,5-bisphosphatidylinositol (Fig. 4.9). Cleavage of the lipid leads to release of **inositol trisphosphate (IP$_3$)** and **diacylglycerol (DAG)**. DAG remains associated with the membrane and is capable of activating **protein kinase C (PKC)**. IP$_3$ moves into the cytosol and binds to specific receptors on the endoplasmic reticulum (ER). Binding of IP$_3$ opens a calcium channel, and the released calcium binds to **calmodulin**, a ubiquitous calcium sensor. Calmodulin undergoes a calcium-induced change in conformation that affects signaling in many

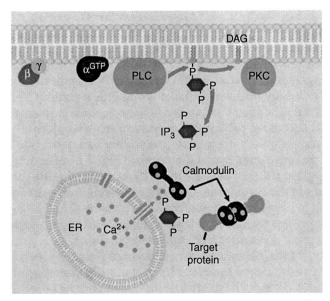

Figure 4.9 Activation of the phosphatidylinositol (PI) signaling system by a G protein. $G_q\alpha$ activates PLC, an enzyme that can cleave the membrane lipid 4,5-bisphosphatidylinositol. Cleavage of the lipid releases two "second messengers," inositol trisphosphate (IP$_3$) and diacylglycerol (DAG). IP$_3$ releases calcium from the endoplasmic reticulum (ER) and triggers calmodulin-regulated pathways. DAG activates protein kinase C (PKC). For details, see Sec. 4.4.1, The phospholipase C (PLC) pathway.

pathways. Thus, activation of G_q initiates two separate signaling pathways that further diverge and converge to integrate cellular events.

As in the case of the cAMP pathway, every step in the signaling pathway involving PLC represents a highly controlled intersection in a complex cellular road map. Because calcium is a critical mediator in this pathway, lowering cytosolic calcium is one of the primary mechanisms for inhibiting signaling through the pathway. Cytosolic calcium is decreased primarily by sequestering it in the ER or by transporting it out of the cell. Sequestering and transport involve energy-requiring calcium "pumps" or **calcium ATPases**. Under resting conditions, pumps maintain cytosolic calcium at a concentration of 0.1 μM which is approximately 1/10,000 of that in the extracellular fluid.

4.4.2 Protein-kinase-associated receptors

Protein-kinase-associated receptors may contain intrinsic kinase activity or the receptors associated directly with a kinase. By definition, **protein kinases** transfer phosphate groups from ATP to specific amino-acid residues in proteins. **Tyrosine kinases (TKs)** phosphorylate tyrosine residues while other kinases (e.g., PKA) primarily target serine and threonine residues. Many receptors ultimately activate protein kinases, including G-protein-coupled receptors (GPCRs) that indirectly activate kinases such as PKA and PKC.

One of the most extensively studied TK-associated receptor is the insulin receptor. The many pathways linked to insulin signaling are beyond the scope of this chapter, but two well-characterized pathways illustrate important aspects of protein-kinase-associated receptors.

The insulin receptor is a heterotetramer comprising two extracellular α-subunits and two transmembrane β-subunits (Fig. 4.10). Binding of insulin to the α-subunits activates *TK activity* associated with the intracellular portions of the β-subunits. Cross-phosphorylation between the β-subunits leads to *phosphorylation of multiple tyrosine (Y) residues* in the insulin receptor. The β-subunits also phosphorylate the **adaptor protein** *insulin-receptor substrate (IRS)* and other proteins.

Phosphorylated tyrosines (phosphotyrosines) serve as docking sites for proteins that bear an amino-acid sequence known as a *Src homology 2 (SH2) domain*. In Fig. 4.10, growth factor receptor-bound protein 2 (Grb2)

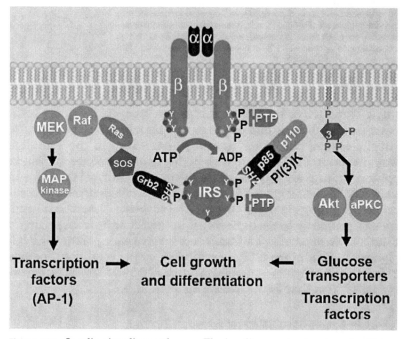

Figure 4.10 Insulin signaling pathways. The insulin receptor is a tyrosine kinase that triggers multiple pathways that lead to glucose uptake, and cell growth and differentiation. Tyrosine phosphorylation (Y-P) of insulin receptor substrate (IRS) by the receptor leads to activation of several pathways. For details, see Sec. 4.4.2.

and phosphatidylinositide 3-kinase [PI(3)K] have been recruited via their SH2 domains to phosphotyrosines on IRS. The pathways initiated by these and other (e.g., **protein tyrosine phosphatases, PTP**) *SH2-containing proteins* diverge and converge to form a highly integrated network.

The pathway initiated by Grb2 contains multiple steps, each of which can serve as a point of amplification and/or integration. *Grb2* is an adaptor (linker) protein that binds via Src homology 3 (SH3) domains to proteins containing proline-rich regions, such as *son of sevenless (SOS)*. SOS is a GEF that activates the small G-protein *Ras*, which in turn recruits the serine/threonine kinase *Raf*. Raf phosphorylates and activates *MAP/ERK (MEK) kinase*, a dual-specificity kinase that phosphorylates tyrosine and threonine residues (Fig. 4.10). MEK in turn activates *mitogen-activated protein (MAP) kinases* and other enzymes that regulate cell growth and differentiation.

The p110 catalytic subunit of PI(3)K catalyzes phosphorylation of phosphatidylinositol (PI) at the 3 position of the inositol ring. Activation of *PI(3)K* results in *multi-phosphorylated phosphatidylinositols* (e.g., PI-3,4,5-P3) that serve as ligands for proteins containing *pleckstrin homology (PH) domains* and other motifs (Fig. 4.10). One protein that binds to PI(3)K-phosphorylated PIs via a PH domain is the serine-threonine kinase *Akt (protein kinase B)*. Akt is activated by binding to the membrane lipid and as a result of phosphorylation by nearby membrane-associated kinases. The activated Akt regulates cell metabolism (e.g., glucose metabolism), cell survival (e.g., inhibition of apoptosis), and cell growth. Phosphorylated PIs also activate *atypical protein kinase C (aPKC)* which also mediates some of the effects of insulin on glucose metabolism and other cellular activities.

Tyrosine kinase signaling provides yet another example of the complexity of cell signaling. In just the cursory description provided above, it is possible to appreciate that activation of multiple, multistep pathways provides incalculable opportunities for interaction and integration.

4.5 Signaling in Apoptosis

Apoptosis or programmed cell death is a normal process in which a cell is disassembled in a highly controlled manner. Intact pieces of the cell or *apoptotic bodies* are removed by macrophages without triggering an inflammatory response. In contrast to apoptosis, *necrosis* involves uncontrolled cell rupture and inflammation. Apoptosis is important for eliminating unwanted cells (e.g., cells forming webbing between our fingers during development) or dangerous cells (e.g., cells that harbor pathogens).

Clearly, apoptosis must be tightly regulated to prevent inappropriate cell death. Apoptosis can be triggered by extrinsic signals from outside the cell (e.g., Fas ligand on a cytotoxic T lymphocyte, CTL), or by intrinsic signals (e.g., DNA damage, high cytosolic Ca^{2+}, oxidative stress, or mitochondrial damage). Pathways triggered by both types of signals result in activation of **caspases** which are aspartate-specific proteases.

The extrinsic apoptotic pathway can be triggered by binding of the Fas ligand on the surface of a CTL to a Fas receptor as juxtacrine messengers (see Sec. 4.1 on Cellular Signaling) (Fig. 4.11). Fas receptors are the best characterized members of the **death receptor** family, which belong to the tumor necrosis factor (TNF) superfamily.

Activation of Fas receptors causes recruitment of adaptors known as *Fas-associated death domains (FADD)* and the subsequent binding of *procaspase 8*. Clustering of procaspase 8 (8' in Fig. 4.11) leads to autocleavage (i.e., proteolysis) and formation of an active caspase composed of a tetramer of two large and two small subunits. Hence, signaling in this system primarily is triggered by proteolysis. Caspase 8 is referred to as an *initiator caspase* because it will activate *effector* or *executioner caspases* that are primarily responsible for cleaving most apoptotic substrates. One effector caspase activated by caspase 8 is

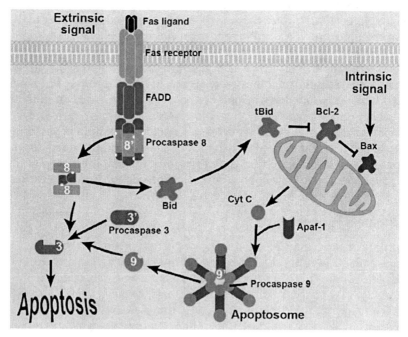

Figure 4.11   Extrinsic and intrinsic activation of apoptotic pathways. A variety of internal and external signals activate caspases (3, 8, 9) that cleave substrates and cause apoptosis. For details, see Sec. 4.5.

caspase 3. This latter caspase is directly involved in events leading to apoptosis.

The intrinsic pathway leading to apoptosis involves the *Bcl-2 family* of proteins. The Bcl-2 family includes proapoptotic proteins (e.g., Bid and Bax) and antiapoptotic proteins (e.g., Bcl-2, $Bclx_L$). An intrinsic apoptotic signal such as DNA damage can cause Bax to translocate from the cytosol to the outer mitochondrial membrane (Fig. 4.11). Insertion of Bax into the membrane promotes release into the cytosol of *cytochrome c (cyt c)* and other mitochondrial proteins. The proapoptotic actions of Bax are inhibited by antiapoptotic proteins such as Bcl-2.

Cytochrome c released from the mitochondrion can associate with the *adaptor protein apoptosis-activating-factor-1 (Apaf-1)* forming a complex **(apoptosome)** that binds procaspase 9 (Fig. 4.11). *Procaspase 9* is activated by its association with the apoptosome and is then capable of activating effector caspases such as caspase 3.

Once activated, caspases can cleave over 100 substrates to cause cell death and promote apoptosis. Targets of degradation include nuclear lamins, DNA repair enzymes, and cytoskeletal proteins. Caspases can also activate other enzymes such as DNAases that produce the characteristic "ladder" of DNA fragments observed using electrophoretic analysis.

As might be expected, programmed cell death is a highly regulated process which involves coordination of the intrinsic and extrinsic pathways of apoptosis and integration with *survival pathways.* One intersection between the intrinsic and extrinsic pathways is activation of the proapoptotic protein Bid by caspase 8. Bid is cleaved to form the active or truncated Bid (tBid) that can interfere with antiapoptotic Bcl-2 proteins and thus promote apoptosis via the intrinsic pathway. Apoptosis can be interrupted by survival factors such as insulin. For example, activation of the protein kinase Akt by insulin can lead to phosphorylation of the proapoptotic proteins Bad and procaspase 9. Phosphorylation of Bad inactivates the protein by sequestration and phosphorylation of procaspase 9 inhibits activation. Hence, as with all signaling processes, regulation of apoptosis is tightly controlled through complex interactions among numerous pathways.

References

Bukhtiar, H. S. and Catt, K. J. (2004). GPCR-mediated transactivation of RTKs in the CNS: Mechanisms and consequences. *Trends Neurosci.* 27: 48–53.

Collingwood, T. N., Urnov, F. D. and Wolffe, A. P. (1999). Nuclear receptors: Coactivators, corepressors and chromatin remodeling in the control of transcription. *J. Mol. Endocrinol.* 23: 255–275.

Hubbard, S. R. and Till, J. H. (2000). Protein tyrosine kinase structure and function. *Ann. Rev. Biochem.* 69: 373–398.

Ji, T. H., Grossmann, M. and Ji, I. (1998). G Protein-coupled Receptors I. Diversity of receptor-ligand interactions. *J. Biol. Chem.* 273: 17299–17302.

Katzenellenbogen, B. S. and Katzenellenbogen, J. A. (2000). Estrogen receptor transcription and transactivation: Estrogen receptor alpha and estrogen receptor beta: regulation by selective estrogen receptor modulators and importance in breast cancer. *Breast Cancer Res.* 2: 335–344.

Katzenellenbogen, J. A., Muthyala, R. and Katzenellenbogen, B. S. (2003). Nature of the ligand-binding pocket of estrogen receptor α and β: The search for subtype-selective ligands and implications for the prediction of estrogenic activity. *Pure Appl. Chem.* 75: 2397–2403.

Lefkowitz, R. J. (1998). G Protein-coupled Receptors III. New roles for receptor kinases and β-arrestins in receptor signaling and desensitization. *J. Biol. Chem.* 273: 18677–18680.

Saltiel, A. R. and Kahn, C. R. (2001). Insulin signaling and the regulation of glucose and lipid metabolism. *Nature.* 414: 799–806.

Shiau, A. K., Barstad, D., Loria, P. M., Cheng, L., Kushner, P. J., Agard, D. A. and Greene, G. L. (1998). The structural basis of estrogen receptor/coactivator recognition and the antagonism of this interaction by tamoxifen. *Cell.* 95: 927–937.

Energy Conversion

James P. Hughes, Ph.D.

OBJECTIVES

- To introduce cellular metabolism and the role of ATP as energy currency
- To describe anaerobic and aerobic cellular respiration
- To present the conversion of light energy to chemical energy in photosynthesis
- To illustrate carbohydrate synthesis in different plants

OUTLINE

5.1 Metabolism and ATP   106
5.2 Anaerobic Cellular Respiration   107
5.3 Aerobic Cellular Respiration   114
5.4 Photosynthesis   126
5.5 Carbohydrate Synthesis   136

The conversion of energy is a feature of life, where photosynthesis and cellular respiration are complementary processes. Plants use the energy of sunlight via photosynthesis to produce biomolecules, which in turn are converted into energy via cellular respiration by all organisms. Understanding of how biological systems produce and utilize energy is helpful for development of novel power and transportation systems and other biology-based technologies that can be applied in medicine and environmental protection plans.

On a sunny day, your brain sends electric signals to your facial muscles with the result that they contract and make you smile. Your facial muscles can contract because they contain chemical energy, which they now convert into mechanical work and heat. The energy of the muscle is derived from the energy of your food, either a plant or an animal that had eaten a plant. The plant has received its energy by converting solar energy into chemical energy in form of carbohydrate molecules. Life, as we know it, is a constant transfer of energy from one system to another. It is the objective of this chapter to introduce the energy transfer in animal and plant cells.

5.1   Metabolism and ATP

Cellular metabolism is simply the sum of all the chemical reactions in the cell. The reactions are parts of highly regulated metabolic pathways that form the basis of all cellular activities. Metabolic pathways can be separated into two basic types, **anabolism** and **catabolism**. Anabolic pathways are energy-requiring (**endergonic**) pathways that result in synthesis of larger molecules from smaller ones (e.g., amino acids to proteins). Catabolic pathways are energy-releasing (**exergonic**) pathways that break down molecules into smaller components (e.g., glucose to pyruvate).

$$\text{Glucose} + 6O_2 \rightarrow 6CO_2 + 6H_2O + \text{energy}$$

$$\Delta G^{\circ\prime} = -686 \text{ kcal/mol}$$

(5.1)

The catabolic process in the cell, in which molecules are oxidized to carbon dioxide and water, is termed **cellular respiration**. The overall reaction for oxidation of glucose is shown above in Eq. (5.1). Cellular respiration includes **anaerobic** pathways (i.e., **glycolysis** and **fermentation**) that do not require molecular oxygen and **aerobic** pathways (i.e., **tricarboxylic acid cycle** and the **electron transport system**) that directly or indirectly require molecular oxygen.

Energy released from cellular respiration has to be conserved in some form that is useful to drive energy-requiring processes in the cell. In other words, the cell must generate some energy "dollar" that is accepted by most processes. In most cases, the energy dollar is adenosine triphosphate (ATP), a molecule consisting of three phosphates attached to the ribose of adenosine (Fig. 5.1). **Hydrolysis** is the process in which a molecule is split by reacting with a molecule of water. Hydrolysis of ATP at the bond between the γ and β (or β and α) phosphates is an exergonic reaction with a **standard free energy** change ($\Delta G^{\circ\prime}$) of -7.3 kcal/mol. This is the energy which is available to do work. For details, see Sec. 1.1. Cellular

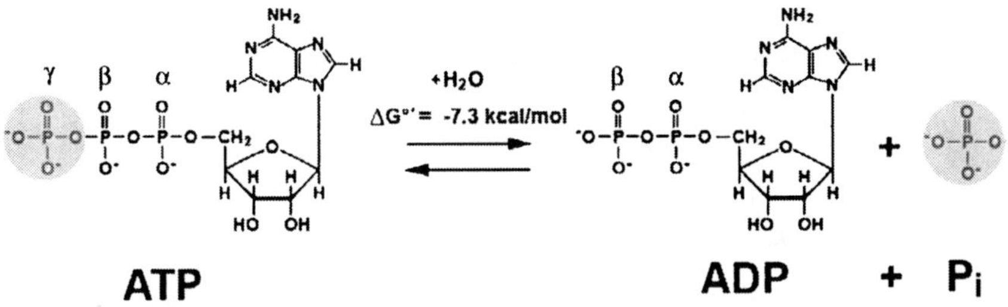

ATP   ADP + P$_i$

Figure 5.1 ATP hydrolysis. Hydrolysis of the γ phosphate from adenosine triphosphate (ATP) to produce adenosine diphosphate (ADP) and inorganic phosphate (Pi) is a highly exergonic reaction ($\Delta G^{\circ\prime} = -7.3$ kcal/mol) often used to drive endergonic reactions.

reactions do not occur under standard conditions; thus, the actual amount of energy released is likely greater than the calculated $\Delta G^{\circ\prime}$. In any case, it is clear that hydrolysis of adenosine triphosphate (ATP) releases a substantial packet of energy that can be used to drive endergonic reactions. To maintain usable energy, the cell must continually resynthesize ATP from adenosine diphosphate (ADP) and inorganic phosphate (Pi). Synthesis of ATP is an energy-requiring process ($\Delta G^{\circ\prime} = +7.3$ kcal/mol) that is fed by the catabolic reactions that comprise cellular respiration. Hence, ATP acts as an intermediary that transfers energy from cellular respiration to the energy-requiring processes in the cell.

5.2 Anaerobic Cellular Respiration

5.2.1 Glycolysis

Glycolysis is a series of cytosolic reactions that converts one molecule of glucose to two 3-carbon pyruvate molecules:

$$\text{Glucose} + 2\text{ADP} + 2\text{Pi} + 2\text{NAD}^+ \rightarrow 2\text{Pyruvate}$$

$$+ 2\text{ATP} + 2\text{NADH} + 2\text{H}^+ + 2\text{H}_2\text{O}$$

(5.2)

In the process, usable energy is stored in two ATP molecules and in two pairs of high-energy electrons passed to **nicotinamide adenine dinucleotide (NAD$^+$)**. NAD$^+$ is a **coenzyme**, which is a small molecule essential for the normal activity of an enzyme (see Chap. 3). Two ATP represent only a small percentage of the total energy that can be obtained from the glucose molecule (i.e., 686 kcal/mol). Nonetheless, glycolysis is a pivotal pathway in cellular respiration because it is the initial pathway for oxidation of glucose, and it generates energy under anaerobic conditions. Anaerobic metabolism can be critically important in a tissue such as skeletal muscle in which oxygen availability may be limited during intense exercise.

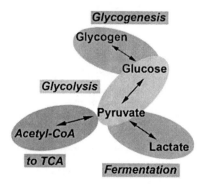

Figure 5.2 Glycolysis and gluconeogenesis overview. Glycolysis is conversion of glucose to pyruvate while the reverse pathway that creates new glucose is gluconeogenesis. Under anaerobic conditions, pyruvate accepts electrons from NADH and is converted to lactate as part of a process called lactate fermentation. If oxygen is available, pyruvate is oxidized to acetyl-CoA and enters the tricarboxylic acid cycle. Glucose not immediately needed for energy is stored in muscle and liver as glycogen.

Molecules produced during **glycolysis**, or the "reverse" process of gluconeogenesis (see Sec. 5.2.3), can enter several different pathways (Fig. 5.2). Pyruvate, the end product of glycolysis, can enter the anaerobic process of fermentation by accepting electrons from NADH, or it can be converted to *acetyl coenzyme A (acetyl-CoA)* and enter aerobic respiration. The fate of pyruvate will be determined by the availability of oxygen. Pyruvate and other substrates also can be converted back to glucose via gluconeogenic pathways that will be discussed in Sec. 5.2.3. Glucose produced by this "reversal" of glycolysis can be released into the blood by tissues such as the liver or stored as glycogen.

Glycolysis is a 10-step pathway (Fig. 5.3) that is tightly controlled by **allosteric** regulators that often are products of the metabolic pathways, and by hormones. Allosteric regulators affect enzyme activity by binding to portions of the enzyme outside the catalytic site (see Sec. 3.3). A few of the most tightly regulated enzymes will be noted in the following discussion. Many of them are **kinases**, which are enzymes that transfer a phosphate group from ATP to the target molecule.

Phosphorylation of the 6-carbon glucose to glucose-6-phosphate by *hexokinase (HK)* (*glucokinase* in liver) initiates glycolysis, and it traps glucose in the cell because phosphorylated glucose does not bind readily to glucose transporters in the cell membrane. *Glucose-6-phosphate* is converted to *fructose-6-phosphate* in step 2. The reaction results in a

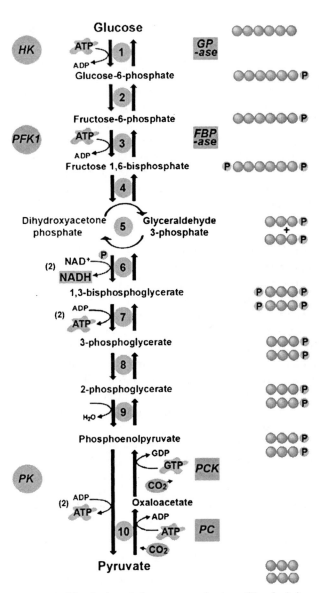

Figure 5.3 Glycolysis and gluconeogenesis steps. Glycolysis is a 10-step pathway that converts glucose to pyruvate. In gluconeogenesis, glucose is generated from substrates (e.g., pyruvate) by reversal of some of the glycolytic reactions and through the use of separate reactions. For details, see Sec. 5.2.3.

hydroxyl on carbon 1 that is phosphorylated in step 3, using another molecule of ATP. *Phosphofructokinase-1 (PFK1)*, which catalyzes step 3, is a key glycolytic enzyme. It is tightly regulated by allosteric factors and hormones and will be discussed in more detail in Sec. 5.2.4. Through step 3 of glycolysis, the cell actually loses usable energy because two molecules of ATP have been consumed to drive early reactions.

Fructose 1,6-bisphosphate is cleaved, yielding the two 3-carbon sugars *dihydroxyacetone phosphate* and *glyceraldehyde 3-phosphate* in step 4 (Fig. 5.3). For glycolysis, dihydroxyacetone phosphate is converted to glyceraldehyde 3-phosphate by a reversible reaction in step 5. In step 6, the two glyceraldehyde 3-phosphate molecules are oxidized to *1, 3-bisphosphoglycerates* when they donate two pairs of electrons to two NAD^+ to form two NADH. If oxygen is available, the NAD^+ needed to continue glycolysis will be regenerated when the electrons carried by NADH are passed across the mitochondrial membrane and enter the electron transport chain. However, in conditions of limiting oxygen, NAD^+ must be regenerated through fermentation as will be discussed later.

In step 7, two ATP are produced when phosphates are transferred from two 1,3-bisphosphoglycerate molecules to two ADP (Fig. 5.3). The direct transfer of a phosphate to ADP from an organic substrate is called **substrate-level phosphorylation**. As will be discussed later, most ATP is produced by **oxidative phosphorylation** that is coupled to electron transport in the mitochondrion (see Sec. 5.3.3). The net result of step 7 is generation of two molecules of ATP and two molecules of 3-*phosphoglycerate*. Hence, glycolysis has paid back the two ATP molecules consumed in earlier reactions.

ATP is again generated at the end of a series of reactions that converts two 3-phosphoglycerate molecules to two pyruvates (steps 8 to 10). First, in step 8, the phosphate in 3-phosphoglycerate is moved to the second carbon to produce 2-*phosphoglycerate*. In step 9, water is removed from 2-phosphoglycerate to produce *phosphoenolpyruvate*, a molecule that contains a high-energy phosphate bond. The enzyme *pyruvate kinase (PK)* catalyses transfer of the phosphates from the two phosphoenolpyruvates to two ADP forming two ATP and two molecules of *pyruvate* (step 10).

The two ATP molecules produced in step 10 constitute the two net ATP generated during glycolysis.

Step 10 is highly exergonic ($\Delta G^{\circ\prime} = -7.5$ kcal/mol) and in effect is irreversible in the cell. Accordingly, conversion of pyruvate back to phosphoenolpyruvate (and then to glucose) in gluconeogenesis requires two reactions to circumvent the substantial energy drop (see Sec. 5.2.3). This final irreversible step is critical in glycolysis; thus, it is not surprising that PK activity, like PFK1 activity, is tightly controlled by allosteric regulators and by hormones.

5.2.2 Fermentation

As indicated above, step 6 in glycolysis requires NAD^+ as an electron acceptor (Fig. 5.3). If oxygen is available in animals, NAD^+ is continually regenerated from NADH as the latter passes electrons, at least indirectly, to the electron transport chain in the mitochondrion. In the absence of oxygen, NAD^+ is regenerated when NADH passes electrons to pyruvate, forming lactate (Fig. 5.4). This process is called lactate **fermentation**. Muscles commonly use lactate fermentation when the demands of strenuous exercise exceed the oxygen supply. In such cases, lactate accumulates in the muscle and then passes to the liver via the blood. In the liver, the lactate can be converted back to glucose via gluconeogenesis and either stored as glycogen or released into the blood. Fermentation is by no means limited to muscle cells, because many cells produce lactate under anaerobic conditions.

Plant and yeast cells also use fermentation under anaerobic conditions but in these cells, the electrons are not passed to pyruvate. Pyruvate is first decarboxylated to acetaldehyde, which is then reduced by electrons from NADH to form ethanol (Fig. 5.4). Because the process yields ethanol, it is called *alcoholic fermentation*. Many industries, including those that produce wine and beer, rely on alcoholic fermentation in yeast.

5.2.3 Gluconeogenesis

In keeping with its name, gluconeogenesis is simply the creation of new glucose. Synthesis of new glucose requires "reversing" the reactions in the glycolytic pathway (Fig. 5.3) and obtaining the substrates needed to create the new glucose. Lactate and pyruvate are common glucose substrates, but amino acids and other sugars such as glycerol are also important.

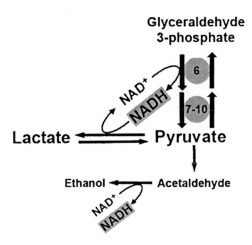

Figure 5.4 Fermentation. Under anaerobic conditions, the NAD^+ required for glycolysis must be regenerated by fermentation. In lactate fermentation (e.g., in animals), NAD^+ is regenerated when NADH transfers electrons to pyruvate to form lactate. In alcoholic fermentation (e.g., in yeast), pyruvate is decarboxylated to acetaldehyde and the latter molecule accepts electrons from NADH. Steps 6,7-10 refer to the 10-step glycolysis pathway as pictured in Fig. 5.3.

Not all reactions in glycolysis can simply be reversed using the same enzyme. Steps 2 and 4 to 9 of Fig. 5.3 are reversible, but steps 1, 3, and particularly 10 require alternate pathways. Step 10 is one of the most difficult to reverse, because it is highly exergonic. The forward reaction is catalyzed by PK whereas the reverse is a two-reaction sequence catalyzed by *pyruvate carboxylase (PC)* and *phosphoenolpyruvate carboxykinase (PCK)* (Fig. 5.3). PC adds carbon dioxide to a 3-carbon pyruvate to form a 4-carbon oxaloacetate. The energy needed to drive the reaction is derived from ATP. PCK then decarboxylates oxaloacetate and adds a phosphate from guanosine triphosphate (GTP) to form phosphoenolpyruvate. In step 3, PFK1 catalyzes the forward reaction whereas the reverse reaction is controlled by *fructose 1,6-bisphosphatase (FBPase*, Fig. 5.3). In step 1, HK catalyzes the forward reaction but *glucose 6-phosphatase (GPase)* catalyzes the reverse reaction (Fig. 5.3). Skeletal muscle expresses little GPase activity; thus, this tissue is unable to dephosphorylate glucose and release it into the blood. Glucose that enters skeletal muscle must be stored as glycogen, used for energy, or released back to the blood as lactate. The enzymes controlling steps 1, 3, and 10 play key roles in glycolysis and gluconeogenesis, and thus are tightly regulated by allosteric factors and hormones.

5.2.4   Regulation of anaerobic respiration

PFK1 and the opposing FBPase are two of the most tightly regulated enzymes in glycolysis and gluconeogenesis. Because PFK1 is a key enzyme in increasing energy production, its activity is inhibited allosterically by ATP and citrate, factors that reflect an abundance of energy in the cell. Conversely, PFK1 activity is increased by ADP and AMP, which are derived from dephosphorylation of ATP and indicate low cellular energy. AMP inhibits FBPase, and thus inhibits gluconeogenesis while increasing glycolysis.

A major activator of PFK1 and inhibitor of FBPase is *fructose 2,6-bisphosphate (F 2,6-P)*. Production of F 2,6-P in the cell is controlled by a single enzyme that can express kinase (*phosphofructokinase-2, PFK2*) or phosphatase (*fructose 2,6-bisphosphatase, F2,6BPase*) activity (Fig. 5.5). Which activity is expressed by this **bifunctional** enzyme depends on its phosphorylation state. Phosphorylation of the bifunctional enzyme by protein kinase A (PKA) or another kinase leads to expression of phosphatase activity and F 2,6-P is converted to fructose 6-phosphate (F 6-P). If the bifunctional enzyme is not phosphorylated or if an existing phosphate is removed, then the enzyme expresses kinase activity and converts F 6-P to F 2,6-P. Thus, the unphosphorylated enzyme or PFK2 favors glycolysis by increasing the production of F 2,6-P, a strong activator of PFK1 (Fig. 5.5).

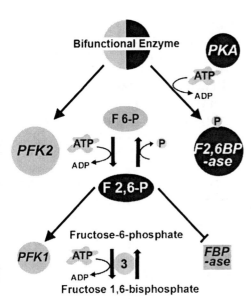

Figure 5.5 Regulation of anaerobic respiration. Fructose 2,6-bisphosphate (F 2,6-P) stimulates the activity of phosphofructokinase-1 (PFK1) and inhibits the activity of fructose 1,6-bisphosphatase (FBPase), key enzymes that catalyze the reactions in step 3 of glycolysis. For details, see Sec. 5.2.4.

The phosphorylation status of the PFK2/F2,6BPase is regulated by hormones. Epinephrine from the adrenal medulla and glucagon from the pancreas bind to membrane receptors and increase intracellular levels of the second messenger cyclic AMP (cAMP). Cyclic AMP activates PKA as part of the "normal" adenylyl cylase-cAMP pathway (see Sec. 4.4.1). PKA phosphorylates the bifunctional enzyme, promoting its F 2,6BPase activity and lowering cellular F 2,6-P. The net result is that epinephrine and glucagon promote gluconeogenesis and the release of glucose into the blood. In contrast, insulin increases PFK2 activity, which promotes glycolysis through increased production of F 2,6-P and subsequent activation of PFK1.

In summary, PFK1 and the opposing FBPase (glycolysis vs. gluconeogenesis) are regulated by cellular allosteric factors that reflect local energy needs and by hormones that signal "global" changes in an organism.

Other key enzymes in glycolysis and gluconeogenesis also are regulated by allosteric factors and hormones. HK is inhibited by glucose 6-phosphate, an example of negative feedback by the product of an enzymatic reaction (Fig. 5.3). PK is inhibited by ATP, acetyl-CoA, and phosphorylation. The first two factors are indicators of abundant energy and result from increased PK activity. Phosphorylation of PK is increased by PKA, which in turn is stimulated by glucagon and epinephrine through increased cAMP. Thus, epinephrine and glucagon inhibit PK activity as well as PFK1 activity. PK activity is increased by upstream generation of fructose 1,6-bisphosphate; thus, a decrease in PFK1 activity inhibits the activity of PK.

PC and PCK, which catalyze the two-step pathway that reverses the reaction catalyzed by PK (Fig. 5.3), also are regulated. Acetyl-CoA stimulates PC activity, and PCK expression is regulated by numerous hormones. Insulin decreases expression of PCK, while epinephrine, glucagon, cortisol, and other gluconeogenic hormones increase its expression. One should bear in mind that the gluconeogenic actions of hormones primarily are expressed in tissues, such as liver, that express GPase activity and are capable of releasing glucose into the blood. In tissues like skeletal muscle that do not express GPase activity, the "gluconeogenic" hormones do not inhibit glycolysis, because inhibition would decrease production of needed energy and prevent release of lactic acid that feeds gluconeogenesis in the liver.

5.3 Aerobic Respiration

Anaerobic fermentation (glucose to lactate) releases only a small portion (~47 kcal/mol) of the total energy (−686 kcal/mol) available in the glucose molecule. However, if oxygen is available as an electron acceptor, glucose and other substrates can be completely oxidized, and a much larger amount of energy can be extracted. Aerobic respiration utilizing oxygen as an electron acceptor takes place in the mitochondrion (for morphology, see Sec. 2.2.1).

It is believed that mitochondria evolved from a bacterium engulfed by an ancestral eukaryotic cell, a theory that is supported by mitochondrial structure (Fig. 5.6). Consistent with a bacterial lineage, the mitochondrion has its own DNA and RNA, and the organelle synthesizes some of its proteins. The mitochondrion consists of two membranes surrounding an inner matrix. The outer membrane is relatively porous because like some bacteria, it contains large membrane channels called porins. In contrast, the inner membrane restricts the movement of solutes, and crossing the membrane usually requires a specific transport mechanism. Infoldings of the inner membrane form **cristae** that increase surface area and, to a certain degree, create an intermembrane

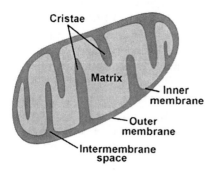

Figure 5.6 The mitochondrion, a double-membrane organelle critical for aerobic respiration. The inner membrane of the mitochondrion forms folds or cristae which act as scaffolds for proteins involved in electron transport. The matrix contains many enzymes involved in the tricarboxylic acid cycle (TCA).

space that is more isolated from the porous outer membrane. The large surface area provided by the folded inner membrane serves as a scaffold for the proteins that participate in electron transport (see Sec. 5.3.3) and oxidative phosphorylation (see Sec. 5.3.4). The matrix of the mitochondrion contains most of the enzymes required for pyruvate oxidation (see Sec. 5.3.1) and the tricarboxylic acid cycle (TCA; see Sec. 5.3.2).

5.3.1 Pyruvate oxidation

$$\text{Pyruvate} + \text{CoA} + \text{NAD}^+ \rightarrow \text{acetyl-CoA} + \text{CO}_2 + \text{NAD} + \text{H}^+$$

(5.3)

$$\Delta G^{\circ\prime} = -7.5 \text{ kcal/mol}$$

If oxygen is available, *pyruvate* produced during glycolysis will be transported into the matrix of the mitochondrion and undergo **oxidative decarboxylation** (Fig. 5.7) before entering the TCA cycle. This reaction is catalyzed by a large multienzyme complex called *pyruvate dehydrogenase (PDH)*. Pyruvate is decarboxylated to acetate, oxidized by losing a pair of electrons to NAD^+, and coupled to coenzyme A (CoA) forming *acetyl-CoA*. Oxidation of pyruvate to acetyl-CoA is a highly exergonic process ($\Delta G^{\circ\prime} = -7.5$ kcal/mol) that essentially is irreversible, and there is no return pathway that can circumvent this reaction. Thus, acetyl-CoA is committed to entering the TCA cycle or to entering pathways such as fat synthesis or ketone formation. Accordingly, acetyl-CoA derived from β-oxidation (breakdown of fatty acids to acetyl-CoA) cannot be used as substrate for gluconeogenesis.

PDH activity is tightly regulated by a number of factors. Enzyme activity is allosterically decreased by NADH, ATP, and acetyl-CoA and increased by NAD^+, AMP, and CoA. Hence, PDH activity is decreased by factors that reflect energy abundance and by feedback inhibition from the immediate product of the reaction it catalyzes. PDH activity is decreased by phosphorylation, a reaction catalyzed by a PDH kinase, and conversely is activated by a PDH phosphatase that dephosphorylates the

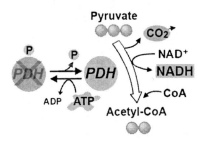

Figure 5.7 Pyruvate oxidation, the reaction that links glycolysis to the TCA cycle. The reaction is catalyzed by pyruvate dehydrogenase (PDH), a large multienzyme complex. Phosphorylation of PDH decreases its activity. In pyruvate oxidation, pyruvate is decarboxylated to acetate, oxidized by passing electrons to NAD^+, and coupled to coenzyme A (CoA).

enzyme. Insulin activates PDH by stimulating the PDH phosphatase, and thus insulin promotes conversion of pyruvate to acetyl-CoA.

5.3.2 TCA cycle

$$\text{Acetyl-CoA} + 2H_2O + 3NAD^+ + FAD + GDP + Pi$$
$$\rightarrow 2CO_2 + 3NADH + 3H^+ + FADH_2 + CoA \qquad (5.4)$$

A summary of TCA-cycle reactions is presented in Eq. (5.4). The TCA cycle is an eight-step cycle that starts with a condensation reaction where the 2-carbon acetate of *acetyl-CoA* is added to a 4-carbon *oxaloacetate* to form a 6-carbon *citrate* (Fig. 5.8). The reaction, which is catalyzed by citrate synthase (Table 5.1), is highly exergonic ($\Delta G^{\circ\prime} = -7.5$ kcal/mol, Table 5.1). Steps 3, 4, and 8 of the cycle are oxidations where electrons are passed to NAD^+ to form NADH. Step 6 also is an oxidation, but the electron acceptor is FAD. This oxidation is catalyzed by succinate dehydrogenase (SDH) (Table 5.1), which is associated with the inner membrane of the mitochondrion, unlike the other TCA enzymes that are in the matrix. SDH and bound FAD form a part of Complex II in the electron transport system, so electrons passed to FAD can be transferred directly to coenzyme Q (CoQ) in the system (see Sec. 5.3.3). Steps 3 and 4 involve decarboxylations with the release of carbon dioxide. The carbons released in these reactions are not the two carbons brought in by the acetate. Nevertheless, loss of the second carbon returns the cycle to a 4-carbon intermediate, which is *succinate* complexed to CoA. Hydrolysis of the bond between succinate and CoA provides sufficient energy for the generation of GTP from GDP and Pi that occurs in step 5 (Fig. 5.8). The terminal phosphate of GTP can be transferred to ADP to form ATP. *Oxaloacetate* regenerated in step 8 can undergo condensation with acetate and begin a new cycle.

As might be expected, the enzymes of the TCA cycle are regulated by allosteric factors. All the dehydrogenases that pass electrons to NAD^+ to form NADH (steps 3, 4, and 8) are inhibited by high levels of NADH. Alpha-ketoglutarate dehydrogenase (step 4) also is inhibited by succinyl-CoA, the product of the reaction it catalyzes (Fig. 5.8 and Table 5.1). Isocitrate dehydrogenase activity (step 3) is increased by ADP.

The only usable energy produced in the TCA cycle from two molecules of acetyl-CoA (derived from one glucose molecule) is that stored in two molecules of ATP. These two ATP are in addition to the two ATP produced by substrate phosphorylation in glycolysis. Most of the energy extracted from the acetyl-CoA is in the form of high-energy electrons in 6 NADH (3 per acetyl-CoA) and 2 $FADH_2$. High-energy electrons also were stored during glycolysis (2 NADH) and pyruvate oxidation (2 NADH).

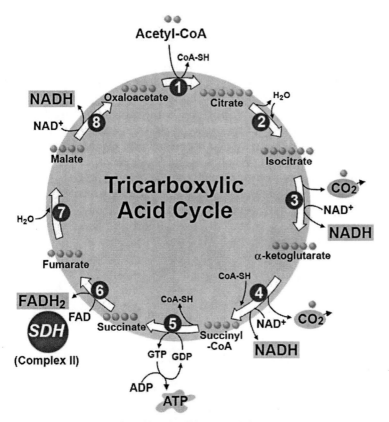

Figure 5.8 Tricarboxylic acid cycle (TCA). The TCA cycle is an 8-step reaction that oxidizes acetate to CO_2. In step 1, the 2-carbon acetate is complexed with the 4-carbon oxaloacetate to form the 6-carbon citrate. In subsequent steps, decarboxylations yield two molecules of CO_2, pairs of electrons are transferred to three NAD^+ and one FAD, and one GTP (ATP) is produced. Succinate dehydrogenase (SDH), which catalyzes reaction 6, is the only enzyme of the cycle associated with the inner mitochondrial membrane. SDH is part of Complex II in electron transport.

TABLE 5.1   TCA Cycle Enzymes

Step	Enzyme	$\Delta G^{\circ\prime}$
1	Citrate synthase	−7.5
2	Aconitase	+1.5
3	Isocitrate dehydrogenase	−2.0
4	α-ketoglutarate dehydrogenase	−7.2
5	Succinyl-CoA synthase	−0.8
6	Succinate dehydrogenase	~0
7	Fumarate hydratase	−0.9
8	Malate dehydrogenase	+7.1

Thus, complete oxidation of one molecule of glucose generates thus far 10 NADH, 2 FADH$_2$, and 4 ATP.

The energy in these stored electrons can be converted to ATP "dollars" in the electron transport system (see Sec. 5.3.3).

The discussion of the various metabolic pathways has, thus far, focused on oxidation of glucose, but it should be remembered that other sugars, lipids, proteins, and nucleic acids can enter these pathways. Triglycerides (lipid) can be disassembled into glycerol and fatty acids. The glycerol can enter glycolysis, and β-oxidation of fatty acids yields acetyl-CoA that can enter the TCA cycle. Amino acids (protein) can be deaminated to produce carbon skeletons that can enter as pyruvate (alanine) or as various intermediates in the TCA cycle. Therefore, virtually all biological molecules can be considered to be energy substrates.

5.3.3   Electron transport

$$NADH + H^+ + \frac{1}{2}O_2 \rightarrow NAD^+ + H_2O$$

$$\Delta G^{\circ\prime} = -52.4 \text{ kcal/mol}$$

(5.5)

$$FADH_2 + H^+ + \frac{1}{2}O_2 \rightarrow FAD + H_2O$$

$$\Delta G^{\circ\prime} = -45.9 \text{ kcal/mol}$$

(5.6)

Electron transport is the first system required to "cash in" the energy associated with electrons transferred to the coenzymes NAD^+ and FAD during the TCA cycle. In the system, electrons from the coenzymes are passed through a series of carriers to oxygen. Most of the carriers are associated with four (I, II, III, and IV) large protein complexes or respiratory complexes in the inner mitochondrial membrane. The transfers of electrons from NADH [Eq. (5.5)] and FADH$_2$ [Eq. (5.6)] to oxygen are highly exergonic processes (Fig. 5.9 and Application Box 5.1) that release enough energy to synthesize multiple molecules of ATP from ADP and Pi. The exact number of ATP molecules produced is unclear, but accepted average values are three ATP from NADH and two ATP from FADH$_2$.

Multiple carriers that undergo reversible reduction-oxidation (redox) reactions allow for a stepwise release of energy from the electrons stored in NADH and FADH$_2$. With each transfer, the electrons lose free energy and as will be discussed in Sec. 5.3.4, the "packets" of released energy are stored in the form of a proton electrochemical gradient. The carriers

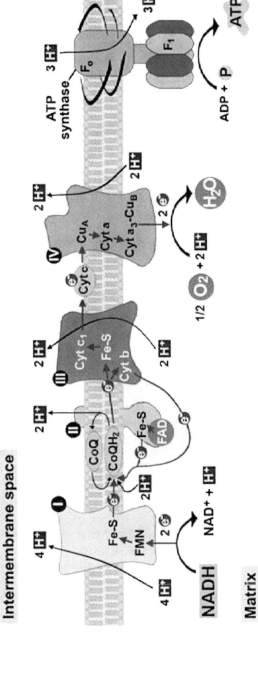

Figure 5.9 Electron transport. Electrons from NADH and FADH$_2$ are transferred to a series of carriers that undergo reversible oxidation-reduction reactions. Most of the carriers are associated with respiratory complexes numbered I, II, III, and IV. For details, see Sec. 5.3.3.

119

APPLICATION BOX 5.1
Reduction-Oxidation (Redox) Potentials, $\Delta G^{\circ\prime}$, and ATP

(1) Redox potentials

The **standard redox potential** (E_0') for a redox couple such as NAD^+-NADH
is determined by the ability of the couple to transfer electrons to (or take
them from) the standard redox couple H^+-H_2 or standard hydrogen electrode.
A strong reducing agent such as NADH coupled to a weak oxidizing agent
(NAD^+) will transfer electrons (electrons have a negative voltage) to the standard couple. A strong oxidizing agent such as O_2 will have a positive voltage
with respect to the standard hydrogen electrode. The half-cell potential of
NADH with respect to molecular oxygen is indicated below by $\Delta E_0'$.

$$NAD^+ + 2H^+ + 2e^- \leftrightarrow NADH + H^+ \quad E_0' \text{ (volts, V)} = -0.320$$

$$\tfrac{1}{2}O_2 + 2H^+ + 2e^- \leftrightarrow H_2O \quad E_0' \text{ (volts, V)} = +0.816$$

$$\Delta E_0' = +0.82 \text{ V} - (-0.32 \text{ V}) = 1.136 \text{ V}$$

Note: E_0' for FAD-$FADH_2$ is $+0.031$.

(2) Calculation of $\Delta G^{\circ\prime}$ from $\Delta E_0'$

$$\Delta G^{\circ\prime} = -nF \, \Delta E_0'$$

where n = number of electrons transferred and F = Faraday's constant
of 23.063 kcal/V mol.

$$\Delta G^{\circ\prime} = -(2)(1.136 \text{ V})(23.063 \text{ kcal/V mol}) = -52.5 \text{ kcal/mol}$$

(3) Efficiency of energy transfer to AT

$\Delta G^{\circ\prime} = +7.3$ kcal/mol for synthesis of ATP from ADP and Pi. However, a
cell doesn't function under standard conditions, so a more realistic value
for ATP synthesis is 10 kcal/mol. If one assumes that three ATP are produced from the energy derived from a pair of electrons in NADH, then the
calculated efficiency for energy transfer to ATP would be about 57%, which
is a high efficiency for a cellular process.

---

include flavoproteins, cytochromes (cyt), iron-sulfur proteins (Fe-S),
copper atoms, and coenzyme Q (CoQ).
   Flavoproteins are polypeptides that bind the prosthetic groups (i.e.,
non-amino acid compounds attached to proteins) flavin mononucleotide
(FMN) or flavin adenine dinucleotide (FAD) (Fig. 5.9). **Cytochromes**
are proteins that contain the heme prosthetic group with an iron atom
that switches between Fe^{2+} and Fe^{3+} oxidation states when it accepts
or donates one electron. *Fe-S proteins* contain iron atoms linked to the

sulfur in cysteines. Although the Fe-S centers may contain multiple iron atoms, they are only capable of donating or accepting one electron. Copper atoms in Complex IV switch between Cu^{2+} and Cu^+ oxidation states and like iron, accept or donate a single electron. CoQ is not a protein. It is a lipid-soluble quinone that can be reduced in two steps to carry two electrons. CoQ does not associate with one of the four respiratory complexes; rather, it serves as a mobile electron carrier in the membrane, shuttling electrons from Complex I or II to Complex III. Reduced CoQ also acquires protons from the matrix and upon oxidation, releases the protons into the intermembrane space (Fig. 5.9). By transferring protons across the membrane, CoQ increases the proton electrochemical gradient used to generate ATP.

As noted above, oxidation of glucose via glycolysis, pyruvate oxidation, and the TCA cycle result in electrons being stored in 10 molecules of NADH and 2 molecules of $FADH_2$. All the reduced coenzymes are in the mitochondrion, except for the two NADH produced during glycolysis, a cytosolic pathway. The electrons associated with the cytosolic NADH most often are transferred to FAD in the inner mitochondrial membrane via the glycerol phosphate shuttle. Hence, electrons in the two molecules of cytosolic NADH usually lose energy as they are transferred into the mitochondrion. In some tissues, electrons associated with the cytosolic NADH can be passed to mitochondrial NAD^+ via the malate-aspartate shuttle.

Two electrons from an NADH molecule enter *Complex I (NADH dehydrogenase complex)*, which is composed of approximately 45 different subunits and contains 6 to 9 Fe-S centers. The electrons pass to an FMN prothetic group, Fe-S centers and then to CoQ (Fig. 5.9). Energy is released with each succeeding oxidation-reduction. The released energy is somehow coupled to translocation of four protons from the mitochondrial matrix to the intermembrane space. **Translocation** creates a proton (electrochemical) gradient between the matrix and the intermembrane space that preserves some of the energy released in the oxidation-reduction reactions.

Complex II (succinate dehydrogenase) transfers electrons from succinate in step 6 of the TCA cycle to CoQ. The complex consists of four polypeptides, two of which comprise succinate dehydrogenase, and three Fe-S centers. Electrons from succinate pass to FAD, Fe-S centers, and then to CoQ (Fig. 5.9). Electron transfer through Complex II is not coupled to proton translocation; thus, in comparison to the electrons in NADH, the electrons in $FADH_2$ contribute less to the electrochemical gradient (i.e., fewer ATP).

A pair of electrons from reduced CoQ ($CoQH_2$) is transferred to *Complex III (cytochrome bc_1)*, which is composed of 11 polypeptides, 2 cyt b, cyt c_1, and an Fe-S center. Transfer of the electrons across

Complex III to cyt c is associated with translocation of four protons into the intermembrane space (Fig. 5.9). Two of the protons are derived from reduced CoQ when it enters the complex, and two additional protons are translocated as part of the Q Cycle.

The **Q Cycle** stems directly from the initial transfer of electrons from CoQ to Complex III (Fig. 5.9). One electron carried by reduced CoQ is passed to the Rieske Fe-S center, which can only accept one electron. The second electron is transferred to cyt b_L. The electron transferred to the Fe-S center passes to cyt c_1 and then to cyt c (Fig. 5.9). However, the electron passed to cyt b_L is transferred to cyt b_H (only one cyt b is shown in Fig. 5.9) and then back to a CoQ. After CoQ receives a second electron from cyt b_H, it acquires two protons from the matrix and prepares to transfer the electrons to Complex III. Upon transfer of the electrons to the Fe-S center and cyt b_L, the second pair of protons is released to the intermembrane space.

Cyt c carries electrons to *Complex IV (cytochrome oxidase)*, which is composed of 13 polypeptides, including cyt a, cyt a_3, Cu_A (dimeric copper center), and Cu_B. The electron from cyt *c* is initially transferred to Cu_A. Reduced Cu_A then passes its single electron to the heme group (Fe) in cyt a. From cyt a, the electron is transferred to the binuclear redox center consisting of the heme group (Fe) of cyt a_3 and the copper atom of Cu_B (Fig. 5.9). Two electrons are required to reduce the cyt a_3-Cu_B binuclear center. Molecular oxygen then binds to the center and accepts the electrons. The binuclear center essentially holds the oxygen in place until it is fully reduced to two H_2O by acquiring two additional electrons and four protons. Release of partially reduced oxygen species that are extremely reactive leads to cellular damage (see Clinical Box 5.1). For each pair of electrons that passes through Complex IV, two protons translocate (are "pumped") from the matrix to the intermembrane space. In addition, two protons are removed from the matrix when they bind to oxygen to form water. Both actions contribute to the electrochemical gradient between the matrix and the intermembrane space.

5.3.4   Chemiosmosis and ATP synthesis

Peter D. Mitchell (1920–1992), the British biochemist and Nobel laureate, first proposed the chemiosmotic coupling model, which suggested that most energy required for ATP synthesis is derived from an electrochemical gradient across the inner mitochondrial membrane. As described in the previous sections, the electrochemical gradient primarily is created by coupling energy released during electron transport to "pumping" of protons into the intermembrane space. Binding of matrix protons to oxygen to form water also contributes to the gradient. The chemical component of the gradient is due to the higher concentration of hydrogen ions (protons) in the intermembrane space relative to

---

CLINICAL BOX 5.1
Free Radicals and Aging

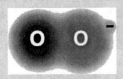

Superoxide ion

A free radical, such as the **superoxide** radical shown on the left, is an unstable molecule that has an unpaired electron in its outer shell. These highly reactive molecules can oxidize and damage proteins, nucleic acid and lipids, triggering oxidative stress and cell death. For example, mitochondrial damage caused by free radicals might trigger apoptosis. Formation of free radicals can increase for many reasons, including immune cell activation, inflammation, ischemia, infection, cancer, or simply in response to radiation from sunlight (i.e., skin damage). Antioxidants include enzymes (e.g., superoxide dismutase, SOD), cellular reducing agents (e.g., glutathione, α-lipoic acid), and nutrients (e.g., vitamin E).

It has been proposed that aging results partly from accumulated damage caused by free radicals. Most of the support for this theory comes from studies of mice and rats. It has been shown that mice lacking mitochondrial SOD survive only for a short period and that mice over-expressing mitochondrial catalase (i.e., the enzyme that detoxifies hydrogen peroxide) have increased life spans. These studies support a role for free radicals in aging and also point to the mitochondrion as a primary source of the destructive molecules. There also are some hints that the long life span of humans is related to control of free radicals. For example, it appears that in comparison to mouse cells, human cells have a greater ability to manage damage by free radicals. Another line of evidence that links longevity to decreased free radicals is that rodents and other animals placed on very low calorie diets live longer and show decreased production of superoxides and hydrogen peroxide. Unfortunately, there is no solid evidence that humans on low-calorie diets experience a similar, or any, increase in life span. Furthermore, recent studies have suggested that the increased life span in response to a low-calorie diet may be due to increased expression of a repair gene called Sirt1 rather than to a direct suppression of free radical production.

Is it time to go on a thousand-calorie diet composed mainly of vitamin E? Free radicals may play some role in the aging process, but they are only one part of a much larger picture. Hence, in most cases there is no way to predict whether the benefits of trying to lower free radical production outweigh the risks.

---

that in the matrix (ΔpH). The electrical component is the membrane potential (V_m) in volts that results from the separation of charge carried by the protons. The energy present in the electrochemical gradient can be expressed as the **proton motive force (pmf)**. The pmf essentially is the force that drives the protons in the intermembrane space across the inner mitochondrial membrane and into the matrix.

To couple the energy in the electrochemical gradient to ATP synthesis, protons are forced to flow through **ATP synthase**, an ATP-synthesizing "motor." ATP synthase is composed of an F_1 head linked to the F_0 base in the inner mitochondrial membrane (Fig. 5.10). The head is composed of nine subunits; 3α, 3β, γ, δ, and ε. As shown in Fig. 5.10, the α- and β-subunits alternate in the *F_1 head*. Each of the three β-subunits contains a site for ATP synthesis; thus, each F_1 head is capable of concurrently catalyzing the synthesis of three ATP molecules. The γ-subunit, in association with the ε-subunit, essentially forms a rotating camshaft that extends from the *F_0 base* into the F_1 head. At each point in a single rotation, the γ-subunit interacts differently with each of the three β-subunits, causing the catalytic site on each subunit to adopt a different conformation (i.e., open, loose, or tight). Hence, each catalytic site undergoes cyclical changes (open →loose→tight) as the γ-subunit rotates, and these changes are responsible for ATP synthesis, as will be discussed later in this section.

Rotation of the γ-subunit (camshaft) is in turn controlled by the F_0 base. At a minimum, the base is composed of 1 a, 2 b, and 10 to 14 c subunits (not detailed in Fig. 5.10). The b subunits in conjunction with the δ-subunit of the F_1 head may help fix the position of the α- and β-subunits. The c subunits (approximately 12) form a rotating ring in the inner mitochondrial membrane that is attached to the γ-subunit. Rotation of the ring is driven by protons moving from the intermembrane space to the matrix. Binding of a proton from the intermembrane space to a c subunit in the F_0 base causes the subunit to undergo a conformational change that

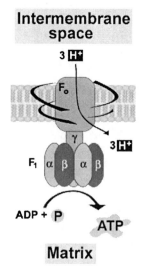

Figure 5.10 ATP synthase. This enzyme is a multi-subunit complex composed of two main structures, an F_0 base and an F_1 head. Protons in the intermembrane space flow through the synthase to the matrix. The flow of protons down an electrochemical gradient is coupled to ATP synthesis. For details, see Sec. 5.3.4.

results in partial rotation of the ring. Presumably, sequential binding of protons to each of the c subunits (12 total) results in one complete rotation. A proton from the intermembrane space may gain access to a c subunit through a channel in the a subunit. After a complete rotation of the ring (sequential binding of 11 more protons), the first proton is able to move into the matrix via a separate channel through the a subunit.

Based on the assumptions above, passage of 12 protons through ATP synthase drives one rotation of the γ-subunit (camshaft), which causes the catalytic site in each of three β-subunits to complete one ATP-synthesizing cycle (open→loose→tight) (Fig. 5.11).

As discussed in the first third of this section, ATP synthesis is driven by γ-subunit-induced changes in the conformations of the catalytic sites in the β-subunits of the F_1 head. The changes for a single β-subunit are shown in Fig. 5.11. In the "open" conformation, the catalytic site of the β-subunit accepts new ADP and Pi. Rotation of the asymmetric γ-subunit (camshaft) in response to binding of protons from the intermembrane space to the c subunits of the F_0 base causes the catalytic site to adopt a loose conformation in which ADP and Pi are loosely bound. Further rotation of the γ-subunit induces the catalytic site to adopt a tight conformation with increased affinity for ADP and Pi. The tight conformation greatly changes the energy requirements for synthesis of ATP from ADP and Pi. In an aqueous solution, synthesis of ATP is highly endergonic with a $\Delta G^{\circ\prime} = +7.3$ kcal/mol. However, the environment provided by the tight conformation of the catalytic site reduces $\Delta G^{\circ\prime}$ to approximately 0 kcal/mol. Hence, the energy in the electrochemical gradient is used to drive the change in the conformation of the catalytic site that creates an environment for spontaneous ATP formation.

5.3.5 Usable energy

The theoretical maximum ATP yield for complete oxidation of a molecule of glucose is 38 ATP (Table 5.2).

If one assumes that hydrolysis of ATP in the cell yields approximately 10 kcal/mol, then oxidation of glucose provides 380 kcal/mol of usable energy.

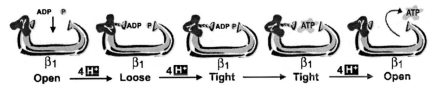

Figure 5.11   Electron flow from the mitochondrial intermembrane space to the matrix. During this process, the γ-subunit of the ATP synthase rotates causing the catalytic site in a β-subunit (β_1) to assume three different conformations designated open, loose, and tight. In the tight conformation, the catalytic site creates an environment that promotes spontaneous formation of ATP at a $\Delta G^{\circ\prime}$ close to zero. For details, see Sec. 5.3.4.

TABLE 5.2   Maximum ATP Yield

Glycolysis	2 ATP	2 ATP
	2 NADH	6 ATP
Pyruvate oxidation	2 NADH	6 ATP
TCA cycle	2 ATP	2 ATP
	6 NADH	18 ATP
	2 FADH$_2$	4 ATP
Total		38 ATP

As noted in Eq. (5.1), the $\Delta G^{\circ\prime}$ for oxidation of glucose is 686 kcal/mol. Based on the assumptions of 10 kcal/mol released upon hydrolysis of ATP and a maximum yield of 38 mol of ATP, the efficiency of energy conservation in aerobic respiration would be about 55.4% (380/686 $\times$ 100). Of course, the assumptions are unrealistic. First, electrons from cytosolic NADH (i.e., NADH from glycolysis) usually are passed to FAD in the mitochondrion, which lowers the ATP yield. Second, the energy in the electrochemical gradient is used for processes in addition to oxidative phosphorylation. For example, movement of pyruvate across the inner mitochondrial membrane is driven by cotransport with a proton.

Nevertheless, aerobic respiration is still an efficient process even if one assumes an average yield of 30 to 32 ATP.

5.4   Photosynthesis

5.4.1   Conversion of light energy to chemical energy

The core of photosynthesis is conversion of light energy to chemical energy or **energy transduction**. Conversion occurs when the energy in a **photon** of light is transferred to an electron in a light-absorbing molecule (e.g., **chlorophyll**). Electrons excited in this first energy-transduction step (**photosystem II**) are then passed through multiple carriers that undergo reversible oxidation-reduction reactions. Multiple reactions allow for a stepwise release of energy from the electrons, and as in electron transport, the released energy is coupled to synthesis of ATP. The electrons are then reexcited by absorption of light energy in a second energy-transduction step (**photosystem I**). The reexcited electrons can be transferred to the coenzyme **nicotinamide adenine dinucleotide phosphate (NADP$^+$)** or used for other purposes. Eventually, much of the energy stored as ATP and high-energy electrons stored in NADPH will be used to convert carbon dioxide to carbohydrates (i.e., sugars) and various other organic compounds.

$$CO_2 + 2H_2O \xrightarrow{\text{light}} CH_2O + O_2 + H_2O \qquad (5.7)$$

The general reaction for photosynthesis, where water is the electron donor, can be written as in Eq. (5.7). Extraction of electrons from water yields molecular oxygen. Therefore, the oxygen that is critical for aerobic respiration is simply a photosynthetic by-product. The low-energy electrons obtained from water are excited by energy from light and stored as high-energy electrons in NADPH. The stored electrons are then used to reduce carbon dioxide, producing a unit of carbohydrate (CH_2O). For a couple of billion years, hydrogen sulfide (H_2S) was the primary source of electrons. Today, H_2S serves as an electron source for only some bacteria.

$$6CO_2 + 6H_2O + energy \rightarrow glucose + 6O_2$$
$$\Delta G^{\circ\prime} = +686 \text{ kcal/mol}$$
(5.8)

Carbohydrate synthesis is an endergonic process that requires a substantial input of energy. The simplified reaction above shows that synthesis of glucose (i.e., the reverse of aerobic respiration) requires 686 kcal/mol. The necessary energy (ATP and NADPH) is generated by energy transduction and energy transfer in the *light-dependent reactions* of photosynthesis. Carbohydrates are then synthesized in the *light-independent* or "dark" reactions of photosynthesis. Carbohydrate synthesis is not truly light independent, because stored energy (ATP and NADPH) is exhausted quickly in the absence of light.

5.4.2 Chloroplasts

The primary photosynthetic organelles in plants and algae are chloroplasts. Like the mitochondrion, the chloroplast is believed to have evolved from a bacterium. Consistent with a bacterial lineage, the chloroplast has its own DNA and RNA, and the organelle synthesizes some of its own proteins. Like some bacteria, the chloroplast has an outer and inner membrane (Fig. 5.12). The outer membrane is relatively permeable,

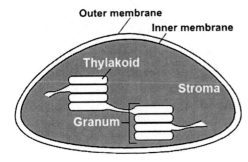

Figure 5.12 Chloroplast. The plant organelle consists of two membranes (inner and outer) that surround an inner stroma. The chloroplast also contains a third membrane system consisting of flattened membranous disks called thylakoids that are arranged in interconnected stacks called grana (one stack is a granum).

because it contains large membrane channels called porins. The inner membrane is much less permeable; therefore, crossing this membrane generally requires specific transporters. Chloroplasts have a third membrane system consisting of membranous disks called **thylakoids**, which are arranged in stacks called grana. The space enclosed by the thylakoid membrane is the lumen. For the sake of discussion, the lumen will be treated as an isolated compartment but in truth, the thylakoids are interconnected. The thylakoid membrane serves as a scaffold for the components (i.e., pigments, photosystem I, photosystem II, and ATP synthase) required for the light-dependent reactions of photosynthesis. Surrounding the thylakoids is the stroma, which contains the components required for carbohydrate synthesis.

5.4.3 Photosynthetic pigments

Chlorophyll is the primary pigment responsible for transducing light energy into chemical energy. The chlorophyll molecule is composed of a porphyrin ring and a hydrophobic phytol tail that can embed in the lipid bilayer of the thylakoid membrane (Fig. 5.13). Recent findings suggest that the properties of the phytol tail make it suitable for use as an adjuvant for vaccines (see Clinical Box 5.2). The alternating single and double bonds in the porphyrin ring provide a cloud of delocalized electrons with a magnesium ion serving as an electron holder/donor. This bond system provides a variety of electron orbitals that can absorb different amounts of energy. Thus, electrons in chlorophyll can effectively absorb energy from photons of a broad **absorption spectrum** (i.e., a broad range of wavelengths of light). The absorption spectrum of chlorophyll is further broadened by interaction of the molecule with different binding proteins and by the presence of "a" and "b" forms of the molecule. Accessory pigments such as carotinoids and phycobilins absorb energy at wavelengths that are not effectively utilized by chlorophyll. These accessory pigments expand photosynthesis' **action spectrum**, which indicates the relative efficiency of photosynthesis at different wavelengths of light. The action spectrum for photosynthesis in green plants (primarily chlorophylls a and b) is maximal in the blue-violet region (wavelengths ~400 to 600 nm) and high in the orange-red region (wavelengths ~600 to 700 nm) (Fig. 5.14).

Chlorophyll and other pigments are associated with large pigment-protein complexes called **photosystems** or with light-harvesting complexes (LHCs). Most of the pigments simply serve as antenna pigments or light-gathering molecules that transfer photon energy to nearby pigments.

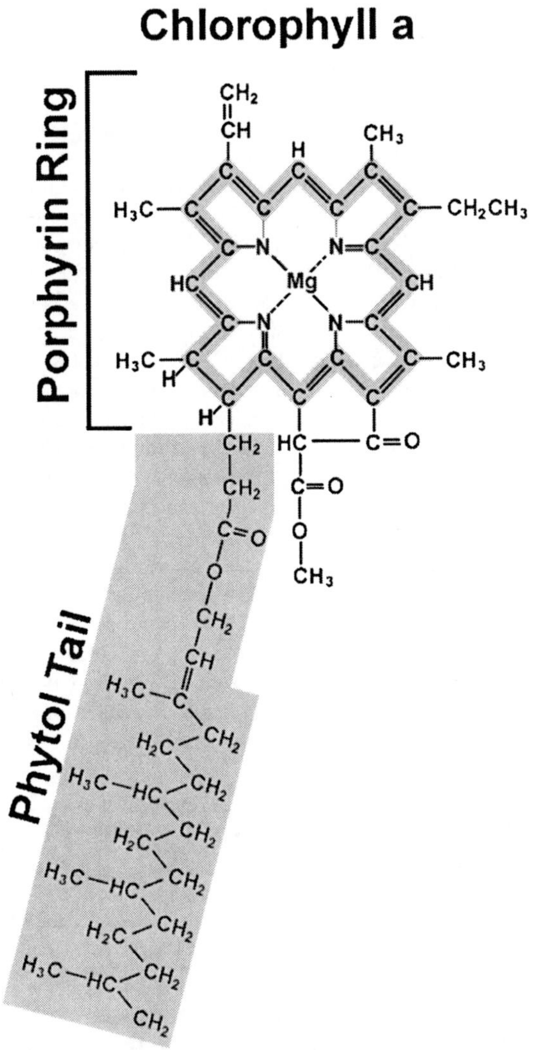

Chlorophyll a

Figure 5.13 Chlorophyll consisting of a porphyrin ring and a hydrophobic phytol tail. The alternating single and double bonds in the porphyrin ring provide a cloud of delocalized electrons with a magnesium ion serving as an electron holder/donor.

Eventually, the photon energy is passed to a **reaction center** chlorophyll in a photosystem. Reaction centers receive energy collected by approximately 250 antenna pigments. Thus, the antenna pigments (like a satellite dish) increase the efficiency of energy reception by the reaction center

CLINICAL BOX 5.2
Phytol and Vaccines

Adjuvants increase vaccine efficacy

In immunology, an **adjuvant** is a substance that enhances the ability of a vaccine to stimulate an individual's specific and nonspecific immune responses. Adjuvants can increase the efficacy of a vaccine by a variety of actions, including (1) retaining the vaccine in the body or at the site of injection, (2) reducing degradation of the vaccine, and (3) recruiting macrophages and other cells to augment immune responses to a pathogen or associated antigens. Many adjuvants that are presently available can enhance the efficacy of a vaccine, but the adjuvants may also cause severe adverse inflammatory reactions. Consequently, the only adjuvant currently accepted by the Food and Drug Administration (FDA) for use in humans is aluminum hydroxide (alum).

Recently, Lim and colleagues at Indiana State University have discovered that the hydrophobic phytol tail of chlorophyll and other phytol derivatives are effective and relatively safe adjuvants. This is an exciting discovery because it has the potential to improve vaccine effectiveness. Furthermore, the discovery highlights the importance of naturally-occurring substances in medicine.

Lim, So-Yon (2005). Environmental Factors in Autoimmunity: Assessment of immunotoxicity of phthalates in the induction of Lupus-like anti-DNA antibodies. Dissertation, Indiana State University.

Ghosh, Swapan K. (2005). Novel phytol derived immunoadjuvants and their use in vaccine formulations. U.S. Patent Pending #20050158329.

(receiver) (Fig. 5.15). A LHC does not contain a reaction center, so the energy gathered by this mobile satellite dish must be passed to a reaction center through antenna pigments associated with a photosystem.

The reaction center consists of two chlorophyll-a molecules that are responsible for converting light energy to chemical energy. The energy passed to the reaction center causes an electron to move from an inner to a high-energy outer orbital. This high-energy state is not stable. One option is that the excited electron can return to its original orbital, releasing its energy as heat and light. The light emitted will be at a longer wavelength (fluorescence) that contains less energy. The second option is that the electron can be transferred to a high-energy orbital in an acceptor molecule. This step essentially completes the conversion of light energy to chemical energy.

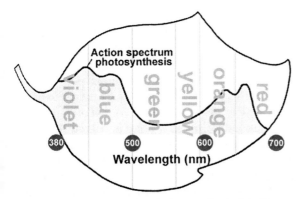

Figure 5.14 Action spectrum of photosynthesis. It indicates the relative efficiency of photosynthesis at different wavelengths of light. The action spectrum for photosynthesis in green plants (primarily chlorophylls a and b) is maximal in the blue-violet region or wavelengths of light that are ~400 to 600 nm.

5.4.4 Z-scheme

In the light-dependent reactions of photosynthesis, electrons are passed from water to photosystem II, to photosystem I, and then to NADPH. As shown in Fig. 5.16, electrons are photoexcited twice in the pathway, first at an absorption maximum of 680 nm in photosystem II, and subsequently at an absorption maximum of 700 nm in photosystem I. This electron flow has been referred to as the *Z-scheme*, because the pathway can resemble a "Z" when electron flow is plotted as a function of **redox potential**, which is a measure of electron energy.

Passage of electrons from water to photosystem II is a very complex procedure for two reasons. First, water is a very stable molecule; thus,

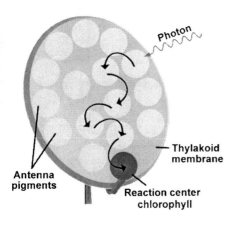

Figure 5.15 Energy absorption in plants. Light energy absorbed by antenna pigments is transferred from one pigment to another until it is "received" by a reaction-center chlorophyll. Like a satellite dish, antenna pigments increase the efficiency of energy reception by funneling energy to a single reaction center.

splitting water is normally a highly endergonic reaction. Second, the splitting of two molecules of water to form molecular oxygen yields four electrons, but the P680 reaction center can only accept one electron at a time. Splitting of water during photosynthesis (photolysis) is possible because the oxidized form of P680 (P680$^+$) has a redox potential of approximately +1.17 (Fig. 5.16). Thus, P680$^+$ is a strong enough oxidant to steal electrons from water, which has a redox potential of only +0.816. The problem of transferring one electron at a time is solved by interposing four manganese (Mn) atoms in the electron pathway between water and photosystem II. Each Mn atom passes a single electron through a tyrosine residue (Tyr) to the P680$^+$. The P680 is reoxidized to P680$^+$ by a photon of light and is ready to accept an electron from the next Mn atom. Once all four Mn atoms have been oxidized, water is split and four electrons pass to the Mn cluster.

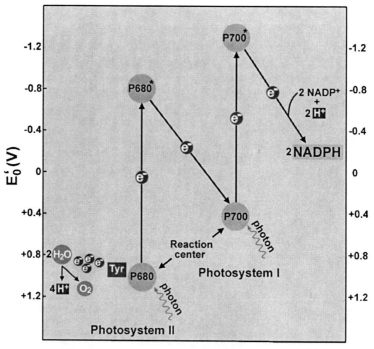

Figure 5.16 Electron flow during the light-dependent reactions of photosynthesis. Electron flow from water to NADP$^+$ is referred to as the Z-scheme, because the pathway resembles a "Z" when flow is plotted as a function of redox potential (E_0', V). Electrons that flow from water to the P680 reaction center pass through a tyrosine (Tyr) residue. For details, see Sec. 5.4.4.

5.4.5 Electron flow through the photosystems

An electron in the P680 reaction center is excited by energy from a photon of light (Fig. 5.17). The energy is transferred to the reaction center from an antenna pigment associated with *photosystem II*. The photoexcited electron must be passed to an acceptor or the electron will simply release its energy and return to its ground state. *Pheophytin (Ph)*, which is a modified molecule of chlorophyll a, serves as the acceptor (Fig. 5.17). *Oxidized P680 (P680$^+$)* then steals a replacement electron from one of the four Mn atoms in the *Mn cluster*.

Ph passes an electron to *plastoquinone A (Q_A)*, which is associated with *protein D2* in photosystem II (Fig. 5.17). Q_A passes two electrons, one at a time, to *plastoquinone B (Q_B)*. Reduced Q_B dissociates from protein D2, picks up two protons from the stroma, and becomes the mobile electron carrier *plastoquinol (PQH_2)*. The role of PQH_2 in photosynthesis is similar to that of CoQ in electron transport (see Sec. 5.3.3). PQH_2 transfers its electrons to the *cyt b_6f complex* which is similar in structure and function to the cyt bc_1 complex in electron transport. The protons carried by PQH_2 are released into the thyakoid lumen. Transfer of electrons to the cyt b_6f complex involves a Q cycle as described for electron transport (see Sec. 5.3.3). Therefore, four protons are translocated from the stroma to the thylakoid lumen for each pair of electrons passing through the cyt b_6f complex. As in electron transport, translocation of protons creates an electrochemical gradient that can be used to drive the synthesis of ATP.

Electrons pass from *cyt f* to a copper ion (Cu^{2+}) in *plastocyanin (PC)* (Fig. 5.17). PC is a water-soluble, mobile electron carrier located on the luminal side of the thylakoid membrane. PC carries the electron to the oxidized *reaction center (P700$^+$)* in *photosystem I*.

An electron in the P700 reaction center is excited by energy from a photon of light. As in photosystem II, the energy is transferred to the reaction center from an antenna pigment associated with the photosystem. The photoexcited electron is passed to an acceptor named A_0, which is a separate chlorophyll molecule. The electron is then transferred to *phylloquinone (A1)* and subsequently to three iron-sulfur clusters *(F_X, F_A, F_B)*. The electron leaves photosystem I and passes to the iron-sulfur protein *ferredoxin (Fd)*.

Electrons passed to ferredoxin can be used to reduce NADP$^+$ or inorganic acceptors, or the electrons can return to the cyt b_6f complex as part of cyclic photophosphorylation (Fig. 5.17; see Sec. 5.4.6). Reduction of NADP$^+$ to NADPH requires the stromal enzyme *ferredoxin-NADP$^+$ reductase (FdNR)*. Ferredoxin can pass only a single electron; thus, two reduced ferredoxins are required. Reduction of NADP$^+$ removes two protons from the stroma, an action that increases the electrochemical gradient across the thylakoid membrane. Some electrons

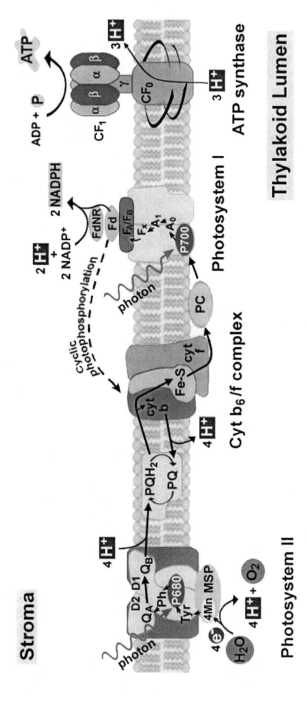

Figure 5.17 Electrons flow through photosystem II, cyt b₆f complex, and photosystem I in the thylakoid membrane of the chloroplast. It creates a proton gradient that can be used to drive ATP synthesis by ATP synthase. For details, see Secs. 5.4.5 to 5.4.7.

passed to ferredoxin are used to reduce nitrate (NO_3^-) to ammonia or to reduce sulfate (SO_4^{2-}) to a sulfhydryl ($-SH$). These reduced chemical groups are critical for synthesis of amino acids and other biological molecules.

5.4.6 Cyclic photophosphorylation

Electrons passed to ferredoxin can be passed back to the cyt b_6f complex (Fig. 5.17). Energy released from the electrons as they pass through the cyt b_6f complex is coupled to translocation ("pumping") of protons (approximately 2 H^+/electron) from the stroma to the thylakoid lumen. The exact mechanism is unknown but may involve plastoquinol. The protons translocated into the thylakoid lumen contribute to the electrochemical gradient that can be used to synthesize ATP. Thus, cyclic photophosphorylation provides additional ATP needed for carbohydrate synthesis and for other cell processes.

5.4.7 ATP synthesis

Synthesis of ATP in the chloroplast follows the same basic mechanism that was described in more detail in Sec. 5.3.3 on electron transport and Sec. 5.3.4 on ATP synthesis. Briefly, the electrochemical gradient created by "pumping" protons into the thylakoid lumen is responsible for the proton motive force (pmf) that drives protons to flow across the thylakoid membrane and into the stroma. Protons can flow across the membrane only by passing through ATP synthase, an ATP-synthesizing "motor" composed of a *CF₁ head* linked to a *CF_O base* anchored in the thylakoid membrane (Figs. 5.17 and 5.10). Rotation of the CF_0 base in response to proton flow leads to cyclical changes in conformations of ATP catalytic sites in the CF_1 head. The cyclical changes in the catalytic sites are responsible for ATP synthesis.

5.4.8 Summary of light-dependent reactions

$$2\ H_2O + 2\ NADP^+ + ?\ ADP + ?\ Pi \xrightarrow{8\ photons} O_2 +$$

$$2\ NADPH + ?\ ATP$$

(5.9)

The primary purpose of the light-dependent reactions is to provide energy and a strong reducing agent for the synthesis of carbohydrate [Eq. (5.9)]. In the initial reactions, two molecules of water are split yielding four electrons, four protons, and molecular oxygen. Photolysis of water is possible because $P680^+$ in photosystem II is a strong oxidizing agent. Each electron is excited in photosystem II and again in photosystem I. Hence, eight photons of light are required to produce four high-energy electrons capable of reducing two $NADP^+$ to two NADPH. Three steps in the light-dependent reactions of photosynthesis help create a proton (electrochemical) gradient across the thylakoid membrane: (1) photolysis of two molecules of water releases four protons into the thylakoid lumen; (2) transfer of two pairs of electrons from photosystem II through the cyt b_6f complex (including the Q cycle) translocates eight protons from the stroma to the lumen; and (3) reduction of two $NADP^+$ removes two protons from the stroma.

The energy in the electrochemical gradient is used to synthesize ATP from ADP and Pi. Additional ATP can be produced by photophosphorylation. The net result is that NADPH and ATP are available for carbohydrate synthesis.

5.5   Carbohydrate Synthesis

5.5.1   C₃ plants

Carbohydrate synthesis requires **carbon fixation** which is incorporation of carbon dioxide into an organic compound. In C_3 plants, carbon fixation occurs in the initial step of the **Calvin cycle**, a cycle that occurs in the stroma of the chloroplast. As described in Sec. 5.4.2, the stroma also contains the ATP and NADPH required for carbohydrate synthesis. The carbon dioxide required for carbohydrate synthesis enters a leaf through pores called **stomata** (Fig. 5.20). It then diffuses into mesophyll cells and undergoes fixation in the stroma of the chloroplast. A continuous supply of carbon dioxide is available only if stomata remain open. However, open stomata can also lead to water loss, so a C_3 plant has to balance its needs for water and carbon dioxide.

In the Calvin cycle, carbon dioxide is fixed and reduced to form a 3-carbon carbohydrate. The first step in the cycle (fixation) involves linking carbon dioxide to the 5-carbon molecule *ribulose 1,5-bisphosphate (RuBP)* to form a *6-carbon intermediate*. The intermediate, probably still bound to enzyme, rapidly splits into two molecules of *3-phosphoglycerate (PGA)*. In Fig. 5.18, three molecules of carbon dioxide are fixed to three molecules of RuBP, producing six molecules of PGA. Fixation of carbon dioxide is catalyzed by *ribulose bisphosphate carboxylase (rubisco)*. Rubisco is an inefficient enzyme in that it catalyzes fixation of only three molecules of carbon dioxide per second. Moreover, the fixation (carboxylation) catalyzed by the enzyme is subject to competitive inhibition by oxygen. As a consequence, plants are required to produce large amounts of rubisco to maintain adequate rates of carbohydrate synthesis.

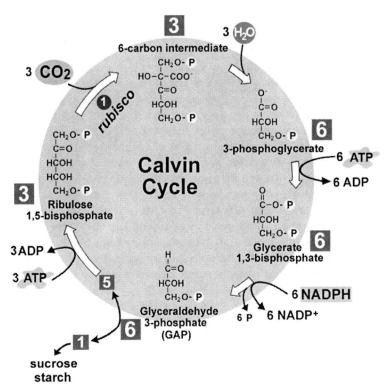

Figure 5.18   The Calvin cycle for converting carbon dioxide into carbohydrate. Three molecules of carbon dioxide are fixed to three molecules of ribulose 1,5-bisphosphate (RuBP) in a reaction catalyzed by the enzyme ribulose bisphosphate carboxylase (rubisco). The three 6-carbon intermediates rapidly split into six molecules of 3-phosphoglycerate (PGA) that are reduced to six molecules of glyceraldehyde 3-phosphate (GAP). One GAP is used for synthesis of sucrose or starch. The numbers in the squares indicate the number of molecules of each intermediate. For details, see Sec. 5.5.1

PGA is reduced to the 3-carbon sugar *glyceraldehyde 3-phosphate (GAP)* using the NADPH and ATP produced during the light-dependent reactions. As shown in Fig. 5.18, entry of three carbon dioxide molecules (i.e., a total of 3 carbons) into the Calvin cycle leads to formation of six molecules of GAP, but only one GAP (i.e., a total of 3 carbons) is used to synthesize *starch* or *sucrose*. Starch, which is a glucose polymer, is synthesized and stored in the chloroplast. In contrast, the disaccharide (glucose plus fructose) sucrose is synthesized in the cytosol and transported to various structures in the plant where it serves as a source of energy.

The remaining five GAP molecules (i.e., a total of 15 carbons) are used to regenerate the three 5-carbon RuBPs (i.e., a total of 15 carbons) that served as acceptors for the carbon dioxide. Several separate reactions (not shown) involving aldolases, transketolases, sedoheptulose bis-phosphatase, and other enzymes are required to convert the five GAP molecules into three RuBP molecules. Regeneration of RuBP may be an important control point in the Calvin cycle because increased expression of these enzymes increases the rate of carbohydrate synthesis. Additional ATP is then required to phosphorylate the RuBP.

Calvin-cycle reactions often are described as being light-independent or "dark" reactions. These descriptors are misleading, because the Calvin cycle is dependent upon factors produced by the light-dependent reactions. For example, the activities of three enzymes in the cycle (GAP dehydrogenase, sedoheptulose bisphosphatase, and ribulose 5-phosphate kinase) are indirectly dependent upon electrons from ferredoxin. Ferredoxin transfers electrons to thioredoxin, which then activates enzymes in the Calvin cycle by reducing key disulfide bonds. The activity of the Calvin cycle also is increased by high levels of magnesium, ATP, and NADPH in the stroma and by high stromal pH. All of these factors are dependent upon ongoing light-dependent reactions.

5.5.2 Photorespiration

Photorespiration in plants refers to light-enhanced uptake of oxygen and release of carbon dioxide (Fig. 5.19). The increased utilization of oxygen occurs because rubisco has some oxygenase activity in addition to its carboxylase (carbon fixation) activity. As discussed above, the carboxylase activity of rubisco catalyzes fixation of carbon dioxide to the 5-carbon RuBP and the subsequent formation of two molecules of the 3-carbon PGA. The oxygenase activity of rubisco catalyzes addition of oxygen to RuBP and the subsequent formation of one 3-carbon PGA and one *2-carbon 2-phosphoglycolate (PG)*. Phosphoglycolate cannot serve as a direct intermediate in the Calvin cycle, but it can enter the glycolate pathway which returns a usable 3-carbon glycerate to the chloroplast. The glycolate pathway is an important salvage pathway, but it is costly because in the pathway, one previously fixed carbon is lost as carbon dioxide and energy is consumed. Under some conditions of high light

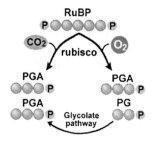

Figure 5.19 The reactions of photorespiration. Rubisco also can act as an oxygenase (as well as a carboxylase) and split RuBP into a 3-carbon PGA and a 2-carbon 2-phosphoglycolate (PG). PG can enter the glycolate pathway and return to the Calvin cycle as PGA. For details, see Sec. 5.5.2.

intensity, photorespiration can cause C_3 plants to lose up to 50% of newly fixed carbon dioxide, which can clearly hinder plant growth. When carbon dioxide levels are lowered artificially to 50 ppm (normally 381 ppm in atmosphere), the rate of carbon dioxide loss via photorespiration equals the rate of carbon dioxide fixation. This artificial situation has real-life correlates, because the carbon dioxide available for fixation decreases substantially when a leaf has to close its stomata to prevent water loss.

Photorespiration and water loss become problematic for plants in environments that are hot and dry. A C_3 plant must keep its stomata open to obtain carbon dioxide, but open stomata lead to water loss. If a C_3 plant closes its stomata to conserve water, then photorespiration causes net carbon fixation to approach zero as carbon dioxide levels in the leaf approach 50 ppm. To circumvent this problem, C_4 plants (see Sec. 5.5.3) and crassulacean acid metabolism (CAM) plants (see Sec. 5.5.4) have developed mechanisms that minimize the amount of time that stomata must remain open. Both types of plants use *phosphoenolpyruvate (PEP) carboxylase* to fix carbon dioxide (Fig. 5.21), and ultimately produce a 4-carbon malate molecule that serves as a reservoir of carbon dioxide. However, C_4 and CAM plants have different mechanisms for using the stored carbon dioxide.

5.5.3 C_4 plants

Plants which use C_4 metabolism include sugarcane, corn, and sorghum. In C_4 plants, photosynthetic tissues differ structurally and functionally from those in C_3 plants. For example, *bundle sheath* cells in C_4 plants are effectively shielded from the atmosphere by a ring of mesophyll cells (Fig. 5.20). This arrangement allows the mesophyll cells to regulate

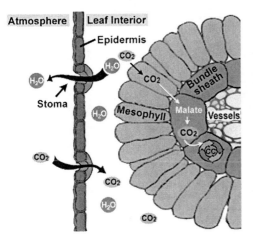

Figure 5.20 Leaf anatomy in a C4 plant. In the leaf of a C_4 plant, mesophyll cells form a cylinder around the bundle sheath cells, isolating them from the atmosphere. Carbon dioxide enters the leaf through a stoma (plural is stomata), diffuses into a mesophyll cell, is incorporated into malate, and is delivered to the Calvin cycle (CC) in bundle sheath.

the gases (e.g., carbon dioxide) that reach the bundle sheath cells. Another difference between C_3 and C_4 plants is that photosynthesis primarily occurs in the mesophyll cells in C_3 plants whereas photosynthesis occurs in both the mesophyll and bundle sheath cells in C_4 plants. Moreover, in C_4 plants, the Calvin cycle is restricted to chloroplasts in the bundle sheath cells (Fig. 5.20).

The cellular arrangement in C_4 plants and a new carboxylation/decarboxylation pathway allow these plants to decrease photorespiration by increasing the carbon dioxide concentration in the bundle sheath cells. The carboxylation/decarboxylation pathway is called the C_4 **pathway** or **Hatch-Slack pathway**. The pathway begins in the mesophyll cells where the cytosolic enzyme PEP carboxylase catalyzes carboxylation of PEP to form *oxaloacetate* (Fig. 5.21). Fixation of carbon dioxide (actually bicarbonate) by PEP carboxylase is a key step, because this enzyme is effective at low concentrations of carbon dioxide and does not exhibit the oxygenase activity of rubisco. As a result, net carbohydrate synthesis in C_4 plants continues until carbon dioxide levels reach 1 to 2 ppm versus 50 ppm in C_3 plants. Hence, C_4 plants can keep stomata closed for longer periods, preventing excessive water loss. Attempts have been made to introduce into C_3 plants some of the desirable properties of C_4 plants (see Application Box 5.2).

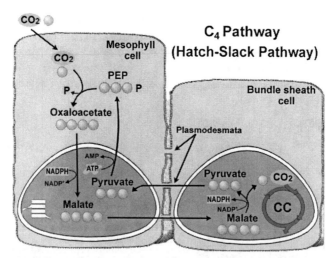

Figure 5.21 C_4 pathway (Hatch-Slack pathway) in a C_4 plant. In mesophyll cells, carbon dioxide is fixed to a 3-carbon molecule of phosphoenolpyruvate (PEP) to form oxaloacetate. The oxaloacetate is reduced to malate in the chloroplast and the malate is transported through plasmodesmata to the bundle sheath cells. Decarboxylation releases carbon dioxide that can enter the Calvin cycle (CC).

APPLICATION BOX 5.2
C_4 Pathways in C_3 Plants

Rice, wheat, and other important food crops are C_3 plants in which photosynthetic rates decrease in hot, dry environments with high light intensities. As might be expected, scientists have attempted to use genetic engineering to introduce parts of the C_4 pathway into these plants. Attempts also have been made to modify C_3 enzymes.

Cartoon of rice-corn hybrid

Corn genes coding for phosphoenolpyruvate (PEP) carboxylase, pyruvate orthophosphate dikinase, and $NADP^+$ malic enzyme have been cloned into rice. The **transgenic** rice plants can express high levels of these enzymes, but it is unclear whether the enzymes substantially increase photosynthesis or plant growth. Equivocal results are not surprising since the genetically modified C_3 plants do not contain all the C_4 enzymes and do not have the specialized leaf anatomy of C_4 plants. Both issues will have to be addressed in future studies.

Attempts to modify C_3 enzymes have met with some limited success. Many mutations have been introduced into the ribulose biphosphate carboxylase (rubisco) gene but thus far, the attempts have not increased rubisco activity. Indeed, most mutations have decreased the catalytic activity of the enzyme. Overexpression of sedoheptulose bisphosphatase, one of the enzymes responsible for regenerating ribulose 1,5-bisphosphate (RuBP) from glyceraldehyde 3-phosphate (GAP), increases the photosynthetic rate in tobacco. Accordingly, sedoheptulose bisphosphatase and other enzymes in the regenerating system may be attractive targets for genetic engineering.

Scientists also have engineered plants to express ictB, a gene involved in bicarbonate (i.e., carbon dioxide) accumulation in cyanobacteria. Plants expressing this gene show faster photosynthetic rates in conditions of limiting carbon dioxide, presumably due to the increased availability of bicarbonate for fixation.

Genetic engineering that can increase carbon dioxide availability or carbon fixation in plants has the potential to increase crop yields. Such engineering will be critical to keep up with the increasing demand for food.

Oxaloacetate produced by carboxylation of PEP is transported into the chloroplast where it is reduced by electrons from NADPH to *malate* in a reaction catalyzed by NADPH-dependent malate dehydrogenase (Fig. 5.21). The malate is then transported through plasmodesmata to adjacent bundle sheath cells. In these cells, $NADP^+$ malic enzyme catalyzes decarboxylation of malate, releasing carbon dioxide that can be fixed to RuBP by rubisco. The high concentration of carbon dioxide generated in bundle sheath cells by decarboxylation of malate decreases photorespiration and facilitates carbohydrate synthesis. Decarboxylation of malate also is accompanied by storage of electrons in NADPH—electrons needed for reduction of intermediates in the Calvin cycle. The net result is that C_4 plants are more resistant than C_3 plants to hot, dry environments with high light intensities.

The C_4 pathway has some costs. *Pyruvate* produced in bundle sheath cells by decarboxylation of malate has to be converted back to PEP in mesophyll cells. Regeneration of PEP is catalyzed by pyruvate orthophosphate dikinase (PPDK), an enzyme unique to the C_4 pathway. The reaction requires hydrolysis of ATP to AMP (Fig. 5.21), an energy loss which is the equivalent to conversion of two ATP to ADP. Because the C_3 pathway avoids this energy requirement, C_3 plants have an advantage in cooler, less arid, temperate regions (i.e., higher latitudes) where light intensity is lower.

5.5.4   CAM plants

Crassulacean acid metabolism (CAM) plants include cacti and a wide variety of succulent plants that are found in deserts and other arid regions. Like C_4 plants, CAM plants avoid opening their stomata during the day by increasing the availability of carbon dioxide through carboxylation/decarboxylation reactions. However, unlike C_4 plants, CAM plants do not partition carboxylation and decarboxylation reactions between two different types of cells. Instead, CAM plants run carboxylation and decarboxylation at different times in a single cell.

In CAM plants, carboxylation is carried out at night when cooler temperatures decrease water loss. Stomata open to allow entry of carbon dioxide, and PEP carboxylase catalyzes fixation of carbon dioxide to PEP, forming oxaloacetate and then malate. During the night, the malate is stored in the central vacuole to avoid decreasing the pH of the cytosol. During the day, the stomata close to prevent water loss, and malate moves into the cytosol from the vacuole. Decarboxylation of malate releases carbon dioxide which can be fixed to RuBP by rubisco. Photorespiration is low because the closed stomata prevent oxygen entry. Because CAM plants primarily open their stomata at night, they can attain water use efficiencies three- to sixfold higher than those of C_4 and C_3 plants.

Suggested Reading

Häusler, R. E., Hirsch, H. J., Kreuzaler, F. and Peterhänsel C. (2002). Overexpression of C_4-cycle enzymes in transgenic C_3 plants: A biotechnological approach to improve C_3-photosynthesis. *J. Exp. Bot.* 53: 591–607.

Jagendorf, A. T. (2002). Photophosphorylation and the chemiosmotic perspective. *Photosynth. Res.* 73: 233–241.

O'Keefe, D. P. (1988). Structure and function of the chloroplast cytochrome bf complex. *Photosynth. Res.* 17: 189–216.

Raines, C. A. (2006). Transgenic approaches to manipulate the environmental responses of the C3 carbon fixation cycle. *Plant Cell Environ.* 29: 331–339.

Rider, M.H., Bertrand, L., Vertommen, D., Michels, P. A., Rousseau, G. G. and Hue L. (2004). 6-Phosphofructo-2-kinase/fructose-2,6-bisphosphatase: Head-to-head with a bifunctional enzyme that controls glycolysis. *Biochem. J.* 381: 561–579.

Shin, M. (2004). How is ferredoxin-NADP reductase involved in the NADP photoreduction of chloroplasts? *Photosynth. Res.* 80: 307–313.

Stryer, L. (2002). *Biochemistry.* 5th ed., W.H. Freeman & Company, New York.

Weber, J. and Senior, A. E. (2003). ATP synthesis driven by proton transport in F1F0-ATP synthase. *FEBS Lett.* 545: 61–70.

Chapter

6

Cellular Communication

Taihung Duong, Ph.D.

OBJECTIVES

- To introduce the receptors on the surface and inside of cells
- To introduce representative neurotransmitters
- To explain the pathway of secretory molecules from production to exocytosis
- To highlight developmental interactions between transmitters and receptors

OUTLINE

READ
6.1 Membrane Receptors   147
6.2 Nuclear Receptors   153
WRITE
6.3 Neurotransmitter   158
6.4 Cell Secretion   168
READ and WRITE
6.5 Synaptic Interactions during Development   174

The human body consists of about 100 trillion (100 × 10^{12}) cells and the communication between them makes the whole organism more than the sum of its parts. Understanding how cells communicate with one another can shed light on the mechanisms of diseases resulting from lack of or miscommunication between cells. It is at the heart of tissue engineering to gain insight into cellular communication for medical applications such as transplantation, gene and drug delivery.

In this chapter, the molecules and mechanisms used in cell communication will be discussed. The main objective is to build on the reader's
understanding of the cell biology presented in other sections of the book
by focusing on the all-important cellular function, communication. This
overview is by no means exhaustive since new methods of cellular communication are uncovered frequently. It is meant to familiarize the
reader with common mechanisms of cellular communication and to
present examples of the myriad ways in which cells can regulate this
function. The emphasis of examples will be on nerve cells or neurons,
since they specialize in communication. In an analogy to a computer
receiving input information and processing it into output information,
this chapter is organized into two sections, the READ, or input section,
and the WRITE, or processing and output section. Each nerve cell in a
neural circuit "reads" incoming information received from thousands to
tens of thousands nerve inputs and "writes" outgoing information to
the next nerve cells in the circuit. This is achievable because each nerve
cell contains a large protein signaling machinery.

The READ Part of the Signaling Machinery

The cell membrane serves as the interface between the outside and
inside of the cell. Embedded in the cell membrane are a number of specialized proteins, known as ion channels, that enable inorganic ions
such as sodium (Na^+), potassium (K^+), calcium (Ca^{2+}), or chloride (Cl^-)
to cross the cell membrane. Inorganic ions carry a positive or negative
electric charge. As the ion channels transport the inorganic ions from
outside to inside the cell, the accumulation of a specific type of inorganic
ions inside the cell will change the membrane potential to reflect the
charge of that ionic species: for example, an increase in chloride (Cl^-)
ions will result in a more negative membrane potential. This change in
membrane potential will result in a physiological response from the
cell. In another example, if all the sodium channels on the cell membrane
open up all at once and allow sodium (Na^+) ions to rush into the cell,
the membrane potential will be driven toward a positive charge. In neurons, such an event will result in a nerve impulse or action potential
propagating to other nerve cells.

Ion channels are characterized by two common properties: they are
ion selective and they are gated. The first implies that sodium channels
will only conduct sodium ions and no other ionic species across the cell
membrane. The second means that, similar to a gate, they open or close
briefly in response to a specific stimulus.

At the present time, ion channels can be classified into three general
groups: voltage-gated channels, mechanically gated channels, and
ligand-gated channels.

Voltage-gated channels open or close with a voltage change in the membrane potential. Thus, they transduce an electrical signal into a physiological response.

Mechanically gated channels are associated with mechanoreceptors, that is, receptors that transduce a mechanical stimulus into a physiological response. Hair cells in the cochlea of the inner ear are examples of mechanoreceptors (see Fig. 2.9). Mechanically gated channels thus open under the influence of a mechanical force to allow flow of inorganic ions.

Ligand-gated channels are further subdivided into nucleotide-gated, ion-gated, or transmitter-gated channels. These channels will open when their specific ligand (nucleotide, ion, or transmitter) binds to them. Although they convert a chemical signal (the ligand) into an electrical response (change in membrane potential due to ionic influx), they tend to be insensitive themselves to changes in the membrane potential. The size of the physiological response elicited by the ligand-gated channels depends largely on the amount and availability of the ligand. In this chapter, the emphasis will be placed on transmitter-gated channels since they are common in nerve cells and they are the targets for psychoactive drugs used in the treatment of neurological conditions.

6.1 Membrane Receptors

The cellular machinery responsible for the READ function resides in the cell surface **receptors**, which are proteins embedded in the cell membrane. For information on the signaling of nuclear and membrane receptors, refer to Secs. 4.3 and 4.4, respectively. In this section, the emphasis is on the structure of neuronal receptors and their mechanisms involved in cellular communication.

Membrane receptors are classified into three types: ionotropic receptors, G-protein-coupled receptors (metabotropic), and protein kinase-associated receptors. The main difference between the three types of receptors is due to their mechanism of action.

6.1.1 Ionotropic receptors

Ionotropic receptors are located in the cell membrane and are protein complexes formed by multiple subunits. These subunits combine to form a channel through the cell membrane (Fig. 6.1). In the inactive state, the channel is closed. When activated by the binding of a ligand to the receptor, the channel opens and admits an influx of ions specific for the channel. This ion influx results in a change in the membrane current, affecting the functional state of the cell.

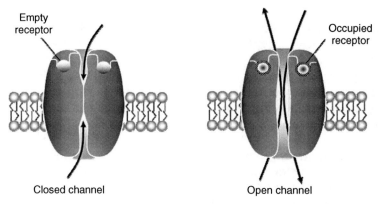

Figure 6.1 Ionotropic receptor. The ion channel is closed in the inactive state (left) and only opens when a specific ligand binds to the receptor (right).

Activation of the channel in the ionotropic receptor requires binding of receptor-specific molecules. Pharmacological agents that trigger the opening of the channel when bound to the receptor are termed **agonists**. Those that inhibit the opening of the channel, and thus the function of the receptor, are **antagonists**. The discovery of agonists and antagonists was a major advance in understanding the function of receptors.

In many cases, the receptor is named after the pharmacological agonist initially used to study its function. For example, the most common excitatory transmitter in the brain is glutamate, which binds to several types of glutamatergic receptors. One type of glutamate receptor is named the NMDA receptor. NMDA stands for *N*-methyl-*D*-aspartate, a pharmacological agent which binds only to a specific type of glutamate receptors. Another type of glutamate receptor is the AMPA receptor, due to its specific binding with AMPA (amino-3-hydroxy-5-methylisoxazoleproprionic acid). Finally, kainate or quisqualate receptors are other glutamatergic receptors named after their agonists. For convenience, the latter three receptors are classified as *non-NMDA receptors*.

In the brain, ionotropic receptors mediate communication using classical **neurotransmitters**: acetylcholine (nicotinic acetylcholine receptor), γ-aminobutyric acid (GABA$_A$ receptor), serotonin (serotonin 5-HT3 receptor), and glutamate (NMDA and non-NMDA receptors).

Ionotropic receptors are formed from the assembly of subunits. For example, the nicotinic cholinergic receptor, which binds the neurotransmitter acetylcholine, is formed by five subunits named α, β, γ, and δ and contains two α subunits. Each of the subunits is further divided into several subtypes with differing physiological properties. The large number of possible combinations between the different subunit subtypes creates a great diversity of ionotropic receptors, which can respond to the same ligand. In a simple analogy, the master password (ligand) can unlock many restricted applications (receptors) in a computer. By this strategy, the synthesis of only a small number of ligands is necessary to elicit a wide variety of physiological responses.

6.1.2 G-protein-coupled receptors (GPCRs)

G-protein-coupled receptors (GPCRs) or metabotropic receptors are a very diverse group of over one thousand identified receptors that have been classified into three subfamilies: the *rhodopsin-adrenergic receptors*, the *secretin-vasoactive intestinal peptide receptors*, and the *metabotropic glutamate receptors*. They share the common mechanism that binding of the ligand to the receptor activates a G-protein, which, in turn, activates secondary messengers leading to physiological effects in the cell (Fig. 6.2). For details on the GPCR signaling cascade, see Sec. 4.4.1.

Because of the intermediate steps required in activation of the secondary messengers, the effects in the cell are slow in onset but they last longer. In the brain, classical neurotransmitters as well as neuropeptide transmitters can activate G-protein. Unlike classical neurotransmitters, which consist of small, low molecular weight compounds, neuropeptides are larger molecules built from several amino acids (see Sec. 5.3.2).

The ability to bind to both types of receptors increases the degree of control of the classical neurotransmitters, which can elicit fast responses over seconds by way of the ionotropic receptors and slower responses over tens of seconds through the GPCRs. Communication between nerve

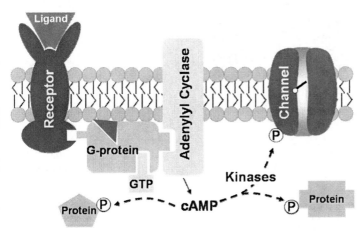

Figure 6.2 G-protein-coupled receptor. Binding of the specific ligand (transmitter) to the receptor results in activation of the G-protein, which in turn activates a secondary messenger (cAMP) triggering eventually a physiological function (gate opening in the ionic channel and phosphorylation of proteins).

cells can thus be modulated with greater precision through the action of the two different types of receptors.

GPCRs exist for acetylcholine (muscarinic acetylcholine receptors), the catecholamines (dopaminergic receptors and adrenergic receptors), serotonin (5-HT receptors types 1, 2, 4-7), glutamate (metabotropic glutamate receptors), γ-aminobutyric acid (GABA$_B$ receptor), and for the neuropeptide transmitters (e.g., neuropeptide Y receptors types 1-5).

A GPCR is a single polypeptide shaped into a 7-transmembrane protein with three intracellular and three extracellular loops (Fig. 6.3). The N-terminus of the receptor protein is located extracellularly and the C-terminus, intracellularly. The neurotransmitter-binding portion of the receptor exists as a pocket in the middle of the seven transmembrane segments (Fig. 6.4). The receptor domain, which interacts with the G-protein appears to be within the second and third intracellular loops. This segment is responsible for the efficiency of coupling and the specificity of the G-protein. The diversity of the cellular response is enhanced by the ability of the GPCRs to interact with different G-proteins: for example, in the case of the α2-adrenergic receptor, even by binding to the same ligand, it can activate four different G-proteins and thus four different pathways linked to these G-proteins within the cell.

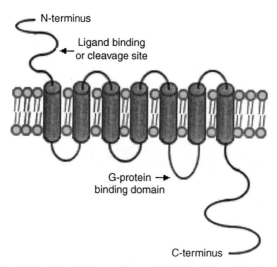

Figure 6.3 G-protein-coupled receptor. Note the typical structure of 7-transmembrane proteins with three extracellular and three intracellular loops. (Reproduced with permission from Bushell, R. and Thompson, University of Strathclyde, Glasgow, Department of Physiology and Pharmacology, (http://spider.science.strath.ac.uk/PhysPharm/showPage.php?u=TBushell&includePage=staffDetails.php)

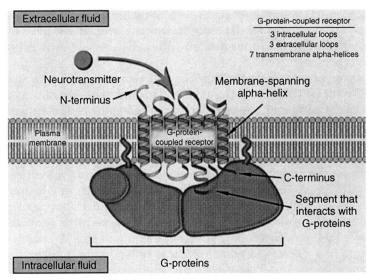

Figure 6.4   G-protein-coupled receptor. Note that the neurotransmitter binds to the pocket formed by the 7-transmembrane proteins of the G-protein-coupled receptor.

Once activated, the G-protein will trigger secondary messengers. The response of the cell is determined by which secondary messenger is activated. For example, in the α2-adrenergic receptor, the G-protein can either activate the enzyme phospholipase C (PLC), or PLC plus another enzyme phospholipase A_2, further increasing the complexity of the cell response.

Within the ligand-binding segment of the receptor protein, specific sets of amino acids will bind to different agonists or antagonists. Agonists, when bound to the receptor, tend to stabilize the activated state of the receptor. Further, the **affinity** of the GPCR for its agonist is increased when it is coupled to the G-protein in a positive feedback mechanism (for receptor affinity, see Sec. 4.2). Antagonists exert their effect in two different ways. Either they can bind to the inactive form of the receptor and are then termed **negative antagonists**, or they can bind to both the active and inactive forms, preventing transition of the receptor into the activated form and are then called **neutral antagonists**.

Binding of a ligand to the receptor will result in activation of the G-protein, leading to a physiological response by the cell. The complexity of responses available to the cell expressing GPCRs is increased in the following ways. Recall that a GPCR is a 7-transmembrane protein (Fig. 6.3). Now consider that contiguity between the different transmembrane segments is not necessary for function: it has been shown that the complex formed by the transmembrane segments one to five and the complex formed by segments six to seven only have to be present in the same cell for binding of the ligand. These complexes do not have to be physically located next to one another, although they both have to be present intracellularly.

Another layer of complexity within the cell response is due to the ability of different GPCRs within the same cell to bind to each other, thus forming different combinations of receptors. These new combinations allow the binding of different combinations of ligands, which consequently allow the cell a wider range of responses.

Modulation of the response is further accomplished in several ways. The cell can increase the magnitude of the response by recruiting or synthesizing more GPCRs. An increased number of GPCRs results in a higher probability for an encounter between the ligand and the receptor. Conversely, the cell can also lower the magnitude of the response through a process called **desensitization**: a constant amount of neurotransmitter will result in a decreased response, thus reducing the dangerous effects of overstimulation. This can be accomplished by two different mechanisms. First, the receptor can be phosphorylated by protein kinase A (PKA), protein kinase C (PKC), or G-protein receptor kinase (GRK). This process is fast, occurs in a few seconds, and leads to quick desensitization of the GPCR for its ligand. A second slower mechanism, which happens over tens of seconds, involves removal of the receptors from the membrane by **internalization** or by **downregulation**. Internalization is a reversible process because the receptors are only removed from the membrane and can be reinserted into it when needed. Downregulation is irreversible because the receptors are dismantled.

Finally, the complexity of responses afforded by the diversity of the separate ionotropic receptors and GPCRs is further augmented by the recent observation that these two different types of receptors can also associate with each other and produce functional units with a greater range of responses.

6.1.3   Protein kinase-associated receptors

These receptors are also activated by extracellular binding of a ligand, but instead of opening a channel or activating a G-protein, the binding of the extracellular signal regulates the catalytic activity of receptor-associated protein kinases. These enzymes (often of the tyrosine kinase types) phosphorylate intracellular target proteins which trigger signaling cascades (see Sec. 4.4.2). Members of this receptor family are often receptors for growth factors such as the **neurotrophins**. In the brain, the family of neurotrophins binds to protein kinase-associated receptors. This leads, depending on the signaling pathways, to cell survival, growth of nerve cell processes, or modulation of neuronal plasticity.

6.2   Nuclear Receptors

Endocrine glands secrete products called hormones into the circulatory system and these are eventually brought into contact with their target tissues. Certain types of hormones, for example steroid hormones, can freely pass the plasma membrane of the target cell and then act by binding to intracellular receptors. The complex of hormone and receptor is then translocated into the nucleus of target cells, where it affects the transcription of genes by binding to certain DNA sequences called **promoters**. By regulating gene transcription, hormones can thus have a profound effect on physiological functions.

Recent advances in imaging techniques, allowing for *in vivo* monitoring of protein trafficking within live cells, have made it possible to study the events following the binding of hormones to their receptors. These events are described in Sec. 6.2.1 with the example of the hormone cortisol and its receptors, but the characteristics of the cortisol-receptor interactions can be generalized to other steroid and nuclear hormone receptors.

Overall, there have been at least 48 nuclear receptors identified in the human genome. They are classified into type I or type II receptors according to their mechanism of action, or subdivided into seven subfamilies based on their amino acid sequence. The subfamilies include receptors for steroid hormones such as cortisol and receptors that are activated by nonsteroidal compounds such as retinoids, thyroid hormones, and vitamin D. Many of the nuclear receptors are still known as orphan receptors, because their *in vivo* ligands are not yet identified.

6.2.1   Steroid hormone receptors

Steroid hormones are all derived from cholesterol. Hence, they are all lipid-soluble and diffuse from the blood through the cell membranes of the target cells, where they act as transcription factors and initiate important cellular responses. Cortisol is a steroid hormone secreted from the adrenal gland, and it plays a significant role in processes as diverse as human development, metabolic response to stress, energy metabolism, and aging. Cortisol acts through two intracellular receptors: the mineralocorticoid receptor and the glucocorticoid receptor (Nishi and Kawata, 2006). Both receptors can be found in the cytoplasm of target cells and are rapidly translocated into the nucleus when bound to their ligand, cortisol. However, some studies have also reported the presence of these corticosteroid receptors in the nucleus even in the absence of its ligand.

In the cytoplasm, unbound glucocorticoid receptor or mineralocorticoid receptor (*R*, Fig. 6.5) are part of a protein complex that includes the *heat shock protein 90 (hsp90)*, an immunophilin protein of FK506 family,

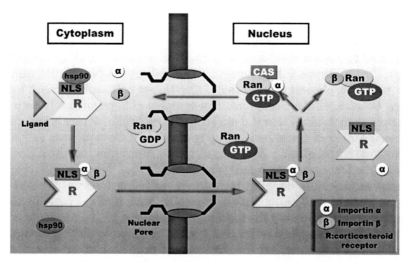

Figure 6.5 Nuclear translocation of the corticosteroid receptor. The ligand cortisol (triangle on the left) binds to the corticosteroid receptor (R), which also contains nuclear localization signals (NLS). The transport mechanism (importin α and β) binds the NLS and docks with the nuclear pore complex, which then translocates the whole corticosteroid receptor into the nucleus. (Reproduced with permission from Waxman, S.G., Sage Publications Inc.; Nishi, M. and Kawata, M. (2006). Brain corticosteroid receptor dynamics and trafficking: Implications from live cell imaging, *Neuroscientist*. 12:119–133.)

and the protein p23 (not shown). These proteins will dissociate from the glucocorticoid receptor when it becomes activated. During the process of synthesis and maturation, the receptors are also transiently associated with other proteins, termed molecular **chaperones**, such as heat shock protein 70 or Cyp40. It was previously thought that the association of the receptors with their chaperone protein complex in the cytoplasm prevented translocation into the nucleus. However, recent results have shown that some unbound corticosteroid receptors are also assembled into protein complexes in the nucleus. Thus, the location of the complex formed by association of the steroid receptor with their chaperone proteins appears to be specific for cell types, rather than intracellular functions.

In experimental cell types such as the COS-1 cells (a transformed cell line from fibroblasts derived from the normal kidney of the African green monkey), the mineralocorticoid receptor has about a tenfold greater affinity for cortisol than the glucocorticoid receptor. This means that at low cortisol concentration [10^{-9} Molar (M)], the mineralocorticoid receptor is activated for binding to cortisol, while the glucocorticoid receptor only becomes activated at higher levels of the ligand (10^{-6} M or a thousand times higher). Two different receptors are thus used for the same hormone depending on its concentration, with the mineralocorticoid receptor operating at physiological concentrations of cortisol and the glu-cocorticoid receptor participating at high cortisol levels. Interestingly, this difference in affinity for the ligand by the corticosteroid receptors is not present in cultured nerve cells from the hippocampus of the temporal lobe of the brain, suggesting that mechanisms for nuclear translocation of bound corticosteroid receptors differ between cell types.

When cortisol binds to either the glucocorticoid receptor or the mineralocorticoid receptor, it induces a loss of associated chaperones such as hsp90 from the protein complex, which holds the receptor in the cytoplasm, triggering a change in the protein shape or conformation of the receptor. This conformational change exposes signal proteins known as *nuclear localization signals* (*NLS*, Fig. 6.5).

The transport mechanism recognizes the NLSs and binds to the receptor complex. In the case of the glucocorticoid receptor and the mineralocorticoid receptor, the transport mechanism is formed by proteins of the importin family. *Importin α* binds to the nuclear localization signals as well as to *importin β*. The latter docks the *nuclear pore complex*, which then mediates the translocation into the nucleus. It remains unclear as to how the bound steroid receptor complex is shuttled from cytoplasm to the nucleus. Microtubules, which are cytoskeletal elements for intracellular transport of other proteins (see Sec. 2.3.2), do not seem to be involved.

The nuclear localization signals are important in mediating the translocation process, not only because they are recognized and bound by importin α, but also because different types of nuclear localization signals exist with different properties. For example, the glucocorticoid receptor contains two nuclear localization signals named NLS-1 and NLS-2. NLS-1 is involved in rapid transport into the nucleus with a half-time of 5 to 15 min, inducible by binding of either agonist or antagonist to the glucocorticoid receptor. This rapid transport is active when the glucocorticoid receptor is expressed at high levels in the cell. NLS-2, on the other hand, induces a slow nuclear transport with a half-time of 45 min, and this occurs only with agonist binding. Thus, depending on the environmental and cellular conditions, one or the other nuclear localization signal will be favored, increasing the versatility of the cell response.

The corticosteroid receptors also interact with molecules other than their specific ligands: cyclic AMP (cAMP) is a signaling molecule, which, when present, can activate mineralocorticoid receptors located in the nucleus, even in the absence of the ligand. This action of cAMP results in low levels of transcription activity by the cell. Transcriptional activity is increased when the ligand is present and more mineralocorticoid receptors are translocated from the cytoplasm to the nucleus. Thus, cAMP and the ligand both act in promoting transcription activity but at different levels.

Once in the nucleus, the translocated corticosteroid receptors have been observed to cluster in certain nuclear regions, although these clusters do not always colocalize with areas of transcription activity in the nucleus. The clusters are formed by single receptors as well as receptors bound to each other. Both the glucocorticoid receptor and the mineralocorticoid receptor can bind with receptors of their own type to form **homodimers** through interface in a particular protein region known as the **zinc finger region**. Interestingly, the zinc finger region of glucocorticoid receptors and that of mineralocorticoid receptors have the same amino acid sequence. Studies have shown that the glucocorticoid receptor and the mineralocorticoid receptor can also bind to one another to form **heterodimers**, and that this reaction depends on the level of cortisol. As discussed above, at low levels of the ligand, mineralocorticoid receptors have a greater affinity for cortisol than glucocorticoid receptors, and homodimer formation between mineralocorticoid receptors is favored. At higher concentrations of the ligand, when the glucocorticoid receptors are activated, the probability of heterodimer formation with mineralocorticoid receptors is increased. The usefulness of heterodimer formation lies in the increased versatility for regulating gene expression, by having not just two types of corticosteroid receptors affecting transcription, but also homo- and heterodimers of these receptors. Surprisingly, *in vitro* studies have demonstrated both synergistic and inhibitory regulation of transcription by heterodimer formation of corticosteroid receptors. The precise role of heterodimer formation of steroid receptors remains to be elucidated.

Within the nucleus, the corticosteroid receptors also interact with cofactors known as steroid receptor coactivators or SRCs. These SRCs act as regulators of corticosteroid receptors-induced transcription by recruitment of other coactivators or by modification of chromatin structure. The presence of SRCs is required for transcription, and the amount of SRCs present in the nucleus can limit the level of activation: low levels of SRCs result in low level activation and higher levels increase the level of activation.

It must be noted that the physiological functions mediated by the nuclear transport pathway are not confined solely to the transport of hormones from the cell body to the nucleus. In nerve cells, importin α and β play diverse roles in the retrograde transport of proteins for processes extending from the cell body back to the nucleus. For example, transport from the tip of the growing axon (termed the **axonal growth cone**) to the nucleus during cell development, from the dendrites to the nucleus, and also from an area of axonal injury to the nucleus has been observed. Studies have demonstrated an association between the activities of the nuclear transport and the processes that induce long-term changes in communication at the **synapse** (the contact point between nerve cells). Thus, the nuclear transport proteins such as the importins play an important role in signaling between the synapse and the nucleus in nerve cells. Furthermore, the implication that factors produced by synaptic activity and carried to the nucleus can alter nuclear transcription suggests that communication between nerve cells may have profound effects not just at the level of the synapse but also at the transcription level in the nucleus.

The WRITE Part of the Signaling Machinery

The most prevalent method of communication is chemical transmission, illustrated in this section by the communication between nerve cells.

Nerve cells or neurons have special shapes: they have a cell body or **perikaryon** from which processes extend such as dendrites and an axon. In Fig. 6.6 drawn by the famous neuroanatomist Ramon Y. Cajal, there are two types of nerve cells: the large cells are *Purkinje cells* (*A*, Fig. 6.6) and the smaller ones are *granule cells* (*B*, Fig. 6.6). The dendrites of the Purkinje cells ramify densely above the cell body, whereas the thin axon originates below the cell body. In the granule cell, the dendrites are the three or four short processes ending in a tuft, and the axon is the thin, long structure stretching upward to meet with the Purkinje cell dendrites. Thus, the dendrites remain close to the cell body whereas axons can project over a long distance to make contact with other cells. For example, neurons that control muscles in the foot have their cell bodies located in the spinal cord about half way down the back. To reach the foot, their axons project from this location along the rest of the back and the whole length of the lower limb, a distance of over 1 m (over 3 ft) in an individual of average height. In general, the dendrites and cell body receive signals from previous nerve cells in the connecting circuitry and the axon transmits signals to the next nerve cells. The neuron is described as polarized since signals travel in one direction: from the dendrites to the cell body and then into the axon to reach the terminal endings, which interact with other neurons or muscle cells.

The area of interaction between the transmitting (presynaptic) and the receiving (postsynaptic) nerve cells is a small intercellular gap called the synapse (Fig. 6.7). At the synapse, the presynaptic neuron, when stimulated, releases a transmitting chemical or **neurotransmitter**. The latter crosses the synaptic gap by diffusion and binds to cell surface receptors, which are specialized proteins on the membrane of the postsynaptic neuron. Neurotransmitter-receptor interactions result in activation or inhibition of the postsynaptic neuron. These events constitute synaptic chemical transmission between a signal-emitting neuron and a signal-receiving neuron.

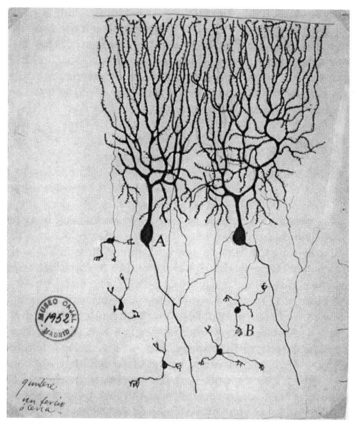

Figure 6.6   Differing morphologies of neurons. Note the large cerebellar Purkinje neurons (A) with extensive dendritic architecture and compare to the small granule cells (B) with sparse dendrites. (Public domain figure produced by Santiago Ramón y Cajal, 1899. Instituto Santiago Ramón y Cajal, Madrid, Spain, and displayed at *http://www.answers.com/topic/ purkinjecell-jpg.*)

6.3   Signaling Molecules

Neurotransmitters vary greatly in their composition. Acetylcholine, the biogenic amines, and some amino acids form the so-called classical neurotransmitters (because they were first to be discovered and studied extensively; see Sec. 6.3.1). Other neurotransmitters are the neuropeptides and neurosteroids (see Sec. 6.3.2). Finally, recent research has revealed that neurons can also release gaseous transmitters such as nitric oxide or carbon monoxide to affect nearby nerve cells. They constitute the third class of neurotransmitters called unconventional transmitters.

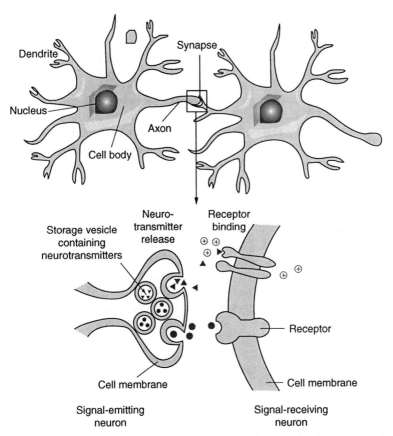

Figure 6.7   The synapse. The axonal terminal from the presynaptic neuron and the dendritic membrane from the postsynaptic neuron form the synapse. The neurotransmitter is released from the presynaptic terminal, diffuses through the synaptic cleft, and binds to receptors located on the membrane of the postsynaptic neuron. (Public domain figure produced by Goodlett, C.R. and Horn, K.H. Mechanisms of alcohol-induced damage to the developing nervous system. *Alcohol Res. Health.* 2001;25(3):175–184, and displayed at National Institute on Alcohol Abuse and Alcoholism of the National Institutes of Health, *http:// www.niaaa.nih.gov/Resources/GraphicsGallery/Neuroscience/synapse.htm.*)

For a substance secreted by a nerve cell to qualify as a neurotransmitter, it must have the following characteristics:

■ The presynaptic neuron must contain the appropriate mechanism for synthesis and secretion of the neurotransmitter.

■ Presynaptic nerve terminals must release the neurotransmitter in a form identifiable by biochemical or pharmacological techniques.

■ The effects of the neurotransmitter on the postsynaptic cell must reproduce the effects obtained by stimulation of the presynaptic neuron.

- Competitive antagonists as well as inhibitors of synthesis must block the effects of the presynaptic stimulation.

- Finally, there must be synaptic mechanisms, such as enzymatic inactivation or reuptake, to terminate the effects of the neurotransmitter.

The process of chemical transmission includes several steps: the synthesis, the storage, and the active release of the neurotransmitter, as well as the interaction of the neurotransmitter with postsynaptic receptors and the termination of this process. These steps are explained in the following discussion of the classical neurotransmitter acetylcholine (ACh) and illustrated in Fig. 6.8. ACh is also the transmitter at the synapse between nerve cells, and cardiac and voluntary muscle cells (**neuromuscular junction**) in mammals. Neurons using ACh as a neurotransmitter are termed cholinergic.

The first step in chemical transmission *involves the synthesis of the neurotransmitter.* As for all classical neurotransmitters, synthesis of ACh takes place in the presynaptic terminal far away from the nerve cell body: the enzyme choline *acetyltransferase* (*ChAt* in Fig. 6.8) transfers the acetyl group from acetyl-coenzyme A to choline. The required

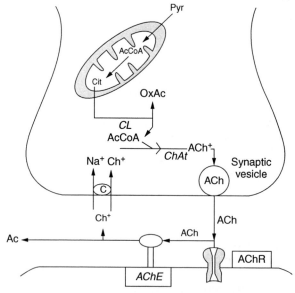

Figure 6.8 Cholinergic synapse. Acetylcholine (ACh) is synthesized by choline acetyltransferase (ChAt) in the presynaptic membrane. After release into the synaptic cleft, ACh binds to the cholinergic receptor (AChR). The effect of ACh is terminated by the enzyme acetylcholinesterase (AChE). Ac = acetic acid; AcCoA = acetyl-coenzyme A; C = choline transporter; Ch^+ = choline; CL = citrate lyase; OxAc = oxaloacetate; Pyr = pyruvate.

agents for this reaction are compartmentalized within the presynaptic terminal: *acetyl-coenzyme A* (*AcCoA*) exists within the mitochondria whereas *choline* (Ch^+) and ChAt are cytoplasmic. Thus, the synthesis of ACh within the presynaptic terminal can be controlled by the rate at which AcCoA is released from the mitochondria or by the availability of Ch^+ within the cytoplasm.

The next step in chemical transmission is *the storage of the neurotransmitter*. The vesicular cholinergic **transporter** is responsible for the translocation of the newly synthesized acetylcholine into presynaptic vesicles where it is protected from degradation by enzymes. As for all classical neurotransmitters, the synaptic vesicles are small in diameter (~50 nm), and they are arranged in a predetermined spatial arrangement in the presynaptic terminal for quick release.

The following step, *the active release of neurotransmitter into the synaptic cleft*, occurs when repetitive firing of the neurons occurs. The transmitter release is regulated by extracellular stimulation of the presynaptic neuron. The presynaptic neuron generates depolarizing signals, which travel along the axon to its presynaptic terminals triggering fusion of the synaptic vesicles with the presynaptic cell membrane and the release of the neurotransmitter into the synaptic cleft (for a discussion on secretion, see Sec. 6.4).

In the next step, *the neurotransmitter interacts with receptors on the postsynaptic as well as the presynaptic membrane*. Membrane receptors include ionotropic receptors and G-protein-coupled receptors (GPCRs) (see Sec. 6.1). In the case of ACh, there are two types of cholinergic receptors: nicotinic and muscarinic. These differ in their mechanism since nicotinic receptors are ionotropic and muscarinic receptors are GPCRs. Nicotinic and muscarinic receptors have different distributions in the body and subserve different functions; furthermore, each type of receptors can be differentiated into different subtypes. For example, the muscarinic cholinergic receptors include five different subtypes numbered M1 to M5, differentiated by function as well as by distribution in the body. M1 cholinergic receptors modulate signaling in brain areas such as the cortex and the hippocampus. M2 cholinergic receptors are involved in tremor and hypothermia. M3 cholinergic receptors mediate gland secretion, contraction of smooth muscle, and pupil dilation. M4 cholinergic receptors modulate the activity of the neurotransmitter dopamine in brain motor functions. Finally, M5 cholinergic receptors control the muscle tone of cerebral blood vessels. Thus, the existence of different receptor types and subtypes increases the versatility of the same neurotransmitter, in our example ACh, in triggering and modulating different physiological functions.

The last step is *the termination of the action of the released neurotransmitter*. This must occur because continuous stimulation of the postsynaptic cell could threaten its survival. Passive and active mechanisms of transmitter termination exist. The passive termination occurs when the neurotransmitter diffuses away from the receptor and becomes diluted into the extracellular fluid to insignificant concentrations. The cell membrane on the pre- and the postsynaptic cells also contain active mechanisms to terminate the action of the released neurotransmitter. In the case of ACh, termination is primarily by the action of the membrane enzyme *acetylcholinesterase* (*AChE* in Fig. 6.8) secreted into the synaptic cleft where it exerts its effect on the neurotransmitter. Acetylcholinesterase breaks down ACh into *acetic acid* (*Ac*) and *choline* (*Ch$^+$*). Choline is transported back into the presynaptic terminal by the action of the *choline transporter* (*C* in Fig. 6.8), to be stored and reused in neurotransmitter synthesis. As its name indicates, the choline transporter is a specialized membrane protein, which binds to choline and shuttles it back into the presynaptic terminal in a sodium-dependent process. Termination of the neurotransmitter action ends the process of synaptic transmission.

Electrophysiological studies of nerve cells have revealed that communication is not limited to active periods (i.e., periods of active stimulation by the presynaptic neuron), but appears to be ongoing at all times. Recordings during inactive periods at the neuromuscular junction show that the synapse is not silent. Instead, small (1 mV) electrical signals termed **miniature end-plate potentials** can be continuously detected (Fig. 6.9). The mechanism that generates miniature end-plate potentials remains unclear. The amount of ACh necessary to generate a miniature end-plate potential is termed a **quantum**. The most commonly accepted model posits that each quantum is the result of the interaction of one synaptic vesicle with a fusion pore located on the

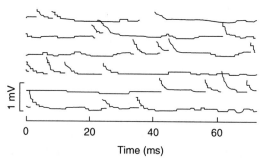

Figure 6.9   Miniature end-plate potentials recordings. Each upward deflection represents one miniature end-plate potential. (Reproduced with permission from Purves, D., Augustine, G.J., Fitzpatrick, D., Hall, W.C., Lamantia, A.-S., McNamara, J.O. and Williams, S.M. *Neuroscience*, 3rd ed., 2004, Sinauer Associates, Inc.)

presynaptic membrane and subsequent release of the content of this single synaptic vesicle.

However, another model for quantal release also exists whereby arrays of synaptic vesicles docked to the nerve terminal membrane by a complex of fusion pores synchronously release ACh to generate a quantum. In Fig. 6.10a, the presynaptic terminal of a frog neuromuscular junction has been processed by the freeze-fracture technique. The electron micrograph image on the left of Fig. 6.10a shows an unstimulated presynaptic terminal. The dots in the picture represent particles that are thought to be calcium channels. The electron micrograph image on the right shows a terminal stimulated by an action potential. The dimple-like structures in the picture represent the fusion pores by which the synaptic vesicles fuse with the presynaptic membrane. Figure 6.10b is the schematic drawing of synaptic vesicles interacting with the presynaptic membrane to form such fusion pores. In the schematic, calcium channels are symbolized as cylinders integrated in the presynaptic terminal membrane. Based on freeze-fracture and other studies, a new model of transmitter release is developed, which proposes that miniature end-plate potentials caused by the simultaneous fusion of several synaptic vesicles may not be dismissed as "background noise" but may constitute a low level of intercellular communication with potential physiological significance. Moreover, the model takes into account the existence of subminiature end-plate potentials (less than 1 mV), which may be the result of partial transmitter release from a single

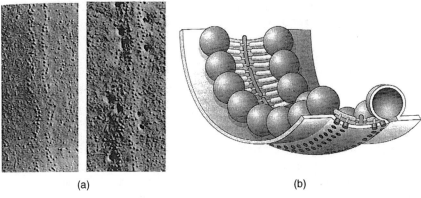

(a)   (b)

Figure 6.10 Fusion pore. (a) The left scanning electron micrograph shows the synaptic side of an inactive presynaptic membrane. After depolarization and transmitter release, fusion pores are visible in the right scanning electron micrograph. (b) The drawing shows how synaptic vesicles interact with the presynaptic membrane to form fusion pores through which neurotransmitter is released. (Left figure reproduced with permission from Heuser, J.E., Reese, T.S., Dennis, M.J., Jan, Y. and Jan, L. (1979). Synaptic vesicle exocytosis captured by quick freezing and correlated with quantal transmitter release. *J. Cell Biology*. 81: 281. Right figure reproduced with permission from Purves, D., Augustine, G.J., Fitzpatrick, D., Hall, W.C., Lamantia, A.-S., McNamara, J.O. and Williams, S.M. *Neuroscience*, 3rd ed., 2004, Sinauer Associates, Inc.)

fusion pore complex representing another degree of low-level intercellular communication.

6.3.1   Classical transmitters

The classical transmitters are ACh, the biogenic amines, or products of intermediary glucose metabolism such as the amino acids γ-aminobutyric acid (GABA), glutamate, or aspartate. The term **biogenic amine** refers to a group of biologically active amines, which serve as neurotransmitters. They include the catecholamines: dopamine, norepinephrine, and epinephrine. The last two names have replaced the older nomenclature of noradrenaline and adrenaline. Another important biogenic amine is 5-hydroxytryptamine or serotonin.

In Sec. 6.3, the formation of ACh by the enzyme choline acetyltransferase (ChAt) was presented. However, synthesis of classical transmitters can also occur by sequential action of different enzymes. For example, the first step in the synthesis of the catecholamines begins with the conversion of the amino acid *phenylalanine* to *tyrosine* by the enzyme *phenylalanine hydroxylase* (Fig. 6.11). Breakdown of proteins eaten during a meal provides for the major source of phenylalanine. Next, tyrosine is accumulated in the nerve cell and converted by the enzyme *tyrosine hydroxylase* to *3,4-dihydroxyphenylalanine (DOPA)*. The latter is in turn rapidly converted to *dopamine* by the enzyme *aromatic amino acid decarboxylase (AADC)*. Thus, nerve cells, which use dopamine as a neurotransmitter, manufacture only the enzymes tyrosine hydroxylase and AADC. These cells are referred to as dopaminergic neurons. Noradrenergic neurons, which utilize norepinephrine as a neurotransmitter also manufacture the enzyme *dopamine-β-hydroxylase (DBH)* to convert dopamine to *norepinephrine*. Finally, adrenergic neurons synthesize the enzyme *phenylethanolamine N-methyltransferase (PNMT)* for the conversion of norepinephrine to *epinephrine*. Thus, the sequential action of enzymes contained within the same nerve cell can synthesize the specific neurotransmitter required for intercellular communication.

Synthesis of classical transmitters occurs in presynaptic axon terminals, near the site of neurotransmitter release. Because of this, characteristic synthetic enzymes for the classical neurotransmitters can be found in axonal terminals. For example, the enzymes tryptophan hydroxylase (converts the amino acid tryptophan to 5-hydroxytryptophan) and AADC (also converts 5-hydroxytryptamine to serotonin) are localized in the cytoplasm near terminal endings. This distribution of synthetic enzymes has been exploited to identify the type of neurotransmitter used by a nerve cell. Figure 6.12 shows dopaminergic neurons that are identified in culture by the presence of fluorescently labeled

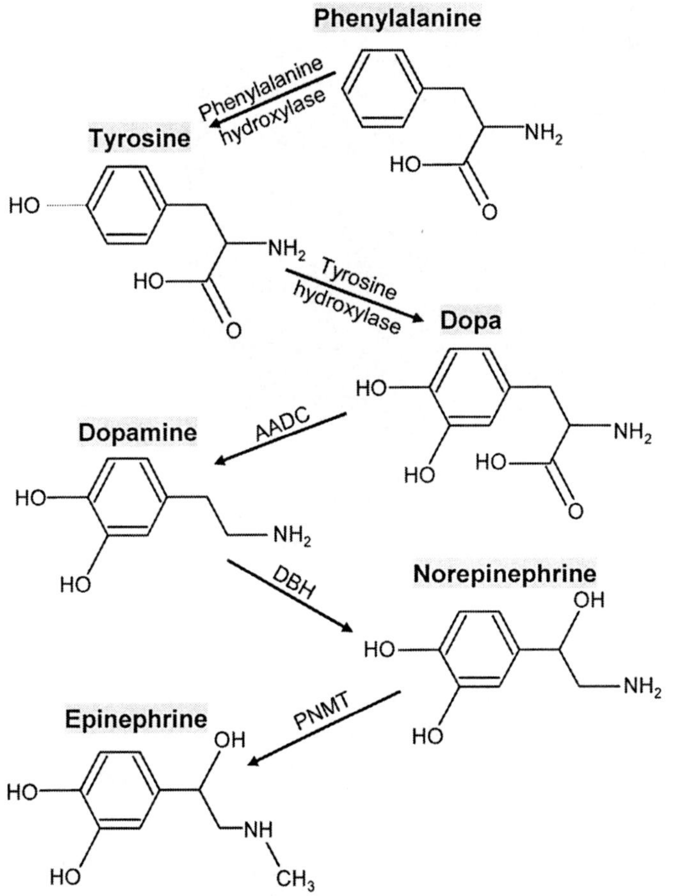

Figure 6.11   Catecholamine synthetic pathway. A series of synthetic enzymes converts phenylalanine to the bioactive catecholamines dopamine, norepinephrine, and epinephrine. AADC = aromatic acid decarboxylase; DBH = dopamine-β-hydroxylase; PNMT = phenylethanolamine N-methyl transferase.

tyrosine hydroxylase, required for the synthesis of DOPA from tyrosine as can be seen in Fig. 6.11. Clinical Box 6.1 presents an example how cells can be labeled *in vivo* with fluorescent markers and the diagnostic value of fluorescent tagging.

Release of classical transmitters is elicited by depolarization of the nerve cell and is calcium-dependent (see Sec. 6.4). Termination of the action of classical neurotransmitters is performed either by the action of enzymes and/or by reuptake mechanisms. The action of AChE, which is the primary mode of transmitter termination for ACh, has already been discussed in Sec. 6.3. For the catecholamines, the enzymes

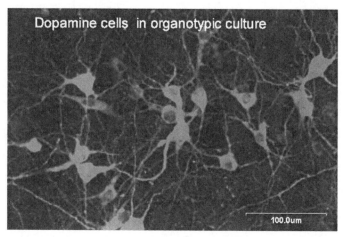

Dopamine cells in organotypic culture

100.0um

Figure 6.12   Dopamine neurons identified by tyrosine hydroxylase labeling and imaged by a red fluorescence tag. (Reproduced with permission from Bannon, M.J. Wayne State University School of Medicine, and displayed at *http://serotonin.med.wayne.edu/mbannon/*.)

CLINICAL BOX 6.1
In Vivo Labeling of Cells

Cells monitor their environment through receptor proteins embedded in the cell membrane. Receptor proteins have several attached molecules of sugar or oligosaccharide chains protruding into the extracellular space. These oligosaccharide chains or glycans serve as binding points for ligands to the receptor protein. Developing the ability to detect glycans specific to a particular type of receptors or cells in the living patient would be a major advance in diagnosis and therapy.

A recent study (Prescher et al., 2004) has reported an important contribution toward this goal. Prescher and colleagues synthesized the precursor of a single sugar molecule or monosaccharide with an attached nontoxic azide group: peracetylated mannosamine. When injected into mice, this precursor was converted by the mouse cellular machinery into azide-bearing sialic acid and transferred to glycans in the cell membrane. Thus, membrane proteins on the mouse cells now contained this artificially manufactured azide-bearing sialic acid. When an appropriate probe, in this case phosphine attached to a peptide named Flag, was injected into these previously treated mice, it bound to the sialic acid. Cells were then harvested from the mice. The complex formed by the artificially induced azide-bearing sialic acid and phosphine attached to Flag was expressed on the cells as a tag, detectable by labeling with a fluorescent antibody specific to the Flag peptide.

This work has important implications because it is a first step in directing the cellular machinery to synthesize artificial tags in cells bearing a particular type of receptor. It has the potential to allow for the artificial labeling of a specific group of cells for detection or therapy in the living. An important step toward this goal would be to uncover the mechanism, which will direct the expression of the artificial tags only in desired tissues or cell types in the living patient. Future applications could focus on developing imaging methods for detecting the tagged cells in living patients for diagnostic purpose; for example, cells in a brain region could be tagged and checked for abnormal growth if a tumor is suspected. Furthermore, therapies can also be developed whereby antibodies specific for the artificial tags can seek out tag-labeled tumor cells and destroy them. This approach would be invaluable in areas difficult to access by surgery such as the brain.

monoamine oxidase (MAO) and catechol-O-methyl transferase (COMT) are responsible for transmitter termination. Although these enzymes play a major role in ending the effects of catecholamine in the bloodstream, they play a secondary role in the brain. The major method of transmitter termination in the brain is by way of catecholamine transporters. These specialized membrane proteins shuttle the catecholamines back into the presynaptic terminal in an energy-dependent process, which requires the presence of the salt ions sodium and chloride. These transporters differ from the vesicular transporters, which load neurotransmitters into synaptic vesicles.

Two transporters specific to nerve cells have been characterized for the catecholamines: the dopamine transporter and the norepinephrine transporter. The existing nomenclature is somewhat misleading because the transporters are not specific and will bind to either dopamine or norepinephrine. The catecholamine transporters are located outside of the synapse; thus, following release and interaction with the receptors, the catecholamine neurotransmitters are terminated first passively by diffusion away from the synaptic cleft and then actively by their transporters. The arrangement of the transporters outside of the synaptic cleft allows for termination of catecholamines released not only from a nearby synapse, but also from distant sites. This mechanism increases the efficiency of communication between nerve cells by decreasing the likelihood of random neurotransmitter action through extracellular diffusion. Furthermore, it is protective to the cell since overstimulation can deplete metabolic stores within the postsynaptic cell and/or cause overexcitation within a particular neural pathway resulting in cell damage and death. This hypothesis of **excitotoxicity** may be the basis of severe neurological disorders such as strokes or amyotrophic lateral sclerosis.

6.3.2 Neuropeptide transmitters

Short chains of amino acids form neuropeptide transmitters. There is a large number of identified neuropeptide transmitters with new ones discovered on a regular basis. Substance P, vasoactive intestinal peptide, cholecystokinin, somatostatin, and neurotensin are examples of neuropeptide transmitters acting not only in the nervous system but also in the gastrointestinal tract as paracrines and hormones (see Sec. 6.4).

Although they play a similar role in cell communication, neuropeptide transmitters differ from the classical transmitters in some aspects. Synthesis of neuropeptides occurs in the cell body or soma of nerve cells, far away from their site of release. Cell transport mechanisms shuttle the neuropeptides from the cell body to the axonal terminals. Given the constraint of longer transport of the peptide neurotransmitter to the axon terminal and slower transmitter replenishment, the release of the neuropeptide transmitter must be coordinated with its synthesis to avoid shortage.

Although storage of the neuropeptide transmitters takes place in vesicles that are twice as large as the vesicles observed with the classical neurotransmitters, synaptic mechanisms involved in the release of the neuropeptide transmitters appear to be similar to those involving classical transmitters.

Finally, the action of neuropeptide transmitters is terminated passively by diffusion away from the synapse or actively by enzyme degradation. Thus far, no reuptake process via transporters has been identified for the neuropeptide transmitters.

6.4   Cell Secretion

Cells communicate with themselves and with each other by multiple means. Chemical communication can occur in an endocrine (humoral), paracrine, autocrine, or juxtacrine way. These terms were already introduced in Chap. 4 and they are again briefly explained below. In order to communicate chemically, the cell must synthesize a substance intracellularly and release it extracellularly. This process is called secretion which is the broad topic of this section.

Endocrine: The endocrine system is characterized by humoral communication in which a cell synthesizes a hormone and secretes it directly into the circulatory system. The blood transports the hormone to a target tissue where it binds to receptors located on the membrane or inside cells to promote its effects. Application Box 6.1 presents bioengineering attempts to restore in diabetic patients communication between cells that produce the hormone insulin and cells that express insulin receptors.

Paracrine: In paracrine communication, cells release active substances, which affect nearby cells.

Autocrine: A cell releases a substance, which acts upon the same cell.

Juxtacrine: In this type of communication, the messenger acts on neighboring cells without diffusing from the cell where it was produced. For instance, a ligand anchored in the cell membrane binds to a receptor in the membrane of another cell.

APPLICATION BOX 6.1
The Artificial Pancreas of the Future

All cells use glucose to derive energy for their functions. The hormone insulin allows for the passage of glucose from the bloodstream into cells. Without insulin, glucose builds up in the bloodstream and is unavailable for cell use.

In patients with type 1 diabetes (juvenile diabetes), β cells from the pancreas stop secreting insulin. At present, one possible therapy is transplantation of pancreatic tissues from cadavers to restore insulin secretion. However, the disadvantages of this method include having to use two separate pancreas donors for each recipient because only a fraction of the transplanted tissue will become functional; further, the rate of independence from injected insulin in these patients decreases over time: from 80% at 1 year to 65% by 2 years. A powerful potential therapy eliminating the need for donors would be the use of artificial β cells.

Bioengineering has contributed enormously to the development of improved treatments of diabetes. Implantable insulin pumps have been continuously improved since their first introduction in the early eighties. The pumps are nowadays connected with a miniature glucose sensor, and at the current writing, the first real-time monitoring systems have just been approved for use in humans. It implies that for diabetics, the widespread use of implanted closed-loop insulin delivery systems that mimic the functions of the human pancreas may not be in the too distant future.

There are other experimental approaches to the treatment of diabetes in which bioengineering design may be important. For instance, bioengineers aim at developing an implantable membrane device that entraps β cells isolated from animals or human donors and grown in culture. The membrane has to be designed to be freely permeable to glucose and insulin. But the membrane must also protect the entrapped β cells from immune factors such as antibodies and cytokines from the implanted patient, since they can elicit an immune reaction against the implanted device and interfere with its function. Further research is needed to develop such a valuable device.

Though most attention has been focused on chemical transmission as the generally accepted mode of cell to cell communication, electrical communication exists as well, in the form of gap junctions and ephaptic information exchange.

Gap junctions: Communication occurs through protein channels that form in the apposed membranes of adjacent cells and allow the exchange of small molecules and ions between the cells. The flow of ions allows electric impulses to be transmitted directly from one cell to the other.

Ephaptic: Cells communicate by apposed cell membranes with regions of low resistance flanked by regions of high resistance. Electrical signals flow across membranes at regions of low resistance.

Cellular secretion is a complex and highly regulated task. It first involves *manufacturing* of the substance to be secreted (see Sec. 6.4.1). The secretory cell has to synthesize two types of products: those destined for internal use and those destined for secretion, that is, release to the external environment. Next, the synthesized products must be packaged for efficient transport to their destination and also to prevent them from acting prematurely. These processes are presented under *packaging* in Sec. 6.4.2. The packages must have specific labels because they will be delivered by different systems to their respective destinations. The processes involved in *sorting and delivery* are crucial aspects of secretion and explained in Sec. 6.4.3. Finally, the products destined for secretion can only be secreted when there is a need for them and therefore secretion is highly regulated as presented in Sec. 6.4.4. The fusion of the secretory vesicle with the cell membrane and the release of their content extracellularly is called *exocytosis*, and presented in Sec. 6.4.5.

6.4.1 Manufacturing

Peptides or proteins formed on free ribosomes are usually destined for intracellular processes. Proteins destined for secretion are manufactured in ribosomes bound to the endoplasmic reticulum. The latter is called rough endoplasmic reticulum (RER) because of its high content of ribosomes and its appearance in electron micrographs. RER can be seen in Chap. 2 in the schematic of Fig. 2.1 and in the electron micrograph of Fig. 2.5. The process of synthesizing peptides and proteins from RNA is called translation and is described in Sec. 7.5.

6.4.2 Packaging

As proteins leave the RER, they are shuttled to the Golgi apparatus by way of the *endoplasmic reticulum Golgi intermediate compartment*

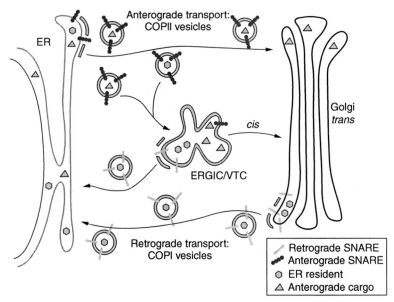

Figure 6.13   Endoplasmic reticulum Golgi intermediate compartment (ERGIC). The ERGIC, also called vesicular tubular cluster (VTC), shuttles synthesized proteins from the endoplasmic reticulum (ER) to the Golgi apparatus for processing and packaging. (Reproduced, with permission from the *Annual Review of Cell and Developmental Biology* 20, 2004, by Annual Reviews *http://www. annualreviews.org.*)

(ERGIC), also called *vesicular tubular cluster (VTC)* (Fig. 6.13). During this process, the secretory proteins are wrapped in transport vesicles which bud from the RER exit sites. These transport vesicles, known as *COPII-coated vesicles*, will move by way of interconnected tubules in the ERGIC or along microtubules to the *cis*-face of the Golgi apparatus.

The Golgi complex is organized into the *cis*-Golgi network (entry site) and the *trans*-Golgi network (exit site), which is shown in Fig. 2.6 as an electron micrograph and in schematic form. The *cis* and *trans* sites are both sorting stations: the *cis* site separating proteins destined for return to the RER from those continuing into the Golgi complex, and the *trans* site triaging proteins destined for the cell membrane from those with intracellular destinations. Within the Golgi complex, proteins are further modified by glycosylation, that is, attachment of carbohydrate entities, as they are shuttled toward the *trans*-Golgi network, the exit point of the Golgi complex. In case vital proteins are needed again in the RER, the Golgi complex has a mechanism for trapping them and sending them back to the RER via retrograde transport in so-called *COPI vesicles* (Fig. 6.13).

At the Golgi exit site, the secretory proteins will be packaged, initially into pieces of membrane from the *trans*-Golgi network, which loosely wrap themselves around the aggregates of synthesized secretory proteins. The resulting structures are called immature secretory vesicles. The signals directing the proteins into aggregates and the signals packaging the aggregates into secretory vesicles remain unknown. However, they may be common to all secretory cells because cells that normally do not synthesize a particular protein will still package that protein appropriately if forced to express the gene for that protein by artificial methods.

A number of processes take place further in the interior of the secretory vesicles as they mature. The vesicles concentrate their content by continuous removal of portions of their membrane, which are then recycled back to late endosomes or to the Golgi apparatus. A progressive acidification of the interior of the secretory vesicles also occurs due to an increase in ATP-dependent H^+ pumps in the vesicle membrane. Finally, the protein content of the secretory vesicles may be processed from an inactive precursor form to an active form by cleavage. For example, the secretory vesicle may contain a single precursor polyprotein, which will be cleaved off into multiple end-products. All these steps result in a more efficient packaging system for delivery.

6.4.3 Sorting and delivery

Secretory vesicles once formed must move toward their target membrane for eventual fusion and release of their content. It remains unclear as to how cells label the secretory vesicles for specific destinations. The delivery is performed by a system of microtubules, which shuttle the vesicles, and once they arrive, the vesicles become tethered to their target membrane, mainly by a group of GTP-binding proteins named Rabs. The specificity of the secretory vesicles for their target membrane is determined by the type of Rabs. It is believed that this tethering process is the earliest step in the fusion of the secretory vesicle with its target membrane.

The process by which the secretory vesicle interacts with the target membrane is known as docking, and it is performed by the interaction of two different categories of proteins known as SNAREs. The latter are formed by the strong binding of three synaptic proteins shown in Fig. 6.14: the vesicle-associated membrane protein (VAMP) or *synaptobrevin*, *syntaxin*, and the *synaptosomal-associated protein of 25 kDa (SNAP-25)*. The binding of these three synaptic proteins form the SNAP receptor complex also known as SNARE. SNAREs confer further specificity to the secretory vesicle for the fusion process to its target membrane. Secretory vesicles contain v-SNAREs, incorporated into their membrane during their budding from the *trans*-Golgi network. Target membranes contain t-SNAREs, interacting to form complexes with the approaching v-SNAREs, pulling the lipid bilayers of the secretory vesicle and target membrane into contact for fusion.

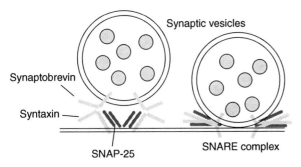

Figure 6.14   Docking mechanism of the synaptic vesicle to the presynaptic membrane. The docking mechanism involves binding of synaptobrevin, syntaxin, and SNAP-25, which form the SNARE complex.

6.4.4   Regulation of secretion

Since secretion occurs only on demand, upon reception of specific signals, interaction of the SNAREs is by itself not sufficient for membrane fusion. For example, in nerve cells, the v- and t-SNARE complexes remain in an inactive conformation due to the binding of complexin, and the secretory vesicle remains docked with its target membrane, awaiting the activating signal for fusion. This triggering signal in most cells will be an increase in calcium ion concentration.

Once fusion occurs, the SNAREs will be removed by the binding of a protein named *N*-ethylmaleimide sensitive factor (NSF) or *N*-ethylmaleimide sensitive fusion protein. NSF first binds a cofactor called α-soluble NSF attachment protein (α-SNAP), and this complex then binds to the SNARE. The latter will be reduced to its component proteins in an energy-dependent process by hydrolysis of adenosine triphosphate (ATP) and returned to its dissociated state.

6.4.5   Exocytosis

Exocytosis is a common process of many cells, but has been studied best at synapses, the contact points between nerve cells or between nerve cells and their target tissues. As an electric impulse arrives at the nerve terminal, an influx of calcium ions is triggered and the SNARE complex pulls the lipid bilayers of the secretory vesicle and its target membrane together (Fig. 6.14). In nerve cells, this fusion is regulated by synaptotagmin, a calcium-binding protein. It has been shown by *in vitro* experiments that the pulling action of SNAREs is enough to trigger fusion of the membranes. During this process, the cytosolic face of the secretory vesicle (that portion of the membrane of the secretory vesicle which is turned toward the outside of the vesicle) becomes part of the cytosolic surface of the cell membrane. The luminal surface of the vesicle (that is the portion of the membrane of the secretory vesicle turned toward the inside of the vesicle) becomes part of the outer surface of the cell membrane. The end result is the formation of a "fusion pore" illustrated in Fig. 6.10. The fusion pore rapidly dilates to release the content of the secretory vesicle.

Interactions between READ and WRITE of the Signaling Machinery

6.5   Synaptic Interactions during Development

Much is known about the molecular interactions between pre- and post-synaptic structures, the READ and WRITE mechanisms in our model of signaling machinery. A large part of this knowledge is acquired from studies that investigate the synaptic interactions during development, specifically of the neuromuscular junction, the interface between a voluntary nerve fiber and a muscle cell (Kummer et al., 2006). The voluntary nerve fibers are derived from **motoneurons** in the spinal cord and their terminals will interact with cholinergic receptors of the nicotinic types (see Sec. 6.3) on the muscle membrane.

During development, a pre-pattern of dense clusters of nicotinic cholinergic receptors forms on the muscle cell membrane. These clusters are expressed well in advance of the arrival of the motoneuron axon to the muscle cell. It was originally assumed that the presence of the motoneuron axon is necessary for expression of receptor clusters on the muscle cell membrane. But recent results have shown that the muscle cell can express this pre-pattern of nicotinic cholinergic receptors on its own. In fact, tissue culture studies have shown that muscle cells can form complex clusters of cholinergic receptors, which resemble neuromuscular junctions. These complex clusters, named "aneural pretzels" are shaped under the influence of a muscle-derived protein named LL5β. Thus, the shape of the mature neuromuscular junction is the result of interactions between muscle-derived and nerve-derived factors.

As the axon terminals from the motoneuron contact the muscle cell membrane, some of the pre-pattern clusters disappear whereas others are incorporated into the neuromuscular junction. The pre-pattern receptor clusters appear to play an important role in attracting and stabilizing the incoming axon terminals. Once neuronal contact has been established, the presence of the axon terminals is required for continued existence of the neuromuscular junction: if the axon terminals do not make contact, the nicotinic cholinergic receptor clusters eventually disappear.

Interestingly, the presence of the neurotransmitter acetylcholine (ACh) promotes the loss of the pre-pattern clusters of nicotinic cholinergic receptors and the incoming axon terminals secrete agrin, to counteract this effect. Agrin, a large heparan sulfate proteoglycan, stabilizes the pre-pattern receptor clusters contacted by axon terminals, emphasizing the importance of the presynaptic influence on the receptor distribution pattern in the postsynaptic element.

Additionally, the presynaptic nerve fiber also influences elements other than the postsynaptic structure: presynaptic axon terminals secrete neuregulin, which influences surrounding nonneuronal cells called Schwann cells. These play an important supportive role in maintaining the neuromuscular junction.

Finally, the type of neurotransmitter synthesized by the presynaptic element strongly influences the type of receptors manufactured by the postsynaptic structure. Recent results (Spitzer et al., 2004) have shown that the level of calcium activity inside a nerve cell during development can influence the type of neurotransmitters produced: an increase in calcium activity results in a shift to the production of the inhibitory transmitter γ-aminobutyric acid (GABA), whereas a decrease leads to synthesis of the excitatory transmitter glutamate. Surprisingly, when glutamatergic axon terminals contact muscle cells, which normally express cholinergic receptors, the muscle cells respond by manufacturing glutamate receptors, demonstrating not only the influence of the presynaptic element on the postsynaptic structure, but also the versatility of the postsynaptic element in adapting to new or different stimuli.

In summary, the study of molecular interactions between pre- and postsynaptic elements during development allows for a glimpse of the complex interplay necessary in sculpting and maintaining the communication between cells. It also points to potential approaches in manipulating or designing agents, which can shape or repair synaptic formation, an invaluable therapeutic step in developmental or degenerative disorders.

References

Kummer, T.T., Misgeld, T. and Sanes, J.R. (2006). Assembly of the postsynaptic membrane at the neuromuscular junction: Paradigm lost. *Curr. Opin. Neurobiol.* 16:74–82.

Nishi, M. and Kawata, M. (2006). Brain corticosteroid receptor dynamics and trafficking: Implications from live cell imaging. *Neuroscientist.* 12:119–133.

Prescher, J.A., Dube, D.H. and Bertozzi CR (2004). Chemical remodelling of cell surfaces in living animals. *Nature.* 430:873–877.

Spitzer, N.C., Root, C.M. and Borodinsky, L.N. (2004). Orchestrating neuronal differentiation: Patterns of Ca^{2+} spikes specify transmitter choice. *Trends Neurosci.* 27:415–421.

Cellular Genetics

Michael W. King, Ph.D.

OBJECTIVES

- To present the structure of DNA and chromatin
- To explain the concept of DNA synthesis and repair
- To elucidate the path from DNA reading to protein production

OUTLINE

7.1 DNA Structure   178
7.2 Chromatin Structure   181
7.3 DNA Synthesis and Repair   184
7.4 Transcription: DNA to RNA   193
7.5 Translation: RNA to Protein   199

Our improved understanding of the relationship between genes, cell processes, and disease will have a great impact in twenty-first century research, medicine, and society. The medicine of the near future will include the use of genetic information to determine the predisposition of patients toward diseases and the ability to design individualized preventative medical programs. Genetic engineering has become the gold standard for protein research, and genetically engineered food and drugs are tools in an attempt by our society to fight world hunger and disease.

"I just successfully labeled Van Gogh." Hearing this sentence from a biologist does not necessarily mean that he, or she, is proud of having placed a sticky tag on a painting by a famous artist. Instead, it might mean that the fluorescent labeling of a fruit fly gene has occurred. The recent advances in genetic research have produced the need for the naming of quite a few genes. Although nomenclature committees have been established to define and oversee guidelines for the process of giving names to genes, a lot of them end up with peculiar, often highly entertaining and creative names. The Van Gogh gene received its name since mutations to Van Gogh result in the hairs of a fruit fly wing forming swirling patterns, which resemble the swirls in the Van Gogh painting *Starry Night*. After this short introduction, let's delve into the wonderful world of genes.

7.1   DNA Structure

7.1.1   Composition of DNA in cells

Deoxyribonucleic acid (DNA) is composed of polymers of the four deoxynucleotides whose structures are described in Sec. 1.5. To recall, nucleotides are the compounds that exist of a heterocyclic base, a sugar, and a phosphate group. Utilizing x-ray diffraction data obtained from crystals of DNA, James Watson and Francis Crick proposed a model for the structure of DNA. This model (subsequently verified by additional data) predicted that DNA would exist as a helix of two complementary antiparallel strands, wound around each other in a rightward direction and stabilized by H-bonding between bases in adjacent strands. In the Watson-Crick model (see Fig. 7.1), the bases are in the interior of the helix aligned at a nearly 90° angle relative to the axis of the helix. **Purine** bases form hydrogen bonds with **pyrimidines** in the crucial phenomenon of base-pairing. Experimental determination has shown that, in any given molecule of DNA, the concentration of *adenine (A)* is equal to *thymine (T)* and the concentration of *cytosine (C)* is equal to *guanine (G)*. This means that A will only base-pair with T, and C with G. According to this pattern, known as **Watson-Crick base-pairing**, the base pairs (bp) composed of G and C contain three *hydrogen (H) bonds*, whereas those of A and T contain two H-bonds (see Fig. 7.1). This makes G-C base pairs more stable than A-T base pairs.

The antiparallel nature of the helix stems from the linear orientation of the individual strands. The backbones of the DNA strands are formed between alternating sugars (2-deoxyribose) and phosphates. The phosphates form phosphodiester bonds between the third and fifth carbon atoms (counting clockwise from the oxygen molecule) of adjacent sugar rings. The asymmetric ends of DNA strands are referred to as 5′ (five prime) and 3′ (three prime) ends. While one strand is oriented from left to right beginning with the 5′-end toward the 3′-end, the complementary strand will be oriented in the 3′ to 5′

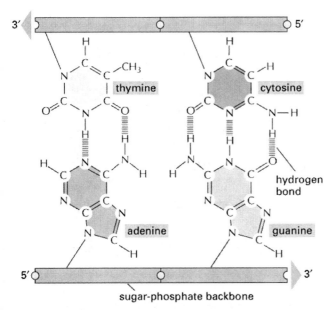

sugar-phosphate backbone

Figure 7.1   DNA base pairs. Orientation of the base pairs formed through hydrogen bonding between adenine (A) and thymine (T) residues and guanine (G) and cytosine (C) residues in a duplex of DNA. (Reproduced with permission from Garland Science, Taylor and Francis Group. *Molecular Biology of the Cell* by Alberts, B. et al., 4th ed., 2002.)

direction (see Figs. 7.1 and 7.2). Both strands typically are right-handed helices (similar to the threads on a standard screw) that wind around each other leaving gaps between them for the base pairs.

On its exterior surface, the double helix of DNA contains two deep grooves between the ribose-phosphate chains. These two grooves are of unequal size and termed the *major* and *minor grooves* (see Fig. 7.2). The difference in their size is due to the asymmetry of the deoxyribose rings and the structurally distinct nature of the upper surface of a base pair relative to the bottom surface.

The double helix of DNA has been shown to exist in several different forms depending upon sequence content and ionic conditions of crystal preparation (see Table 7.1). Panel A of Fig. 7.2 shows a model of the **B-form of DNA**, which prevails under physiological conditions of low ionic strength and a high degree of hydration. Conversely, the **A-form** helix forms under conditions of reduced hydration. Regions of the helix that are rich in stretches of the dinucleotide repeat $(CG)_n$ (where n defines the number of repeats) can exist in a novel left-handed helical conformation termed Z-DNA. CG dinucleotides are more correctly referred to as CpG dinucleotides where the lower case *p* indicates the *phosphodiester bond* between the C and G nucleotides (see Fig. 7.2). The

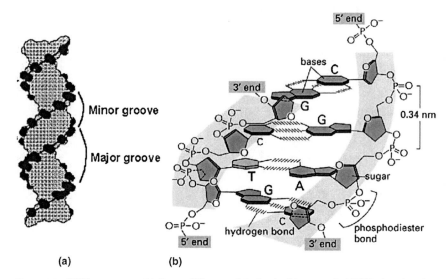

Figure 7.2   DNA structure. (a) Space-filling model of a double-stranded DNA showing the DNA backbone and the major and minor grooves. (b) Depiction of the hydrogen bonding between adjacent Watson-Crick base pairs indicating the near 90° plane of the nucleobases relative to the axis of the DNA helix. (b reproduced with permission from Garland Science, Taylor and Francis Group. *Molecular Biology of the Cell* by Alberts, B. et al., 4th ed., 2002.)

TABLE 7.1   Physical Characteristics of the Three Major DNA Helices

Parameters	A Form	B Form	Z Form
Direction of helical rotation	Right	Right	Left
Residues per turn of helix	11	10	12 base pairs
Rotation of helix per residue (in degrees)	33	36	-30
Base tilt relative to helix axis (in degrees)	20	6	7
Major groove	Narrow and deep	Wide and deep	Flat
Minor groove	Wide and shallow	Narrow and deep	Narrow and deep
Orientation of N-glycosidic bond	Anti	Anti	Anti for Py, syn for Pu
Comments		Most prevalent within cells	Occurs in stretches of alternating purine-pyrimidine base pairs

Z-DNA conformation results from a 180° change in the orientation of the bases relative to that of the more common A- and B-DNA.

7.1.2 Thermal properties of the DNA helix

As cells divide, it is necessary that the DNA be copied in such a way that each daughter cell acquires the same amount of genetic material. This process is called replication and is presented in Sec. 7.3.1. In order for this process to proceed, the two strands of the helix must first be separated in a process termed **denaturation**. Denaturation inside the nucleus occurs as a consequence of the action of adenosine triphosphate (ATP)-dependent helicases that unwind the DNA duplex. The process of denaturation can also be carried out *in vitro*. If a solution of DNA is subjected to high temperature, the H-bonds between bases become unstable and the strands of the helix separate in a process of thermal denaturation.

The base composition of DNA varies widely from molecule to molecule and even within different regions of the same molecule. Regions of the duplex that have predominantly A-T base pairs will be less thermally stable than those rich in G-C base pairs. For this reason, regions of the chromosomes where replication begins are enriched in A-T base pairs since these are unwound more easily than G-C base pairs (see Sec. 7.3).

In the *in vitro* process of thermal denaturation, a point is reached at which 50% of the DNA molecule exists as single strands. This point is the **melting temperature (T_m)**, and is characteristic of the base composition of that DNA molecule. The T_m depends upon several factors in addition to the base composition. These include the chemical nature of the solvent and the identities and concentrations of ions in the solution.

When thermally melted DNA is cooled, the complementary strands will again re-form the correct base pairs in a process termed **annealing**. **Hybridization** and renaturation are equivalent terms to annealing in this context. The time rate of annealing is dependent upon the **nucleotide** sequence of the two strands of DNA.

7.2   Chromatin Structure

The size of eukaryotic genomes is vastly larger than those of prokaryotes. The **genome** of an organism is defined as the total complement of DNA. For example, the genome in human **diploid** cells (i.e., cells other than the egg or sperm which are haploid) contains DNA making up 46 chromosomes. A goldfish has 94 chromosomes and a mosquito has 6 chromosomes. A chromosome is a long strand of DNA that is densely packaged with proteins and other molecules. If a human chromosome would be stretched into one long thread, it would be about 3 m long. Each chromosome has a *centromere*, which divides it into a short arm and a long arm. Chromosomes are an efficient way to store genes; in the case of humans about 25,000 genes (see Sec. 10.2.1). Genes can be as short as 1000 base pairs or as long as several hundred thousand base pairs. Although the different size of the genomes between eukaryotes and prokaryotes is partly due to the complexity of eukaryotes compared to prokaryotes (see Sec. 10.1.1), the size of a particular eukaryotic genome is not directly correlated to the organism's complexity. This is the result of the presence of a large amount of noncoding DNA. Coding DNA refers to the **genes** and noncoding to all the rest of the DNA in the genome. The functions of noncoding DNA are only partly understood. Some noncoding DNA sequences are involved in the control of gene expression while others may simply be present in the genome to act as an evolutionary buffer able to withstand nucleotide mutation without disrupting the integrity of the organism.

One abundant class of DNA is termed **repetitive DNA**. There are two different subclasses of repetitive DNA: highly repetitive and moderately repetitive. Highly repetitive DNA consists of short sequences 6 to 10 bp long, reiterated from 100,000 to 1,000,000 times. **Nonrepetitive DNA** defines the genes (the coding sequences) of the genome since most genes occur, but once, in the genome of an organism. However, it should be pointed out that several genes exist as tandem clusters of multiple copies of the same gene ranging from 50 to 10,000 copies such as the case for the rRNA genes and the genes for histone (for DNA-associated proteins, see Sec. 7.2.1; for RNA types, see Sec. 7.4.1).

In addition to differences in genome size, another characteristic feature that distinguishes eukaryotic from prokaryotic genomes is that eukaryotic genes are composed of regions called exons and introns. **Exons** are the coding regions of genes that remain in the RNA following posttranscriptional processing (see Sec. 7.4). **Introns** are stretches of DNA sequences that separate the coding exons of a gene. The existence of introns in prokaryotes is extremely rare. Essentially all human genes contain introns. Notable exceptions are the histone genes which do not contain introns. In many genes, the presence of introns separates exons into coding regions exhibiting distinct functional domains.

7.2.1 Histones and formation of nucleosomes

Chromatin is a term designating the structure in which DNA exists within cells. The structure of chromatin is determined and stabilized through the interaction of the DNA with DNA-binding proteins. There are two classes of DNA-binding proteins: histones and nonhistone proteins. **Histones** are the major class of DNA-binding proteins involved in maintaining the compacted structure of chromatin. There are five different histone proteins identified as H1, H2A, H2B, H3, and H4. The nonhistone proteins include the various transcription factors, polymerases, hormone receptors, and other nuclear enzymes. In any given cell there

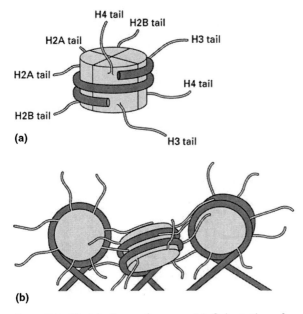

Figure 7.3 Model of a nucleosome. (a) Orientation of a strand of DNA wound around the octamer core of histones (H2A, H2B, H3, and H4) in a nucleosome. (b) Indication of how the tails of the histone proteins interact to stabilize higher order interactions forming the "beads on a string" structure that can be seen using electron microscopy. (Reproduced with permission from Garland Science, Taylor and Francis Group. *Molecular Biology of the Cell* by Alberts, B. et al., 4th ed., 2002.)

are greater than 1000 different types of nonhistone proteins bound to the DNA.

The binding of DNA by the histones generates a structure called the **nucleosome** (see Fig. 7.3). The nucleosome core contains an octamer protein structure consisting of two subunits each of *H2A, H2B, H3,* and *H4* (see Fig. 7.3*a*). Histone H1 occupies the internucleosomal DNA and is identified as the linker histone. The nucleosome core contains approximately 150 bp of DNA. The linker DNA between each nucleosome can vary from 20 to more than 200 bp. These nucleosomal core structures would appear as beads on a string if the DNA were pulled into a linear structure and viewed under an electron microscope (see Figs. 7.3*b* and 7.4).

The nucleosome cores interact through the tails of the histone proteins to compact the DNA, ultimately compacting it into the characteristic X-like structure of **chromosomes** seen in a typical mitotic chromosome preparation (see Fig. 7.4). The protein-DNA structure of chromatin is stabilized by attachment to a nonhistone protein scaffold called the **nuclear matrix**.

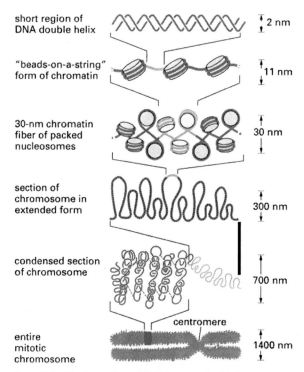

short region of
DNA double helix   2 nm

"beads-on-a-string"
form of chromatin   11 nm

30-nm chromatin
fiber of packed   30 nm
nucleosomes

section of
chromosome in   300 nm
extended form

condensed section
of chromosome   700 nm

                                    centromere
entire
mitotic   1400 nm
chromosome

Figure 7.4   Condensation of DNA into a mitotic chromosome. Interaction of a duplex of DNA with histones forms the nucleosomes which further interact with each other in a progressive condensation yielding the typical structure of mitotic chromosomes. (Reproduced with permission from Garland Science, Taylor and Francis Group. *Molecular Biology of the Cell* by Alberts, B. et al., 4th ed., 2002.)

7.3   DNA Synthesis and Repair

7.3.1   Mechanics and regulation

Replication of DNA is the normal process of doubling the DNA content of cells prior to normal cell division. Because the genetic complement of the resultant daughter cells must be the same as the parental cell, DNA replication must possess a very high degree of fidelity. The entire process of DNA replication is complex and involves multiple enzymatic activities.

The mechanics of DNA replication was originally characterized in the bacterium *Escherichia coli (E. coli),* which contains three distinct enzymes capable of catalyzing the replication of DNA. These have been identified as DNA **polymerase** (pol) I, II, and III. Pol I is the most abundant replicating enzyme in *E. coli* but has as its primary role to ensure the fidelity of replication through the repair of damaged and mismatched DNA. Replication of the *E. coli* genome is the job of pol III.

This enzyme is much less abundant than pol I; however, its activity is nearly 100 times that of pol I.

There have been five distinct eukaryotic DNA polymerases identified: α, β, γ, δ, and ε. The identity of the individual polymerase relates to its subcellular location as well as its primary replicative activity. The polymerase of eukaryotic cells that is the equivalent of *E. coli* pol III is pol-α. The pol I equivalent in eukaryotes is pol-β. Polymerase-γ is responsible for replication of mitochondrial DNA.

The ability of DNA polymerase to replicate DNA requires a number of additional accessory proteins. These accessory proteins include (not ordered with respect to importance):

1. Primases

2. Processivity accessory proteins

3. Single-strand binding proteins

4. Helicases

5. DNA ligases

6. Topoisomerases

The process of DNA replication begins at specific sites in the chromosomes termed origins of replication. The duplex at an origin of replication is unwound to generate single-stranded regions providing access for DNA polymerase. DNA polymerase requires a **nucleic acid** strand bearing a free 3'-OH, called a primer, for DNA synthesis to begin. Replication of DNA proceeds specifically in the 5' to 3' direction on both strands of DNA concurrently and results in the copying of the template strands in a **semiconservative** manner. The semiconservative nature of DNA replication means that the newly synthesized daughter strands remain associated with their respective parental template strands.

Synthesis of DNA proceeds in the 5' to 3' direction through the attachment of the 5'-phosphate of an incoming **deoxynucleoside triphosphate (dNTP)** to the existing 3'-OH in the elongating DNA strands with the concomitant release of *pyrophosphate* (see Fig. 7.5).

Initiation of synthesis, at origins of replication, occurs simultaneously on both strands of DNA. Synthesis then proceeds bidirectionally, with one strand in each direction being copied continuously and one strand in each direction being copied discontinuously. During the process in which DNA polymerases incorporate dNTPs into DNA in the 5' to 3' direction, the polymerases are moving in the 3' to 5' direction with respect to the *template strand* (see Fig. 7.5). In order for DNA synthesis to occur simultaneously on both template strands as well as bidirectionally, one strand appears to be synthesized in the 3' to 5' direction. In actuality, one strand of newly synthesized DNA is produced discontinuously.

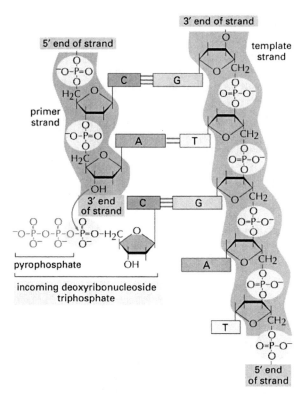

Figure 7.5 Incorporation of deoxynucleoside triphosphates (dNTPs) during DNA replication. Each newly incorporated nucleotide in DNA is attached to the 3'-OH of the preceding nucleotide in the elongating DNA strand. Upon incorporation, the terminal two phosphates of the dNTP are released. (Reproduced with permission from Garland Science, Taylor and Francis Group. *Molecular Biology of the Cell* by Alberts, B. et al., 4th ed., 2002.)

The large size of eukaryotic chromosomes and the limits of nucleotide incorporation during DNA synthesis make it necessary for multiple origins of replication to exist in order to complete replication in a reasonable period of time. At a replication origin, the strands of DNA must dissociate and unwind in order to allow access to DNA polymerase. Unwinding of the duplex at the origin as well as along the strands as the replication process proceeds is carried out by **helicases**. The resultant regions of single-stranded DNA are stabilized by the binding of single-strand binding proteins. The stabilized single-stranded regions are then accessible to the enzymatic activities required for replication to proceed. The site of the unwound template strands is called the **replication fork** (see Fig. 7.6).

In order for DNA polymerases to synthesize DNA, they must encounter a free 3'-OH which is the substrate for attachment of the 5'-phosphate

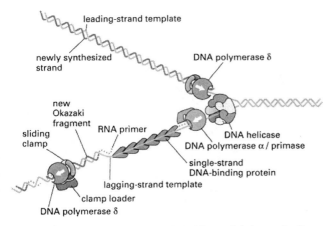

Figure 7.6 Model of a replication fork. The model shows the direction of newly synthesized DNA with associated major activities of replication. For details, see Sec. 7.3.1. (Reproduced with permission from Garland Science, Taylor and Francis Group. *Molecular Biology of the Cell* by Alberts, B. et al., 4th ed., 2002.)

of the incoming nucleotide. During repair of damaged DNA, the 3'-OH can arise from the hydrolysis of the backbone of one of the two strands. During replication, the 3'-OH is supplied through the use of an RNA primer synthesized by the **primase** activity. The primase utilizes the DNA strands as templates and synthesizes a short stretch of RNA generating a primer for DNA polymerase.

The strand of DNA synthesized continuously is termed the *leading strand* and the discontinuous strand is termed the *lagging strand* (see Fig. 7.6). The lagging strand of DNA is composed of short stretches of RNA primer plus newly synthesized DNA approximately 100 to 200 bases long (the approximate distance between adjacent nucleosomes). The lagging strands of DNA are also called **Okazaki fragments** (see Fig. 7.6). The concept of continuous strand synthesis is somewhat of a misnomer since DNA polymerases do not remain associated with a template strand indefinitely. The ability of a particular polymerase to remain associated with the template strand is termed **processivity**. The longer the DNA polymerase associates, the higher the processivity of the enzyme. DNA polymerase processivity is enhanced by additional protein activities of the **replisome** identified as processivity accessory proteins.

How is it that DNA polymerase can copy both strands of DNA in the 5' to 3' direction simultaneously? A model has been proposed where DNA polymerases exist as dimers (two molecules) associated with the other necessary proteins at the replication fork and identified as the replisome (see Fig. 7.7). The template for the lagging strand is temporarily looped through the replisome such that the DNA polymerases are moving along

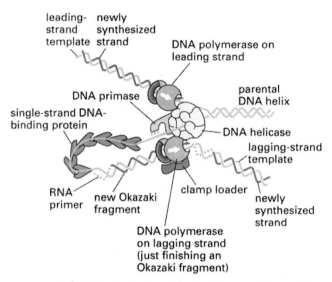

leading- newly
strand synthesized
template strand

DNA polymerase on
leading strand

DNA primase

single-strand DNA-
binding protein

parental
DNA helix

DNA helicase

lagging-strand
template

RNA
primer

new Okazaki
fragment

clamp loader

newly
synthesized
strand

DNA polymerase
on lagging strand
(just finishing an
Okazaki fragment)

Figure 7.7 Model for the looping of the lagging strand through the replisome. This allows for simultaneous replication of DNA on both the leading and lagging strands. For details, see Sec. 7.3.1. (Reproduced with permission from Garland Science, Taylor and Francis Group. *Molecular Biology of the Cell* by Alberts, B. et al., 4th ed., 2002.)

both strands in the 3' to 5' direction simultaneously for short distances, the distance of an Okazaki fragment. As the replication forks progress along the template strands, the newly synthesized daughter strands and parental template strands reform a DNA double helix. This means that only a small stretch of the template duplex is single-stranded at any given time.

The progression of the replication fork requires that the DNA ahead of the fork be continuously unwound. Due to the fact that eukaryotic chromosomal DNA is attached to a protein scaffold, the progressive movement of the replication fork introduces severe torsional stress into the duplex ahead of the fork. This torsional stress is relieved by DNA **topoisomerases**. Topoisomerases relieve torsional stresses in duplexes of DNA by introducing either double- (topoisomerases II) or single-stranded (topoisomerases I) breaks into the backbone of the DNA. These breaks allow unwinding of the duplex and removal of the replication-induced torsional strain. The breaks are then resealed by the topoisomerases.

The RNA primers of the leading strands and Okazaki fragments are removed by the repair DNA polymerases simultaneously replacing the ribonucleotides with deoxyribonucleotides. The gaps (called a "*nick*") that exist between the 3'-OH of one leading strand and the 5'-phosphate of another as well as between one Okazaki fragment and another are repaired by DNA **ligases**, thereby completing the process of replication (see Fig. 7.8).

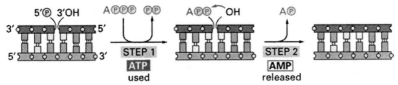

Figure 7.8   Activity of DNA ligase during DNA replication at a nick in the DNA backbone. DNA ligase uses the energy of ATP hydrolysis to generate the phosphodiester bond at nicks in the DNA strand. The 5′ end of the nick is activated first by attachment of AMP with release of pyrophosphate (PP$_i$). In the second step, the phosphodiester bond is formed and AMP is released. (Reproduced with permission from Garland Science, Taylor and Francis Group. *Molecular Biology of the Cell* by Alberts, B. et al., 4th ed., 2002.)

APPLICATION BOX 7.1
The Polymerase Chain Reaction

The **polymerase chain reaction (PCR)** is an *in vitro* technique involving multiple rounds of DNA synthesis with an enzyme that is resistant to heat denaturation. Heat-resistant DNA polymerases are isolated from thermophilic bacteria that are found growing in deep ocean thermal vents or geyser pools. The first thermostable DNA polymerase used was isolated from the bacterium *Thermus aquaticus* and is therefore called *Taq polymerase*. The PCR is used for the amplification of millions of copies of a specifically selected region of DNA. During the PCR, the products of the previous round of DNA synthesis serve as templates for each subsequent round, thus the term polymerase chain reaction. The consequences of the PCR are purification of the selected region from the rest of the DNA in any given sample. The power of the PCR is such that a specific sequence of DNA from as little as a single cell can be rendered easily detectable.

The PCR requires that one has knowledge of the DNA sequences of the target DNA to be amplified. This information is then used to design and synthesize single-stranded oligonucleotides (oligos) complementary to one of the two DNA strands. The oligos are generally from 15 to 24 nucleotides long and by virtue of their base-pairing to complementary DNA in the target, provide the 3′-OH required to prime DNA synthesis. Hence, the oligos used for PCR are commonly called "primers." In order to amplify a given region of DNA, there needs to be two primers in the reaction. One primer is designed to base-pair to the 5′-side of the target sequence and is therefore referred to as the 5′ or upstream primer. The other primer is called the 3′ or downstream primer. Except in specialized PCR assays, the primers are generally designed so that they span a region of between 150 and 350 bp.

A typical PCR contains the template DNA, deoxynucleotide triphosphates (dNTPs), priming oligos, and DNA polymerase. The general steps in a

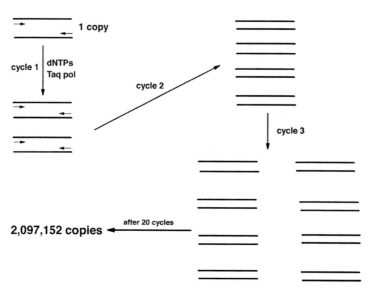

Schematic of the process of the polymerase chain reaction (PCR).

typical PCR involve heating the reaction to 94°C to 95°C in order to thermally denature (melt) the template DNA, lowering the temperature to that which is optimal for primer base-pairing (annealing) to template strands and then heating the reaction to 72°C which is optimal for synthesis by most thermostable DNA polymerases. The annealing temperature is dependent upon the length and nucleotide composition of the priming oligos. As depicted in the figure, following 20 cycles of the PCR, a single copy of a target DNA sequence is amplified to over 2 million copies.

The main enzymatic activity of DNA polymerases is the 5′ to 3′ synthetic activity. However, DNA polymerases possess two additional activities of importance for both replication and repair. These additional activities include a 5′ to 3′ exonuclease function and a 3′ to 5′ exonuclease function. The 5′ to 3′ exonuclease activity allows the removal of ribonucleotides of the RNA primer, utilized to initiate DNA synthesis, along with their simultaneous replacement with deoxyribonucleotides by the 5′ to 3′ polymerase activity. The 5′ to 3′ exonuclease activity is also utilized during the repair of damaged DNA. The 3′ to 5′ exonuclease function is utilized during replication to allow DNA polymerase to remove mismatched bases. It is possible (but rare) for DNA polymerases to incorporate an incorrect base during replication. These mismatched bases are recognized by the polymerase immediately due to the lack of Watson-Crick base-pairing. The mismatched base is then removed by the

3' to 5' exonuclease activity and the correct base inserted prior to progression of replication.

7.3.2 Postreplicative modifications

One of the major postreplicative reactions that modify the DNA is *methylation* (see Fig. 7.9). The sites of natural (i.e., not chemically induced) methylation of eukaryotic DNA is always on cytosine residues that are present in CpG dinucleotides. (Recall from Sec. 7.1.1 that p indicates the phosphodiester bond between C and G.) However, it should be noted that not all CpG dinucleotides are methylated at the C residue. The cytosine is methylated at the 5 position of the pyrimidine ring generating 5-methylcytidine (see Fig. 7.9).

The pattern of methylation is copied postreplicatively by the *maintenance methylase* system. This activity recognizes the pattern of methylated C residues in the maternal DNA strand following replication and methylates the C residue present in the corresponding CpG dinucleotide of the daughter strand.

Methylation of DNA has been shown to be an important regulator of transcriptional activity. A decrease in the methylation state of genes can lead to their transcriptional activation. Conversely, it has been shown that hypermethylation of genes leads to transcriptional silencing. The phenomenon of **genomic imprinting** refers to the fact that the expression of some genes depends on whether or not they are inherited maternally or paternally. Imprinted genes are marked as such by their state of methylation.

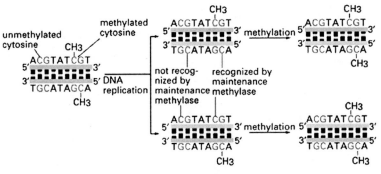

Figure 7.9 Postreplicative methylation of DNA. Following replication, the maintenance methylase enzyme system recognizes the pattern of methylation in the parental strand and incorporates a methyl group into appropriate cytosine residues in the daughter strand. (Reproduced with permission from Garland Science, Taylor and Francis Group. *Molecular Biology of the Cell* by Alberts, B. et al., 4th ed., 2002.)

CLINICAL BOX 7.1
Chemotherapy Targeting Replication

The class of compounds that have been used the longest in chemotherapy as anticancer drugs are the **alkylating agents**. The major alkylating agents are derived from nitrogenous mustards that were originally developed for use by the military. Commonly used alkylating agents include cyclophosphamide (Cytoxan, Neosar), ifosphamide, decarbazine, chlorambucil (Leukeran), and procarbazine (Matulane, Natulan).

Alkylating agents function by reacting with and disrupting the structure of DNA. Some agents react with alkyl groups in DNA resulting in fragmentation of the DNA as a consequence of the action of DNA repair enzymes. Some agents catalyze the cross-linking of bases in the DNA which prevents the separation of the two strands during DNA replication. Some agents induce mispairing of nucleotides resulting in permanent mutations in the DNA. Alkylating agents act upon DNA at all stages of the cell cycle; thus, they are potent anticancer drugs. However, because of their potency, prolonged use of alkylating agents can lead to secondary cancers, particularly leukemias.

Several classes of anticancer drugs function through interference with the actions of the topoisomerases. Two of these classes are the anthracyclines and the camptothecins. The anthracyclins were originally isolated from the fungus, *Streptomyces*. Doxorubicin (Adriamycin, Doxil, Rubex) and daunorubicin (Cerubidine, DaunoXome, Daunomycin, Rubidomycin) have similar modes of action, although doxorubicin is the more potent of the two and is used in the treatment of breast cancers, lymphomas, and sarcomas. The anthracyclins inhibit the actions of topoisomerase II whose function is to introduce double-strand breaks in DNA during the process of replication as a means to relieve torsional stresses. Anthracyclines also function by inducing the formation of oxygen free radicals that cause DNA strand breaks resulting in inhibition of replication.

Another plant-derived anticancer compound that functions through inhibition of topoisomerase II is etoposide (VP-16, Vepesid, Etophos, Eposin). Etoposide is isolated from the mandrake plant and is in a class of compounds referred to as epipodophyllotoxins.

The camptothecins were originally found in the bark of the *Camptotheca accuminata* tree and include irinotecan (Campto, Camptosar) and topotecan (Hycamtin). Camptothecins inhibit the action of topoisomerase I, an enzyme that induces single-strand breaks in DNA during replication.

Anticancer compounds that are extracted from the periwinkle plant, *Vinca rosea*, are called the vinca alkaloids. These compounds bind to tubulin monomers leading to the disruption of the microtubules of the mitotic spindle fibers that are necessary for cell division during mitosis. There are four major vinca alkaloids that are currently used as chemotherapeutics, vincristine (Oncovin, Vincrex), vinblastine (Velbe, Velban, Velsar), vinorelbin (VIN, Navelbine), and vindesine (Eldisine).

The taxanes are another class of plant-derived compounds that act via interference with microtubule function. These compounds are isolated from the Pacific yew tree, *Taxus brevifolia*, and include paclitaxel (Taxol) and docetaxel (Taxotere). The taxanes function by hyperstabilizing microtubules which prevents cell division (cytokinesis). These compounds are used to treat a wide range of cancers including head and neck, lung, ovarian, breast, bladder, and prostate cancers. Taxol has proven highly effective in the treatment of certain forms of breast cancer.

In order for DNA replication to proceed, proliferating cells require a pool of nucleotides. The class of anticancer drugs that has been developed to interfere with aspects of nucleotide metabolism is known as the antimetabolites. There are two major types of antimetabolites used in the treatment of a broad range of cancers: compounds that inhibit thymidylate synthase and compounds that inhibit dihydrofolate reductase (DHFR). Both of these enzymes are involved in thymidine nucleotide biosynthesis. Drugs that inhibit thymidylate synthase include 5-fluorouracil (5-FU, Adrucil, Efudex) and 5-fluorodeoxyuridine. Those that inhibit DHFR are analogs of the vitamin folic acid and include methotrexate (Trexall, Rheumatrex) and trimethoprim (Proloprim, Trimpex).

7.4   Transcription: DNA to RNA

7.4.1 Mechanics

Transcription is the mechanism by which a template strand of DNA is utilized by specific RNA polymerases to generate one of the three different classifications of RNA. These three RNA classes are:

1. **Messenger RNAs (mRNAs):** This class of RNAs is the genetic coding templates used by the translational machinery to determine the order of amino acids incorporated into an elongating polypeptide in the process of translation.

2. **Transfer RNAs (tRNAs):** This class of small RNAs form covalent attachments to individual amino acids and recognize the encoded sequences of the mRNAs to allow correct insertion of amino acids into the elongating polypeptide chain.

3. **Ribosomal RNAs (rRNAs):** This class of RNAs is assembled, together with numerous ribosomal proteins, to form the ribosomes. Ribosomes engage the mRNAs and form a catalytic domain into which the tRNAs enter with their attached amino acids. The proteins of the ribosomes catalyze all of the functions of polypeptide synthesis.

All RNA polymerases are dependent upon a DNA template in order to synthesize RNA. The resultant RNA is, therefore, complimentary to the template strand of the DNA duplex and identical to the nontemplate strand. The nontemplate strand is called the coding strand because its sequences are identical to those of the mRNA. However, in RNA, uracil (U) is substituted for thymine (T).

In prokaryotic cells, all three RNA classes are synthesized by a single polymerase. In eukaryotic cells there are three distinct classes of RNA polymerase: RNA polymerase (pol) I, II, and III. Each polymerase is responsible for the synthesis of a different class of RNA. The capacity of the various polymerases to synthesize different RNAs was shown with the toxin α-amanitin. At low concentrations of α-amanitin, synthesis of mRNAs is affected whereas synthesis of rRNAs and tRNAs remains unaffected. At high concentrations, both mRNAs and tRNAs are affected. These observations have allowed the identification of which polymerase synthesizes which class of RNAs. RNA pol I is responsible for rRNA synthesis (excluding the $5S$ rRNA). There are four major rRNAs in eukaryotic cells designated by their sedimentation coefficient (S), which indicates the relative mass of the molecule. For instance, $28S$ rRNA is heavier than $5.8S$ rRNA in the eukaryotic ribosome. The $28S$, $5S$, and $5.8S$ RNAs are associated with the large ribosomal subunit and the $18S$ rRNA is associated with the small ribosomal subunit. RNA pol II synthesizes the mRNAs and some of the small nuclear RNAs (snRNAs) involved in RNA **splicing**. RNA splicing refers to the modification of genetic information after transcription (see below). RNA pol III synthesizes the tRNAs, the $5S$ rRNA, and some snRNAs.

Synthesis of RNA exhibits several features that are synonymous with DNA replication. RNA synthesis requires accurate and efficient initiation, after which elongation proceeds in the 5' to 3' direction (i.e., the polymerase moves along the template strand of DNA in the 3' to 5' direction). RNA synthesis also requires distinct and accurate termination. Transcription exhibits several features that are distinct from replication:

1. Transcription begins, both in prokaryotes and eukaryotes, from many more sites than replication.

2. There are many more molecules of RNA polymerase per cell than DNA polymerase.

3. RNA polymerase proceeds at a rate much slower than DNA polymerase (approximately 50 to 100 bases/s for RNA vs. nearly 1000 bases/s for DNA).

4. Finally, the fidelity of RNA polymerization is much lower than DNA. This is allowable since the aberrant RNA molecules can simply be turned over and new correct molecules made.

Sequences are present within the DNA templates that act as signals to stimulate the initiation of transcription. These sequence elements are

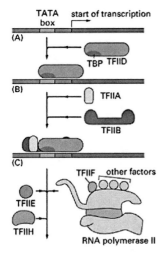

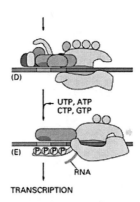

Figure 7.10   Initiation of transcription. Shown are the functions of transcription factors (TFIIA–TFIIH, and others) that bind at the promoter (TATA box) and direct the RNA polymerase II to sites of active transcription. (Reproduced with permission from Garland Science, Taylor and Francis Group. *Molecular Biology of the Cell* by Alberts, B. et al., 4th ed., 2002.)

termed **promoters**. Promoter sequences are so called because they promote the ability of RNA polymerases to recognize the nucleotide at which initiation begins (see Fig. 7.10).

Additional sequence elements are present within genes that act in *cis* to enhance polymerase activity even further. *Cis* is Latin for "on the same side as"; a *cis*-sequence is an RNA region which regulates genes located on the same strand. These sequence elements are termed **enhancers**. Transcriptional promoter and enhancer elements are important sequences used in the control of gene expression.

Elongation involves the addition of the 5′-phosphate of ribonucleotides to the 3′-OH of the elongating RNA with the concomitant release of pyrophosphate (see Fig. 7.11). Nucleotide addition continues until specific termination signals are encountered. Following termination the core polymerase dissociates from the template. The ribosome subunits (core and sigma) can then reassociate forming the holoenzyme (enzyme including all subunits) again ready to initiate another round of transcription.

Eukaryotic RNAs (all three classes) undergo significant posttranscriptional processing. All three classes of RNA are transcribed from genes that contain introns. The sequences encoded by the intronic DNA must be removed from the primary transcript prior to the RNAs being

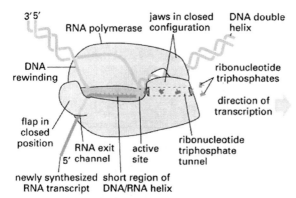

Figure 7.11   Model for the action of RNA polymerases during transcription. The RNA polymerase holoenzyme attaches to DNA at promoter sites as a consequence of the action of numerous promoter-binding transcription factors. The duplex is unwound and RNA polymerase reads the template strand and incorporates the correct ribonucleoside triphosphate (rNTP) into the elongating RNA strand. (Reproduced with permission from Garland Science, Taylor and Francis Group. *Molecular Biology of the Cell* by Alberts, B. et al., 4th ed., 2002.)

APPLICATION BOX 7.2
Genomics—Global Analysis of Gene Activities

Genomics is the term used to describe the comparative analysis of all the genes that are transcriptionally active between any two types of cells or tissues under any given set of conditions. For example, if one wanted to know what genes may be important in the generation of a particular type of cancer, for example, breast cancer, then one could carry out an analysis of all the transcriptionally active genes in normal breast tissue and compare that to those in cancerous breast tissue.

The ability to carry out these massive global transcription analyses was the result of the development of the technology to create gene arrays. Gene arrays are generally of two types: commercial "gene chips" and custom arrays printed on glass slides. Commercial gene chips are created by a sequential process of synthesizing short stretches of DNA corresponding to nearly all of the known genes of an organism onto a solid support. Custom arrays are made by spotting PCR (see figure) generated collections of DNA onto glass slides. The DNA on the gene chip or glass slide is then mixed with mRNAs isolated from the two comparative sources. The mRNAs are modified by the addition of a fluorescent chemical, usually green for one mRNA set and red for the other. The mRNAs **hybridize** to sequences of DNA that are complementary. After washing off the nonspecific interactions, the location of red and green fluorophors are detected. Overlaying the results will yield red or green spots where only one of the two mRNAs is hybridized. It will reveal variable colors between orange and yellow, depending on which mRNA (red or green) is hybridized in greater amount to a given DNA sequence. Computer-generated values for color intensity then allow for accurate assessment of expression level of thousands of genes simultaneously.

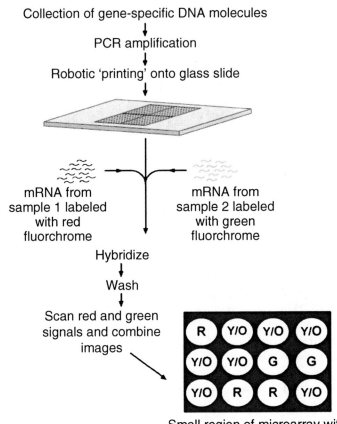

Collection of gene-specific DNA molecules
↓
PCR amplification
↓
Robotic 'printing' onto glass slide
↓

mRNA from sample 1 labeled with red fluorchrome

mRNA from sample 2 labeled with green fluorchrome

Hybridize
↓
Wash
↓
Scan red and green signals and combine images

Small region of microarray with red (R), green (G) and yellow/orange (Y/O) colors

Custom gene array screening. A collection of PCR-generated DNAs from yeast are spotted onto a glass slide and hybridized with mRNAs from two sources. After hybridization and washing, the glass slide is scanned and color levels (**R** red, **G** green, **Y/O** yellow to orange) detected and values assigned by computer analysis of the data. (Modified with permission from Garland Science, Taylor and Francis Group. *Molecular Biology of the Cell* by Alberts, B. et al., 4th ed., 2002.)

biologically active. The process of intron removal is called RNA **splicing** (see Fig. 7.12).

Nuclear mRNA splicing is catalyzed by specialized RNA-protein complexes called **small nuclear ribonucleoprotein particles (snRNPs,** pronounced *snurps).* The RNAs found in snRNPs are identified as U1,

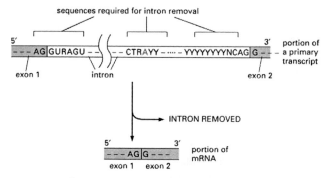

Figure 7.12 Posttranscriptional RNA processing. Shown are the sequences at the ends of exons and within the introns that direct accurate and efficient splicing. (Reproduced with permission from Garland Science, Taylor and Francis Group. *Molecular Biology of the Cell* by Alberts, B. et al., 4th ed., 2002.)

U2, U4, U5, and U6. Analysis of a large number of mRNA genes has led to the identification of highly conserved consensus sequences at the 5′and 3′ ends of essentially all mRNA introns.

The U1 RNA has sequences that are complimentary to sequences near the 5′ end of the intron. The binding of U1 RNA distinguishes the guanine uracil (GU) at the 5′ end of the intron from other randomly placed GU sequences in mRNAs. The U2 RNA also recognizes sequences in the intron, in this case near the 3′ end. The addition of U4, U5, and U6 RNAs forms a complex identified as the **spliceosome** that then removes the intron and joins the two exons together.

Additional processing occurs to mRNAs. The 5′ end of all eukaryotic mRNAs is capped with a unique 5′ to 5′ linkage to a 7-methylguanosine residue (see Fig. 7.13). The capped end of the mRNA is thus protected

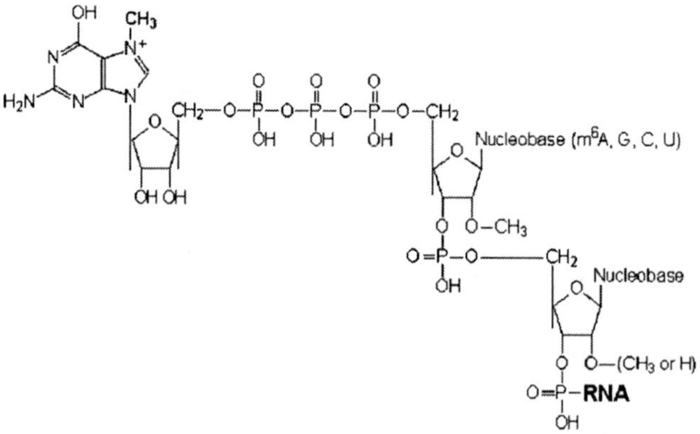

Figure 7.13 Structure of the 5′-cap found on mRNAs. They are capped during posttranscriptional RNA processing. The cap protects mRNA from exonucleases and plays a role during translation.

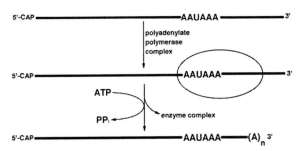

Figure 7.14 Action of polyadenylate polymerase complex. The enzyme complex adds the poly(A) tail $(A)_n$ to mRNAs during posttranscriptional RNA processing. (Reproduced with permission from Garland Science, Taylor and Francis Group. *Molecular Biology of the Cell* by Alberts, B. et al., 4th ed., 2002.)

from exonucleases and more importantly is recognized by specific proteins of the translational machinery.

Messenger RNAs also are polyadenylated at the 3' end. A specific sequence, AAUAAA, is recognized by the endonuclease activity of *polyadenylate polymerase,* which cleaves the primary transcript approximately 11 to 30 bases 3' of the sequence element. A stretch of 20 to 250 A residues is then added to the 3' end by the polyadenylate polymerase activity (see Fig. 7.14).

In addition to intron removal in tRNAs, extra nucleotides at both the 5' and 3' ends are cleaved, the sequence 5'-CCA-3' is added to the 3' end of all tRNAs, and several nucleotides undergo modification. There have been more than 60 different modified bases identified in tRNAs.

7.5 Translation: RNA to Protein

7.5.1 Activation of amino acids

Activation of amino acids is carried out by a two-step process catalyzed by **aminoacyl-tRNA synthetases** (see Fig. 7.15). Each tRNA, and the amino acid it carries, is recognized by individual aminoacyl-tRNA synthetases. This means there exists at least 20 different aminoacyl-tRNA synthetases; there are actually at least 21 since the initiator met-tRNA of both prokaryotes and eukaryotes is distinct from noninitiator met-tRNAs.

First, the enzyme attaches the amino acid to the α-phosphate of *ATP* with the release of pyrophosphate (PP_i, in Fig. 7.15 symbolized by an encircled *P*). This is termed an *adenylated amino acid* intermediate. In the second step, the enzyme catalyzes transfer of the amino acid to either the 2'- or 3'-OH of the ribose portion of the 3'-terminal adenosine residue of the tRNA generating the activated *aminoacyl-tRNA*. Although

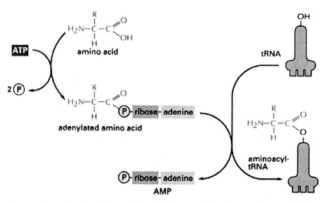

Figure 7.15 Activation and attachment of amino acids to tRNAs during translation. (Reproduced with permission from Garland Science, Taylor and Francis Group. *Molecular Biology of the Cell* by Alberts, B. et al., 4th ed., 2002.)

these reactions are freely reversible, the forward reaction is favored by the coupled hydrolysis of PP_i.

Accurate recognition of the correct amino acid as well as the correct tRNA is different for each aminoacyl-tRNA synthetase. Since the different amino acids have different side chains (R groups), the enzyme for each amino acid has a different binding pocket for its specific amino acid. It is not the **anticodon** (sequence of three nucleotides at the end of a tRNA) that determines the tRNA utilized by the synthetases. Although the exact mechanism is not known for all synthetases, it is likely to be a combination of the presence of specific modified bases and the secondary structure of the tRNA that is correctly recognized by the synthetases.

It is absolutely necessary that the discrimination of correct amino acid and correct tRNA be made by a given synthetase prior to release of the aminoacyl-tRNA from the enzyme. Once the product is released, there is no further way to proofread whether a given tRNA is coupled to its corresponding tRNA. Erroneous coupling would lead to the wrong amino acid being incorporated into the polypeptide since the discrimination of amino acid during protein synthesis comes from the recognition of the anticodon of a tRNA by the **codon** (nucleotide triplet) of mRNA and not by recognition of the amino acid attached to the tRNA.

7.5.2 Initiation

Initiation of **translation** requires a specific initiator tRNA, $tRNA_i^{met}$, that is used to incorporate the initial methionine residue into all proteins. The initiator tRNA is not used for incorporation of methionine into noninitiation locations in the rest of a protein.

The initiation of translation requires recognition of an (adenine-uracil-guanine) AUC codon which is generally, but not always, the first encountered by the ribosome. A specific sequence context surrounding the initiator AUG aids ribosomal discrimination. This context is $^{A/G}$CC$^{A/G}$CCAUG$^{A/G}$ in most mRNAs (where A/G means either nucleotide at that position).

7.5.3 Eukaryotic initiation factors and their functions

The specific nonribosomally associated proteins required for accurate translational initiation are termed **initiation factors (IF)**. In *E. coli* they are IFs; in eukaryotes they are eukaryotic initiation factors (eIFs). Numerous eIFs have been identified with the important ones listed in Table 7.2.

TABLE 7.2   Eukaryotic Initiation Factors (eIF)

Initiation Factor	Activity
eIF-1	Repositioning of met-tRNA to facilitate mRNA binding
eIF-2	Ternary complex formation, composed of three subunits: α, β, and γ
eIF-2A	AUG-dependent met-tRNA$^{i}_{met}$ binding to 40S ribosome
eIF-2B	GTP/GDP exchange during eIF-2 recycling
eIF-3 composed of ~10 subunits	Ribosome subunit antiassociation, binding to 40S subunit
eIF-4F composed of three primary subunits: eIF-4E, eIF-4A, eIF-4G, and at least two additional factors: PABP, Mnk1 (or Mnk2)	mRNA binding to 40S subunit, ATPase-dependent RNA helicase activity, interaction between polyA tail and cap structure
PABP: polyA-binding protein	Binds to the polyA tail of mRNAs and provides a link to eIF-4G
Mnk1 and Mnk2: eIF-4E kinases	Phosphorylate eIF-4E increasing association with cap structure
eIF-4A	ATPase-dependent RNA helicase
eIF-4E	5' cap recognition
4E-BP (also called PHAS) three known forms	Dephosphorylated 4E-BP binds eIF-4E and represses its activity. When phosphorylated, 4E-BP is released from eIF-4E, resulting in increased translational initiation
eIF-4G	Acts as a scaffold for the assembly of eIF-4E and -4A in the eIF-4F complex; interaction with PABP allows 5'-end and 3'-ends of mRNAs to interact
eIF-4B	Stimulates helicase, binds simultaneously with eIF-4F
eIF-5	Release of eIF-2 and eIF-3, ribosome-dependent GTPase
eIF-6	Ribosome subunit antiassociation

7.5.4   Specific steps in translational initiation

Initiation of translation requires four specific steps:

1. A ribosome must dissociate into its $40S$ and $60S$ subunits.
2. A ternary complex termed the preinitiation complex is formed consisting of the initiator, guanosine triphosphate (GTP), eIF-2, and $40S$ subunit.
3. The mRNA is bound to the preinitiation complex.
4. The $60S$ subunit associates with the preinitiation complex to form the $80S$ initiation complex.

The initiation factors eIF-1 and eIF-3 bind to the $40S$ ribosomal subunit favoring antiassociation to the $60S$ subunit. The prevention of subunit reassociation allows the **preinitiation complex** to form.

The first step in the formation of the preinitiation complex is the binding of GTP to eIF-2 to form a binary complex (see Fig. 7.16). The binary complex then binds to the activated initiator tRNA (met-tRNA^{met}) forming a ternary complex that then binds to the $40S$ subunit forming the $43S$ preinitiation complex ($40S$ + 3: eIF-1, eIF-3, and eIF-2-GTP). The preinitiation complex is stabilized by the earlier association of eIF-3 and eIF-1 to the $40S$ subunit.

The cap structure of eukaryotic mRNAs (see Sec. 7.4.1 and Fig. 7.13) is bound by specific eIFs prior to association with the preinitiation complex. Cap binding is accomplished by the initiation factor eIF-4F. This factor is actually a complex of three proteins: eIF-4E, A, and G. The first protein, eIF-4E, is a 24 kDa protein which physically recognizes and binds to the cap structure. The second protein, eIF-4A, is a 46 kDa protein

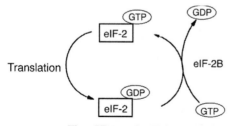

Figure 7.16 The eIF-2 cycle. This process is required for initiation of translation. Following formation of the preinitiation complex, the GTP bound by eukaryotic initiation factor 2 (eIF-2) is hydrolyzed to provide the energy needed to bind and align the activated initiator tRNA (met-tRNA). The GDP is exchanged for GTP through the action of eIF-2B.

which binds and hydrolyzes ATP and exhibits RNA helicase activity. Unwinding of mRNA secondary structure is necessary to allow access of the ribosomal subunits. The third protein, eIF-4G, aids in binding of the mRNA to the 43S preinitiation complex.

Once the mRNA is properly aligned onto the preinitiation complex and the initiator met-tRNA^{met} is bound to the initiator AUG codon (a process facilitated by eIF-1), the 60S subunit associates with the complex. The association of the 60S subunit requires the activity of eIF-5 which has first bound to the preinitiation complex. The energy needed to stimulate the formation of the 80S initiation complex comes from the hydrolysis of the GTP bound to eIF-2. The guanosine diphosphate (GDP) bound form of eIF-2 then binds to eIF-2B which stimulates the exchange of GTP for GDP on eIF-2. When GTP is exchanged eIF-2B dissociates from eIF-2. This is termed the eIF-2 cycle (see Fig. 7.16). This cycle is absolutely required in order for eukaryotic translational initiation to occur. The GTP exchange reaction can be affected by phosphorylation of the α-subunit of eIF-2.

At this stage, the initiator met-tRNA^{met} is bound to the mRNA within a site of the ribosome termed the **P-site**, for peptide site. The other site within the ribosome to which incoming charged tRNAs bind is termed the **A-site**, for amino acid site.

7.5.5 Elongation

The process of elongation, like that of initiation requires specific nonribosomal proteins. In prokaryotes such as *E. coli,* these are **elongation factors (EFs)** and in eukaryotes eukaryotic elongation factors (eEFs). Elongation of polypeptides occurs in a cyclic manner such that at the end of one complete round of amino acid addition, the A site will be empty and ready to accept the incoming aminoacyl-tRNA dictated by the next codon of the mRNA (see Fig. 7.17). This means that not only does the incoming amino acid need to be attached to the peptide chain, but the ribosome must move down the mRNA to the next codon. Each incoming aminoacyl-tRNA is brought to the ribosome by an eEF-1α-GTP complex. When the correct tRNA is deposited into the A site, the GTP is hydrolyzed and the eEF-1α-GDP complex dissociates. In order for additional translocation events, the GDP must be exchanged for GTP. This is carried out by eEF-1$\beta\gamma$ similarly to the GTP exchange that occurs with eIF-2 catalyzed by eIF-2B.

The peptide attached to the tRNA in the P-site is transferred to the amino group at the aminoacyl-tRNA in the A-site. This reaction is catalyzed by peptidyltransferase. This process is termed transpeptidation. The elongated peptide now resides on a tRNA in the A-site. The A-site needs to be freed in order to accept the next aminoacyl-tRNA. The

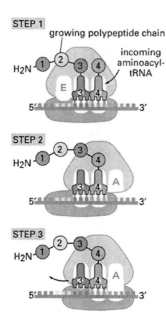

Figure 7.17 Sequence of events during the process of elongation of the polypeptide chain. In step 1, the growing polypeptide resides in the P-site attached to the tRNA that carried the last amino acid to the ribosome (amino acid 3 in step 1). The next amino acid to be incorporated resides in the A-site attached to its tRNA (amino acid 4 in step 1). In step 2, peptidyltransferase transfers the polypeptide to the amino acid and then the ribosome translocates freeing up the A-site and the cycle can begin again. (Reproduced with permission from Garland Science, Taylor and Francis Group. *Molecular Biology of the Cell* by Alberts, et al., 4th ed., 2002.)

process of moving the peptidyl-tRNA from the A-site to the P-site is termed translocation. Translocation is catalyzed by eEF-2 coupled to GTP hydrolysis. In the process of translocation, the ribosome is moved along the mRNA such that the next codon of the mRNA resides under the A-site. Following translocation, eEF-2 is released from the ribosome. The cycle can now begin again.

7.5.6 Termination

Like initiation and elongation, translational termination requires specific protein factors identified as eukaryotic **releasing factors, eRFs.** The signals for termination are specific codons present in the mRNA. There are three termination codons: UAG, UAA, and UGA. The binding of eRF to the ribosome stimulates the peptidyltransferase activity to transfer the peptidyl group to water instead of an aminoacyl-tRNA. The resulting uncharged tRNA left in the P-site is expelled with concomitant hydrolysis of GTP. The inactive ribosome then releases its mRNA, and the $80S$ complex dissociates into the $40S$ and $60S$ subunits ready for another round of translation.

7.5.7 Heme control of translation

Regulation of initiation in eukaryotes is effected by phosphorylation of a serine residue in the α-subunit of eIF-2. Phosphorylated eIF-2 in the

APPLICATION BOX 7.3
Proteomics—Global Analysis of Cellular Protein Composition

Proteomics is the term used to describe the global analysis of all the proteins present in a given cell or tissue at a given time or condition. A determination of how the protein complement of a cell or tissue changes during development, in response to various stimuli or as a consequence of disease states is crucial to our understanding of how these various processes occur. Thus, the ability to detect hundreds and even thousands of protein differences simultaneously greatly enhances our ability to formulate models of the functions of the protein changes that occur under a variety of conditions.

Proteomics involves the use of **mass spectrometry** (MS) to determine the precise mass of intact proteins as well as peptides derived from enzymatic digestion (e.g., trypsin digestion) of protein samples. Traditional proteomics began with the use of high resolution two-dimensional (2D) gel electrophoresis for protein separation followed by isolation of specific protein spots from the gel and MS analysis. The key developments in proteomics were the invention of time-of-flight (TOF) MS and methods of ionization of proteins and peptides that did not destroy the sample. The most common ionization methods are matrix-assisted laser desorption ionization (MALDI) and electrospray ionization (ESI). In using MALDI, the sample(s) are solidified within an acidified matrix that absorbs energy of a specific UV range with the energy being dissipated thermally. The rapid transfer of energy generates a vaporized matrix which ejects the samples into the gas phase where they acquire charge. An electric field between the MALDI plate and the entrance to the MS tube leads to different speeds at which various charged samples reach the MS based upon their mass to charge ratios. The advantage of MALDI-TOF is that sample analysis can be accomplished with high throughput (e.g., 96 samples at one time). ESI allows rapid transfer of samples from the liquid to the gas phase by creating droplets that enter the MS which then go through repetitive solvent evaporation steps until only the charged sample is left in the gas phase. ESI has an advantage over MALDI in that it can be coupled to highly specific separation techniques like liquid chromatography (LC) which in turn allows for high throughput.

absence of eIF-2B is just as active an initiator as nonphosphorylated eIF-2. However, when eIF-2 is phosphorylated, the GDP-bound complex is stabilized and exchange for GTP is inhibited. The exchange of GDP for GTP is mediated by eIF-2B (also called guanine nucleotide exchange factor, GEF). When eIF-2 is phosphorylated it binds eIF-2B more tightly, thus slowing the rate of exchange. It is this inhibited exchange that affects the rate of initiation (see Fig. 7.18).

The phosphorylation of eIF-2 is the result of an activity called *heme-controlled inhibitor (HCI)* which functions as diagrammed in Fig. 7.18.

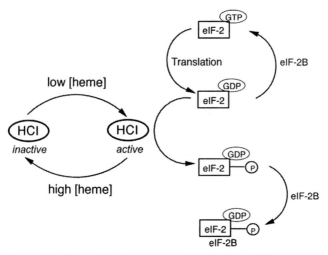

Figure 7.18   Role of the heme-controlled inhibitor (HCI) in regulating the activity of eukaryotic initiation factor 2 (eIF-2) and thus the rate of translational initiation. Shown at the top right is the normal process of the eIF-2 cycle of translation initiation. Shown in the lower right is the effect of HCI phosphorylation of eIF-2 leading to tight binding of eIF-2B with the result that exchange of GTP for GDP is greatly slowed.

HCI is generated in the absence of heme, a mitochondrial product. Removal of phosphate is catalyzed by a specific eIF-2 phosphatase which is unaffected by heme. The presence of HCI was first seen in *in vitro* translation systems derived from lysates of reticulocytes. Reticulocytes synthesize almost exclusively hemoglobin at an extremely high rate. In an intact reticulocyte, eIF-2 is protected from phosphorylation by a specific 67 kDa protein.

7.5.8   Interferon control of translation

Regulation of translation can also be induced in virally infected cells. It would benefit a virally infected cell to turn off protein synthesis to prevent propagation of the viruses. This is accomplished by the induced synthesis of interferons. There are three classes of interferons, which are classified according to their cellular origin: the leukocyte or α-interferons, the fibroblast or β-interferons, and the T lymphocyte or γ-interferons. Interferons are induced by **double-stranded RNAs (dsRNAs)**. This is the form of RNA in some viruses, which has two complementary strands similar to DNA. Interferons themselves induce a specific kinase termed RNA-dependent protein kinase (PKR) that phosphorylates eIF-2, thereby shutting off translation in a manner similar to that of heme control of translation. Additionally, interferons induce the synthesis of

2′-5′-oligoadenylate synthetase, an enzyme that generates the compound pppA(2′p5′A)n (where n refers to multiple 2′ to 5′ linked adenosines), which in turn activates a preexisting ribonuclease, RNase L. RNase L degrades all classes of mRNAs, thereby shutting off host mRNA translation.

Suggested Reading

Alberts, B., Johnson, A., Lewis, J., Raff, M., Roberts, K. and Walter, P. (2002). Basic genetic mechanisms. In: *Molecular Biology of the Cell*, (B. Alberts, A. Johnson, J. Lewis, M. Raff, K. Roberts, P. Walter, eds.). Garland Science, New York, NY, pp. 191–466.

8

Cell Division and Growth

David A. Prentice, Ph.D.

OBJECTIVES

- To present the concept of the cell growth cycle and mechanism of cell division
- To explain stem cells, their clinical potential, and associated controversy
- To outline the processes involved in cell aging
- To introduce cancer cells as cells with unregulated growth control

OUTLINE

8.1	Growth of Cells: Cell Cycle	210
8.2	Mitosis	215
8.3	Stem Cells: Maintenance and Repair of Tissues	222
8.4	Cell Senescence: Cell Aging	229
8.5	Cancer: Abnormal Growth	230

Each cell in your body knows when to grow and divide and when to stop growing. Cell growth relies on an ordered progression of signals to move the cell through its growth cycle and divide. Stem cells are the normal physiological body cells with the capacity to multiply continuously. A better understanding of why some cells proliferate pathologically, with uncontrolled growth, will contribute to the development of new cancer treatments. The knowledge of how to use stem cells to regenerate damaged tissues and organs has the potential to revolutionize medicine. Engineers use engineering principles to put biological knowledge to use for applications in biotechnology and medicine.

Growth of a human being from a single-celled zygote into the complex system of approximately 100 trillion cells of an adult requires an obvious increase in cell number for growth of the organism. But this multiplication of cells must take place in an organized, controlled manner, not only in terms of cellular interactions and specializations but also on an individual cell level. As it grows, each cell must respond to cues from other cells, and adjust its growth according to the needs of the local tissue environment as well as the overall organism. Each new cell has several paths it may follow—continue growth to divide again, pause in the process of growth and division, or begin a pathway of differentiation that leads to a specialized cell type.

8.1   Growth of Cells: Cell Cycle

Historically, the growth cycle of a cell was defined by what was visible through the light microscope. The condensation and segregation of chromosomes was evident, as was the division of the one cell into two, at **mitosis**. The rest of the cell growth cycle seemed rather bland and uneventful through the microscope and was termed **interphase**, denoting that it was the interval between cell division phases. Biochemical evidence later showed that replication of DNA took place during the middle of this interphase of the cell cycle, with the time of DNA synthesis separated from the mitotic phase by gaps in time on either side. Thus, the four major phases of the eukaryotic cell growth cycle were initially defined (Fig. 8.1).

8.1.1   Phases of the cell cycle

Starting immediately upon cell division, the phases of the cell cycle are:

1. **G1 phase** ("gap 1"), when cells respond to various signals for growth or differentiation, as well as make preparation for DNA synthesis,

2. **S phase** ("synthesis"), when the DNA is replicated,

3. **G2 phase** ("gap 2"), when cells prepare for chromosome segregation and cell division, and

4. **M phase** ("mitotic"), when cells separate their chromosomes, ending with cell division into two new daughter cells. M phase itself is divided into a series of stages which will be described later.

Despite their original identification as mere gaps in time between the identified markers of cell division and DNA synthesis, G1 and G2 phases are not idle times for the cell. In particular, G1 phase is the stage where the cell's future is decided, in response to internal and external cues, in terms of whether the cell enters another cell growth

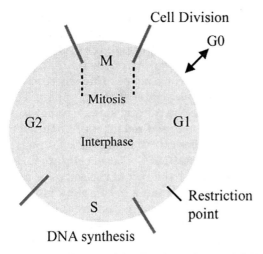

Figure 8.1   The cell cycle. The four phases of the eukaryotic cell cycle are G1, S, G2, (collectively termed interphase), and M (mitosis). DNA synthesis occurs during S phase, and cell division occurs at the end of M phase. G0 is considered a resting phase; the restriction point is a stage that must be passed for the cell to be committed to complete the entire cell cycle.

cycle round, pauses in its growth, or exits the cell growth cycle, including undergoing differentiation. A cell that receives signals to continue growth moves on through G1 phase uninterrupted, becoming committed to finish another cycle and cell division. A cell that receives different signals, or is not stimulated by growth signals, moves into a sort of side loop of the cell cycle termed **G0 phase**, sometimes also called "resting phase." It may move into a lower metabolic state, termed **quiescence**, while it waits for further signals, or it may exit the growth cycle and begin a path of differentiation.

A cell in G1 phase that has received a signal to proceed through the cell cycle is still not committed to complete the cycle until it passes the **restriction point**, a point of no return late in G1 phase. At this point, the metabolic machinery and cell cycle trigger proteins have been activated and accumulated to the extent that the cell is committed to proceed on into S phase and complete an entire cell cycle (usually about 24 hours long for mammalian cells).

8.1.2   Studying cell cycle phases

Typically, within a population of cells under normal conditions, whether in a body tissue or in a laboratory culture dish, cells within the population will be scattered throughout various points in the cell cycle. In working with cells, it is often useful to be able to accumulate an entire population of cells at one point in the cell cycle to study the characteristics of that phase or the progression of cells through the cycle.

Isolation of a population of cells synchronized at one point in the cell cycle usually relies on the use of synchrony agents that specifically target some metabolic event or molecule associated with that phase. For example, microtubules are essential for the mitotic spindle apparatus and separation of chromosomes, but are not required to progress through any other point in the cell cycle, so synchronization of cells at M phase often uses inhibitors of microtubule formation. Cultured cells grown in the presence of such an inhibitor complete the rest of the cell cycle and enter M phase, but are then trapped and unable to complete mitosis, accumulating at that point in the cell cycle. G1 phase progression relies on production of key cell cycle regulatory proteins, stimulated by external growth signals and adequate amino acid availability for protein synthesis, so withholding growth stimulators and/or selective amino acid depletion can block cells in G1 phase. For S phase, DNA synthesis is the obvious choice of a target for inhibition. Currently for most cells no target lends itself easily to G2 blockade, but since cells that have reached S phase are committed to completion of the cycle, synchronization at S phase followed by release and timing of the end of DNA synthesis gives a reliable population of cells all in G2 phase. It should be borne in mind, however, that any metabolic inhibitor that blocks necessary cellular activities can also be toxic to the cells, so careful monitoring of the health of a synchronized cell culture is necessary in synchrony experiments.

A high-tech solution to isolating cells at different points in the cell cycle is the use of a **fluorescence-activated cell sorter (FACS)**. This machine sends a stream of cells through one or more laser beams, and the light scattered from the cells is detected in one or more photomultipliers (Fig. 8.2). Cells can be sorted based on size, which increases throughout the cell cycle. Cells can also be pretreated with a nontoxic fluorescent dye specific for DNA or other molecules and the fluorescence emission excited by the laser can be detected. A flow cytometer analyzes the output of the photodetectors, while the FACS uses the emissions from each cell to sort the cell into collection vessels for further study. In this way an entire population of cells can be analyzed for their position in the cell cycle, for example, based on their DNA content, and sorted into subpopulations for the different cell cycle phases.

8.1.3   Control of cell cycle

Proteins are the triggers for progression of a cell through the different cell cycle phases. In particular, diverse combinations of two different types of protein drive the various phases of the cell cycle. These two protein

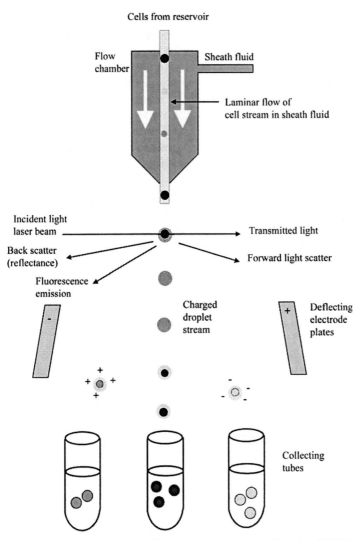

Figure 8.2   Flow cytometry and fluorescence-activated cell sorting (FACS). Individual cells in droplets are passed through a laser beam, exciting fluorescence emission. Photodetectors are used to analyze the emission and light scatter, and the signal can be coupled through a computer to place a charge on the cell droplet to deflect individual cells into different collecting vessels. (Modified from Figure 15.9, *Culture of Animal Cells*, 5th ed., by Freshney, Wiley-Liss.)

families are the cyclins, and the cyclin-dependent kinases (cdks). The **cyclin-dependent kinase (cdk)** proteins are kinases, phosphorylating various target proteins in the cell to activate (or inactivate) them and start a series of events, like knocking over the first domino in a row, that move the cell on to another portion of the cycle. The cdks are usually present throughout a cell cycle. However, they must be complexed with a cyclin for their activity. The **cyclins** are regulatory and targeting proteins for the cdks. When a particular cyclin binds to a cdk, it activates the cdk and also targets its enzymatic activity. Cyclins are transient proteins, synthesized when needed and quickly degraded. When cells are stimulated to grow by external signaling factors and their associated signal transduction, specific cyclin proteins are synthesized as one of the early responses to these growth signals. Different cyclins complex with specific cdks at various points in the cell cycle to drive the cell through each phase of the growth cycle, targeting diverse molecules necessary for each step. Then the cyclin is degraded, deactivating its paired cdk.

Continued movement through the cycle at each step is an ordered progression that requires the synthesis of a different cyclin, its complexing with, activating, and targeting of a cdk. It further requires the cessation of cyclin synthesis and degradation of that cyclin, deactivating the cdk. The different cyclin-cdk pairings for movement through the cell cycle are shown in Fig. 8.3. Starting with a cell "resting" or "waiting" in G0 which receives a growth signal, the first cyclin synthesized is *cyclin D* (and in some cells also cyclin F), which complexes with *cdk4* and *cdk6*. Moving the cell on through G1 and into S phase requires production of *cyclin E*, which complexes with *cdk2*. Once S phase is initiated, *cyclin A* is produced and complexes with *cdk2*. Both cyclin A and newly-produced *cyclin B* complex with *cdk1* to initiate M phase and the events associated with mitosis and **cytokinesis** (the final stage of cell reproduction).

One of the targets of the early G1 phase cyclin D-cdk4/6 complex is an inhibitory protein called *Rb*, the retinoblastoma protein (Fig. 8.4). (The protein is so named because its loss or mutation was initially correlated with development of the retinoblastoma type of tumor.) In early G1, the unphosphorylated Rb protein complexes with transcription factors termed *E2F factors*, inhibiting their activity. The E2F transcription factors

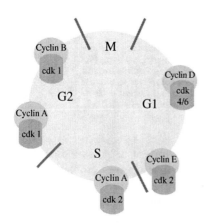

Figure 8.3 Different cyclins and cyclin-dependent kinases (cdks) direct cell-cycle progression. Ordered progression through the cell cycle requires the ordered synthesis of different cyclin proteins, which complex with various cdks to activate and target their phosphorylation of specific molecules at different cell cycle phases. (Modified from Figure 8.2, *Developmental Biology*, 7th ed., by Gilbert, Sinauer Associates, Inc.)

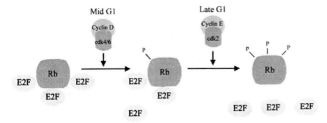

Figure 8.4   Retinoblastoma protein (Rb) release of E2F transcription factor to pass the restriction point. Phosphorylation of Rb by cyclin D-cdk4/6 releases some E2F, which stimulates production of cyclin E and cdk2. The cyclin E-cdk2 complex fully phosphorylates Rb, allowing release of all E2F.

activate genes associated with DNA synthesis, and also stimulate production of the late G1 cyclin E and S phase cyclin A. Phosphorylation of Rb by *cyclin D-cdk4* and *cyclin D-cdk6* releases some of the E2F factors, and these activate production of cyclin E and cdk2 in late G1 (Fig. 8.4). The *cyclin E-cdk2* complex further phosphorylates Rb, completely releasing all E2F transcription factors, and as cyclin E-cdk2 level increases to a critical threshold, it can maintain complete phosphorylation of Rb even when cyclin D is degraded and the early G1 cyclin-cdk complexes are deactivated. This is the primary mechanism that moves the cell past the restriction point and commits it to completing the cell cycle.

Targets for cyclin-cdk complexes later in the cell cycle include DNA pre-replication initiation complexes for the S phase cyclin A-cdk2 complex, and in G2 and M phase the chromatin histone proteins and nuclear laminar proteins for the cyclin B-cdk1 complex. Phosphorylation of histones, especially H1 and H3 (for histones, see Sec. 7.2.1), precipitates chromosome condensation at the beginning of M phase. Phosphorylation of nuclear laminar proteins (protein network lining the inner face of the nuclear membrane) leads to dissociation of the nuclear membrane, allowing separation of the chromosomes into the two daughter cells during mitosis. Moving from one phase to another, and one cyclin-cdk complex to another, requires the cessation of production of each succeeding cyclin and degradation of existing cyclin, primarily due to polyubiquitination of the cyclin targeting it for proteolysis. Polyubiquitination, the binding of **ubiquitin** molecules, is a common process to give proteins the "kiss of death," that is, to label them for proteolytic degradation.

8.2   Mitosis

8.2.1   Stages of mitosis

Mitosis (M phase), though brief in terms of the entire cell cycle, is important because it is the stage where the newly replicated chromosomes

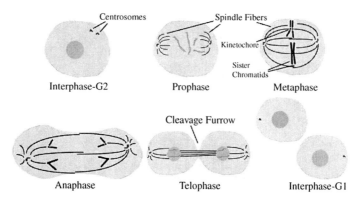

Figure 8.5   Stages of mitosis and cytokinesis. Chromosome condensation and spindle formation begins in prophase. Chromosomes align at metaphase and are separated to opposite poles in anaphase. Chromosomes decondense, the nuclear envelope is reformed, and cell cleavage occurs in telophase. (Modified from Figure 20-29, *Molecular Cell Biology*, 5th ed., by Lodish et al., W.H. Freeman and Co.)

are separated and two new daughter cells are formed from the original cell. Mitosis is divided into four stages: (1) prophase, (2) metaphase, (3) anaphase, and (4) telophase (Fig. 8.5). Mitosis begins with the condensation of the chromosomal material, first appearing as slender threads, and by the end of **prophase,** each chromosome is visible as two identical paired filaments (sister **chromatids**) held together at a constriction called the **centromere.** Also during prophase, two centrosomes take up positions at opposite poles of the cell, and the nuclear envelope fragments into vesicles. Microtubules known as spindle fibers begin to grow from the opposing centrosomes, attaching to the chromosomes at the centromere region at a structure known as the kinetochore.

At **metaphase,** the chromosomes move toward the equator of the cell and align themselves in the equatorial plane. During **anaphase,** the two sister chromatids of each chromosome are separated into independent chromosomes and move toward opposite poles under the influence of the spindle fibers. Finally at **telophase,** the nuclear membrane reforms around each group of chromosomes, the chromosomes decondense, the spindle fibers disappear, and the cell divides (a process known as cytokinesis.)

8.2.2   Mechanics and control of mitosis

Three components are necessary for the alignment and separation of chromosomes into the daughter cells: the centromere, the kinetochore and spindle fibers (explained below), and the cell cycle proteins. As already noted, the cell cycle is driven by cyclin-cdk complexes, and mitosis itself is no different. The complex of cyclin B-cdk1 is the primary driving force for chromosome condensation and dissolution of the nuclear membrane at the beginning of M phase. Phosphorylation of histone proteins in particular begins the chromosome condensation process, while phosphorylation of *nuclear laminar proteins* leads to disaggregation of the protein matrix in the nuclear envelope (Fig. 8.6) and its fragmentation into small vesicles.

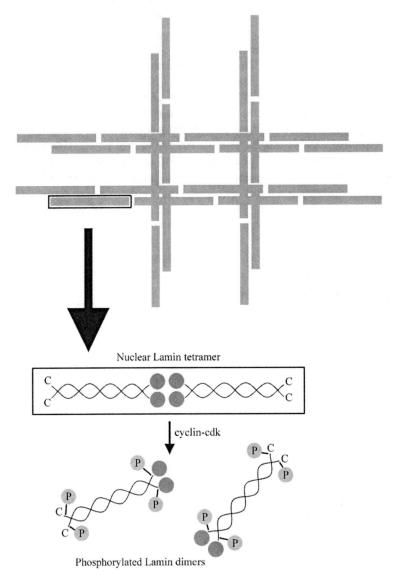

Nuclear Lamin tetramer

cyclin-cdk

Phosphorylated Lamin dimers

Figure 8.6   Nuclear lamina depolymerization is due to phosphorylation by cyclin-cdk. The matrix of lamin tetramers forming the framework for the nuclear envelope disintegrates when phosphorylation of specific serines near the ends of the coiled-coil domains leads to depolymerization into lamin dimers. (Modified from Figure 21-16, *Molecular Cell Biology*, 5th ed., by Lodish et al., W.H. Freeman and Co.)

The mitotic apparatus for alignment and separation of chromosomes is based on microtubular machinery. Microtubules are presented as important cytoskeletal elements in Sec. 2.3.2 and Fig. 2.7. The **centrosome** itself is a protein complex that is a self-propagating organelle (Fig. 8.7, see also Fig. 2.11). The centrosomes facilitate spindle formation, acting as nucleation centers for the assembly of microtubules. The **kinetochore** of each chromosome is a complex, located at the centromeres of chromosomes, that captures spindle fiber microtubules and helps transport chromosomes. The **spindle fibers** themselves are microtubules growing out from the centrosomes. The microtubules are stabilized at their ($-$) end at the centrosomes, but are unstable at the ($+$) ends which are growing into the region of the condensed chromosomes. When the ($+$) end of a spindle fiber microtubule contacts the kinetochore of a chromosome, the kinetochore captures the end of the microtubule. Each sister chromatid has a kinetochore, and the joined pair of sister chromatids is captured by opposing spindle fibers from each pole. In addition, polar microtubules extend from each centrosome and overlap near the center of the cell (see Fig. 8.7). As the spindle fibers continue to grow, the chromosomes are moved to a central mitotic plane at the center of the cell and held in place at that position due to the opposing tension of the microtubules from the opposite poles.

At anaphase, the spindle fibers begin to shorten, pulling the chromatids toward the poles. This is accomplished by disassembly of the microtubules at their ($+$) end. The chromatids remain attached to the

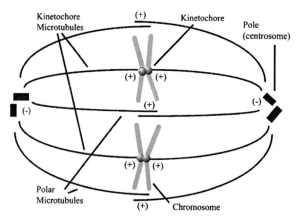

Figure 8.7   Parts of the mitotic apparatus. The mitotic apparatus consists of the centrosomes, from which the spindle fiber microtubules grow, the spindle fiber and polar microtubules, and the kinetochores of the chromosomes. The ($+$) ends of the spindle fiber microtubules contact the kinetochores. (Modified from Figure 20-31, *Molecular Cell Biology*, 5th ed., by Lodish et al., W.H. Freeman and Co.)

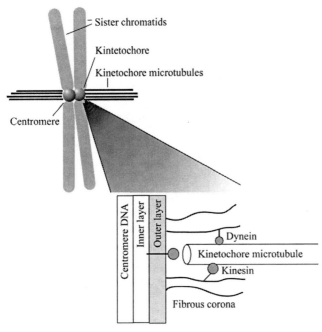

Figure 8.8 Attachment of spindle fibers to the kinetochore. The kinetochore has an inner layer with proteins that bind centromeric DNA, and an outer layer that connects to the (+) ends of kineto-chore microtubules. The outer layer and a fibrous corona around the microtubule ends contain microtubule-binding proteins and motor proteins including dynein and kinesins. (Modified from Figure 20-32, *Molecular Cell Biology*, 5th ed., by Lodish et al., W.H. Freeman and Co.)

microtubules through the interactions of *kinesins*, which maintain contact between the kinetochore and the microtubule (Fig. 8.8). Near the end of anaphase, the centrosomes themselves move further apart, which acts to pull the chromosomes into what will be their respective new cell. This separation is accomplished by elongation of polar spindle fiber microtubules along with the action of the motor protein *dynein* (see Fig. 8.8).

Physical separation of the sister chromatids themselves also requires coordination. Unlinking sister chromatids initiates anaphase. At anaphase, the spindle is in a state of tension, pulling kinetochores toward opposite spindle poles, balanced by forces pushing spindle poles apart. Sister chromatids are held together at the centromeres and multiple positions along chromosome arms by *cohesin* protein complexes (Fig. 8.9). Prior to anaphase, the protein *securin* binds to and inactivates *separase*, a ubiquitous protease. Once all chromosome kinetochores have attached to spindle microtubules, a protein complex called the **anaphase promoting complex (APC)** is activated by its specificity factor, *cdc20*, and

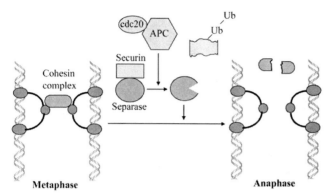

Figure 8.9 Separation of sister chromatids at anaphase. Sister chromatids are held together by cohesion protein complexes. To begin the separation of sister chromatids at anaphase, the anaphase promoting complex (APC) is activated by cdc20. The active APC polyubiquitinates securin, which is then degraded. Securin binds and inactivates the separase protease. With securin removed, separase cleaves the cohesin protein linking sister chromatids, allowing the chromatids to be pulled to opposite poles of the cell. (Modified from Figure 21-19, *Molecular Cell Biology*, 5th ed., by Lodish et al., W.H. Freeman and Co.)

the active APC polyubiquitinates securin, leading to its degradation and the onset of anaphase. With securin degraded, the separase protease is released and cleaves the cohesion protein cross-links between sister chromatids. Poleward force then moves chromatids toward opposite spindle poles. APC also initiates degradation of the cyclin-cdk complexes. Unopposed ubiquitous phosphatase activity then begins to allow chromatin decondensation and reformation of nuclear envelope.

Once the sister chromatids have separated to opposite ends of the cell, telophase begins. The nuclear envelope reforms around the chromosomes as they begin to decondense, and a contractile ring of actin and myosin forms around the center of the cell and constricts to cleave the cell into two new daughter cells (Fig. 8.10). Phosphorylation of myosin by cyclin B-cdk1 inhibits interaction with actin during earlier stages of mitosis. The decrease in cyclin-cdk complexes in late anaphase relieves the inhibition, ubiquitous phosphatase removes the phosphorylation, and this allows formation and action of the contractile ring (Fig. 8.10), initiating cytokinesis to produce the two new daughter cells.

8.2.3 Checkpoints in cell cycle control

Genetic damage can occur if cells progress to the next phase of the cell cycle before the previous phase is properly completed. For example, cells still in S phase can be induced to enter mitosis prematurely, by

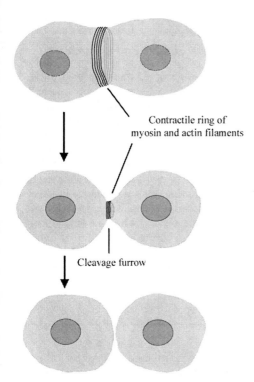

Figure 8.10   Actin filaments are needed for cytokinesis. Decrease in cyclin-cdk during anaphase allows formation of an equatorial contractile ring composed of actin and myosin filaments. The interacting filaments contract, leading to cytokinesis and separation into two new daughter cells. (Modified from page 606, *Cell and Molecular Biology*, 3rd ed., by Karp, G., John Wiley & Sons, New York.)

fusion with a mitotic cell. The chromosomes of the S phase cell will condense under the action of the added mitotic factors including cyclin B-cdk1. However, the replicating chromosomes become fragmented because replication has not been completed. Such premature entry into mitosis in a normal cell cycle would be catastrophic to the cell.

Another example is attachment of kinetochores to microtubules of the mitotic spindle during metaphase. If anaphase is initiated before both kinetochores of the sister chromatids become attached to microtubules from opposite spindle poles, daughter cells will be produced with a missing or extra chromosome. The misallocation of chromosomes to daughter cells is termed **nondisjunction** when it occurs during meiosis. One example of this nondisjunction is seen in Down syndrome (trisomy 21.) In this example, the addition of an extra chromosome is not lethal, but in many instances the loss or gain of a single chromosome is a lethal change (see Clinical Box 8.1).

To minimize the occurrence of mistakes, the cell cycle has several **checkpoints** to assess the cell's progress through the cycle (Fig. 8.11). Control mechanisms that operate at these checkpoints try to ensure that DNA synthesis is properly completed, that the chromosomes are

---

CLINICAL BOX 8.1
Separating Chromosomes Incorrectly: Nondisjunction

Proper attachment of spindle fibers to chromosome kinetochores is necessary for later separation of chromosomes equally between the two daughter cells. If spindle fibers from only one pole capture both sister chromatids, or anaphase segregation of chromosomes starts before both kinetochores are captured, both sister chromatids will be taken into only one of the daughter cells. This nondisjunction results in one daughter cell with an extra chromosome and one daughter cell lacking a chromosome. In most cases this is a lethal defect for the cell, but in some cases the cell survives to pass on the abnormal chromosome number. Nondisjunctions that occur early in development include Down syndrome, where an individual has an extra chromosome 21 (trisomy 21), and Turner syndrome, in which females have only one X chromosome instead of two. An altered chromosome number due to nondisjunction that occurs after birth can also lead to cancer. Cell-cycle checkpoints that monitor proper spindle assembly and chromosome segregation are designed to prevent nondisjunction or stop the cell cycle progress if it occurs, but in some cases the checkpoints fail.

---

intact, and that each stage of the cell cycle is finished before the following stage has begun. One checkpoint for S phase involves maintaining mitotic cyclin B-cdk1 in an inactive state until DNA synthesis is completed. Association of ATR kinase (full name ataxia telangiectasia and Rad3 related kinase) with replication forks activates the kinase, which phosphorylates and activates the Chk1 kinase. Chk1 kinase then inactivates the Cdc25 phosphatase, which normally removes an inhibitory phosphate from the mitotic cdk1. In this way, mitotic cyclin-cdk complexes are kept inactive until replication is completed, at which point the ATR and Chk1 kinases are inactivated, reactivating Cdc25 phosphatase, which can then initiate activation of the mitotic cdk. Cell cycle checkpoints also operate to check for DNA damage and improper replication, as well as spindle assembly and proper chromosome segregation (Fig. 8.11) (Rajagopalan and Lengauer, 2004).

8.3   Stem Cells: Maintenance and Repair of Tissues

8.3.1   The problem of tissue maintenance and turnover

Few cells in the body last a lifetime. Most wear out over time and must be replaced, some cell types more often than others, due to their usage and their exposure to more extreme environments. For example, skin

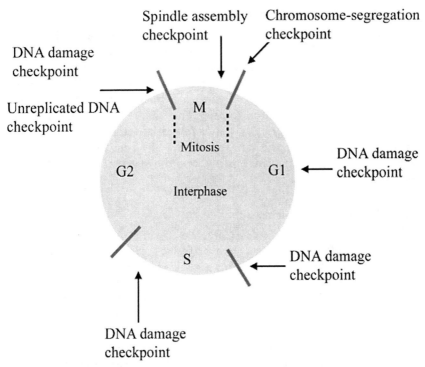

Figure 8.11   Cell cycle checkpoints. The cell cycle contains numerous checkpoints to maintain normal cell cycle progress and ensure production of two healthy daughter cells. For details, see Sec. 8.2.3.

cells are exposed to the external environment and are continually shed and replaced with new layers, and the cells lining the digestive tract must also turn over at a rapid rate. The mechanism of replacement throughout the lifetime of the organism must account for the rate of turnover of cells and need for replenishment, and must also do so with fidelity to ensure that the particular tissue or organ has a continual supply of cells for its needs. Some tissues accomplish this through simple mitosis of existing cells within the tissue, with replenishment regulated usually by slow constant cell division, and the rate of division and replenishment dependent on signals (usually diffusible) within the tissue. However, many specialized cell types are unable to replenish their numbers mitotically, due to the nature of their specialized morphology and function. For example, muscle is formed from fusion of individual muscle cells into long fibers, making it impossible to divide into new individual muscle cells. Human red blood cells and the squamous epithelial cells of the skin contain no nucleus and have filled themselves with specialized proteins. These tissues and many others in the body rely on replenishment through stem cells, unspecialized precursors or progenitors that can produce more of the specialized, differentiated cells.

8.3.2   Tissue stem cells (traditional view)

The historical view of a stem cell, particularly for a tissue, is that it resides within a tissue as a partially differentiated cell. For example, the **hematopoietic** (blood-forming) **stem cell** in the bone marrow does not express the specialized genes for red or white blood cells, but is specialized to produce all the various types of blood cells. In this respect, the tissue stem cell is limited to producing more cells for the tissue in which it is located. Historically, only a few tissue stem cells were recognized, including those in the bone marrow for blood cell formation, as well as those in the skin, intestinal lining, and the spermatozoan-producing stem cells of the testis.

8.3.3   Regenerative medicine with stem cells

A newer view of stem cells and their utility to replenish tissues has been emerging since the late 1990s. This view is that some stem cells have a greater potential to form many differentiated cell types, making them potentially useful to repair tissue damaged by disease or injury, regenerating the cells of the tissue or organ and thereby treating a patient suffering from a medical condition. This idea of "regenerative medicine" holds great hope for millions of patients with degenerative diseases and injuries. Repair of damaged organs and tissues using stem cells could potentially address the needs of these patients, encompassing most of the leading causes of death in the industrialized nations. However, the emotional appeal of stem cells and the political debate in which the science is embroiled have clouded much of the actual results in this area.

8.3.4   Sources of stem cells

Any stem cell has two chief characteristics: (1) a continued ability to proliferate so that a pool of cells is always available for further use and (2) the ability to respond to appropriate signals by differentiating into one or more differentiated cell types (Fig. 8.12a). Various sources of human stem cells exist (Fig. 8.12b), though much of the focus has been on two sources: (1) **embryonic stem cells**, taken from the inner cell mass of young (5 to 7 day post-conception) embryos, and (2) **adult stem cells**, taken to mean any postnatal tissue source and including umbilical cord blood and the solid umbilical cord matrix, placental tissue, and most or all body tissues. The "plasticity" of a stem cell, that is, its ability to form differentiated cell types, ranges from **unipotent** (limited to forming only one differentiated cell type), to **multipotent** (with the ability to form multiple cell types), to **pluripotent** (having the ability to form most or all postnatal tissues), to **totipotent** (with the ability to form all postnatal tissues as well as extraembryonic tissues such as

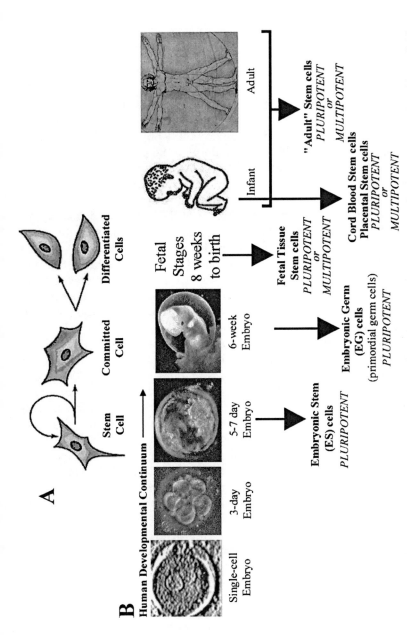

Figure 8.12 Stem cell characteristics and sources. (A) Stem cells continue to proliferate, maintaining a pool of readily-available cells, and respond to signals for differentiation into various specialized cell types. (B) Sources of stem cells include the early embryo, fetal germ cells, other fetal tissues, umbilical cord and cord blood, placenta, and postnatal tissues.

placenta, essentially able to regenerate a complete new embryo and full organism).

8.3.5 Current and potential stem cell uses and points of controversy

Embryonic stem (ES) cells from mice were first successfully grown in the laboratory in 1981, while human ES cells were first successfully cultured in 1998. Isolation of ES cells requires the disaggregation and destruction of the young embryo; hence in terms of the isolation of human ES cells, this has led to the ethical debate over destruction of young human life. ES cells are considered to be pluripotent stem cells with the ability to proliferate extensively in culture and based on their normal function during development or results from reinsertion into another embryo, with the potential to form any tissue. While this potential is attractive for potential treatment of degenerative diseases, initial results in animal models have been modest and have highlighted several scientific hurdles to overcome before ES cells could be applied in clinical situations. These hurdles include the directed differentiation into specific, functional differentiated cell types that can successfully integrate into organs and tissues to function along with the host tissue, potential immune rejection of transplanted ES cells, and the potential for tumor formation by ES cells.

The best examples of potential success in animal models have been in relation to neurodegenerative diseases and injuries, especially in animal models of spinal cord injury and Parkinson disease. ES cells have shown some success at ameliorating spinal cord injury in animals, including improvement in locomotor activity and remyelination of damaged spinal cord. Using animal models of Parkinson disease (degenerative disorder of the central nervous system), ES cells have been successfully transplanted and achieved dopamine (one of the classical neurotransmitters; see Sec. 6.3.1) secretion, alleviating some of the behavioral symptoms in monkeys and rats; however, in many of these experiments the implanted ES cells have yet to show long-term survival, and tumor development has ranged from 20% to 100% in ES cell-treated animals. Tumor formation presents a significant safety concern for the potential clinical use of ES cells—the uncontrolled growth of native or ES-derived progenitor cells is one factor that would preclude their approval for treatment of patients. Some animal studies have also indicated the potential ability of ES cells for cardiac repair or formation of insulin-secreting cells and diabetes treatment. However, at this point there are conflicting reports on the ability of ES cells to form true insulin-secreting cells. It will be important in clinical use of ES cells to demonstrate that functional cells can be formed and that the cells show physiological integration into the target tissues (see Application Box 8.1).

APPLICATION BOX 8.1
The Challenges of Repairing Damaged Tissues with Stem Cells

The concept of regenerative medicine with stem cells is simple: replace dead or damaged cells with newly generated cells for that tissue. If you've had a heart attack, add new heart cells; if the dopamine-secreting cells are missing, leading to Parkinson disease, add back that type of brain cell. But the application is not as easy as the theory. While both embryonic stem cells and adult stem cells have shown the flexibility to change into other cell types, tissue engineering and repair is not so straightforward. For one thing, producing the desired cell type in the laboratory is still inefficient and mostly guesswork in terms of the proper growth factors, combinations, and timing needed. A fully differentiated cell may not even be the desired endpoint for treatment—it may be more effective to start a stem cell on the path to specialization, but use it while it still has some flexibility and let the body's normal signals direct its final differentiation. Once in the tissue, the stem cell must not only form the desired cell type, but must integrate into the tissue structure and function normally with its neighbors. It must also maintain its specialized function over time and not revert to a stem cell, or worse, a cancer cell. There is still a long road to the routine use of stem cells in treatment of disease and injury.

Several possibilities have been proposed for overcoming potential immune rejection of ES cells, including:

- genetic engineering of major histocompatibility antigen genes so that ES cells are recognized as "self" in the host (for graft versus host disease, see Clinical Box 2.1),

- cotransplantation with ES-derived hematopoietic and immune cells so that tolerance of ES-derived tissue cells is developed,

- establishing "banks" of ES cell lines to match all of the potential genetic diversity needed by potential recipients, and

- somatic cell nuclear transfer (SCNT, so-called "therapeutic cloning").

Theoretically, **somatic cell nuclear transfer (SCNT)** would create an embryonic clone of the patient, from which genetically matched embryonic stem cells would be harvested to generate patient-specific cell lines that should not be rejected. While some published evidence exists for the other proposed methods to overcome transplant rejection of ES cells, there has yet to be convincing evidence even in animal models for SCNT cloning, and the research area has been clouded by fraudulent reports of the cloning of human embryos.

Adult stem cells, as mentioned previously, have traditionally been thought to be few in number, difficult to isolate and grow in culture, and extremely limited in their capacity to generate new cell types, being limited to forming more cells from their tissue of origin. However, numerous publications have been overturning this dogma and showing that such postnatal tissue stem cells have a much greater flexibility than previously thought. Several examples now exist of certain adult stem cells with pluripotent flexibility, including cells isolated from bone marrow, peripheral blood, umbilical cord blood, nasal mucosa, amniotic fluid, and placental membrane. Relevant to their potential use in clinical therapies, adult stem cells have also shown effectiveness in treating animal models of disease as well as improving patient symptoms in early clinical trials. While the first successful bone marrow transplant occurred in 1968, and the first cord blood stem cell transplant in 1988, there is a growing realization that adult stem cells have a greater potential for tissue repair beyond reconstitution of blood cell production. In stroke (brain ischemic attack) models, adult stem cells have provided therapeutic benefit. Interestingly, in some experiments the cells showed an ability to "home in" to the site of tissue damage. While this phenomenon is still not completely understood, it provides an intriguing possibility for targeting of regenerative stem cells. For spinal cord injury, adult stem cells have promoted neuronal growth and therapeutic benefit in rodent models, and the published report of initial clinical trials in Portugal has demonstrated improvement in the first seven patients treated. In animal models of Parkinson disease, adult stem cells have shown effectiveness at stimulating dopamine secretion and decreasing behavioral symptoms, and one patient has received a transplant of his own neural stem cells, resulting in decreasing Parkinson symptoms for several years. Other studies designed to stimulate endogenous adult brain stem cells resulted in an average 61% decrease in symptoms for a small number of Parkinson patients.

Adult stem cells have also shown effectiveness at ameliorating retinal degeneration in animal models, raising hopes for possible treatments for diabetic retinopathy (damage to the blood vessels in the retina) and age-related macular degeneration (hardening of the arteries that nourish the retina). Regarding diabetes, several examples now exist showing generation of insulin-secreting cells from various adult stem cells. One research group has achieved permanent disease reversal in mice, and now has approval of the U.S. Food and Drug Administration (FDA) to begin human trials for juvenile (Type I) diabetes (diabetes with minimal or no insulin production by the body).

Use of adult stem cells from bone marrow or mobilized into peripheral blood has become relatively common as an adjunct for cancer chemotherapy to replace the patient's hematopoietic system, or for anemias. Similar techniques to replace or "reboot" the immune system are now being tested with some success in patients with various autoimmune conditions such as scleromyxedema (skin disease), multiple sclerosis (inflammatory disease of the central nervous system), and Crohn disease (inflammatory bowel disease). Similar treatments have also shown promising results for metabolic disorders such as Krabbe disease (abnormal buildup of substances in nerve cells). Adult stem cells have also been used in bone and cartilage repair protocols, and in stimulating repair of cardiac damage in patients in early clinical trials (Prentice, 2006; Prentice and Tarne, 2007).

The mechanism for these regenerative results is still unclear. Adult stem cells in some cases appear capable of interconversion between different tissue types, known as **transdifferentiation**, while in some tissues adult stem cells appear to fuse with the host tissue and take on that tissue's characteristics, facilitating regeneration. And in some studies, the adult stem cells do not directly contribute to the regenerating tissue, but instead appear to stimulate the endogenous cells of the host tissue to begin repair.

8.4   Cell Senescence: Cell Aging

8.4.1   Cellular aging theories and telomerase

Aging is common for us at the organismal level, but it is also a common experience at the cellular level. Normal cells in culture show growth for a period of time, then experience a process of senescence in which cell division ceases, even though the cell remains metabolically active. This cessation in cell growth and division accompanies a loss of response to external stimuli by the senescent cell. There are several theories about cellular senescence and how it might relate to aging of the organism, as well as to cancer. One set of theories is that as our cells age, they accumulate mutations and genetic damage that eventually inactivate vital mechanisms associated with cell division. The mutations may accumulate in vital metabolic genes, the protein synthesis machinery, or in the energy-generating mitochondria, and finally lead to a "catastrophe" in the controls for metabolic machinery and cell division, shutting down cell replication. Another aspect of senescence involves **telomeres**, the sequences at the ends of chromosomes. Because our chromosomes are linear, there is an increasing shortening at the ends as replication occurs, due to the need for a primer for DNA replication and the removal of that RNA primer. Telomeres are repetitive DNA sequences at the ends of chromosomes, and these end sequences use a special DNA polymerase called **telomerase** for replication. The telomerase enzyme brings its own primer, allowing it to extend the ends of the chromosome and overcome the shortening at the ends that would normally be seen. However, the activity of telomerase goes down with age, and less and less of the telomere is replicated. Possibly, senescence may involve shortening of the chromosome ends to the point that vital genes at the ends are not copied. In addition, shortened telomeres may also be perceived as damaged DNA, activating the DNA damage checkpoint mechanism and stopping replication and cell growth.

8.4.2 Cell cycle breakdown

Another aspect of cellular senescence involves the cell cycle regulatory proteins. A senescent cell is unresponsive to external signaling that should stimulate cell growth. This could result from absence or loss of function of a signal transduction protein that would normally stimulate production of the cyclin-cdk complexes. It is also very likely that the cell cycle proteins themselves are mutated or missing, making it impossible for the cell to initiate passage through the cell cycle. This would still allow the cell to continue metabolic activity, but would preclude a cell from completing the entire cell cycle because of a missing actor in the ordered series of proteins needed to progress through the cycle.

8.5 Cancer: Abnormal Growth

8.5.1 Characteristics of cancer

Cancer (**oncogenesis**) and cell aging (**senescence**) are in many respects opposite sides of the same coin. Whereas senescence results in a lack of cell growth irrespective of external stimuli, cancer is the uncontrolled and continuous growth of the cell, again without response to the normal external and internal control mechanisms limiting cell division. Cancers can be characterized in various ways, and can encompass virtually any tissue. In general, a cancer cell can be characterized by uncontrolled growth, altered morphology, and may also include the ability of the aberrant cell to migrate to a site distant from its original tissue and invade other tissues. A **benign tumor** or cancer will proliferate out of control but not show invasiveness and migration, while a **malignant tumor** shows invasiveness and spreads to other tissues even distant from its original tissue, a characteristic termed **metastasis**.

Cancer cells tend to go through various stages of development, usually starting with some alteration in the genetics of the cell that increases its proliferative capacity. **Hyperplasia** (increase in cell numbers) results with excessive growth of the cells. As the cells continue to change, **dysplasia** (abnormal growth) is next, with overproliferation accompanied by alterations in cell morphology. If the cancer cells begin invasion and migration, it is termed **metaplasia**.

8.5.2 Mechanisms of oncogenesis

Mutation figures prominently in cancer. Many cancers have been linked to **oncogenes**, genes that transform a normal cell into a cancerous cell.

Of the many known oncogenes, the vast majority are derived from normal cellular genes, termed **proto-oncogenes**. These genes are normally associated with cell proliferation in various ways, especially as signal transduction proteins and regulators of cell growth. There are various ways in which an oncogene can be turned on or produced in a cell.

Viral infection may introduce an oncogene into a cell. The introduced oncogene is usually altered in such a way that it is overproduced and/or continues to signal cell growth without normal inactivation mechanisms.

Mutation of a proto-oncogene can result in a mutated signaling protein that does not respond to inactivation signals, so that the signal for cell growth is always expressed.

Gene amplification may produce multiple copies of a proto-oncogene, which are then no longer under normal control in terms of amounts and times of expression in the cell. The amplified signal then cannot be shut off.

Insertional mutagenesis (promoter insertion) can occur when some viruses infect a cell. Insertion of the viral genome into the host DNA brings a strong viral promoter, stimulating synthesis not of viral products but also other nearby genes, potentially proto-oncogenes, which would now be expressed at inappropriate times in the cell cycle and cause overexpression of the products.

Chromosomal rearrangement can occur when chromosomal breakage results in loss of small segments that are then placed onto other chromosomes. The misplaced genes can be brought into proximity of strong transcriptional activators, resulting in continuous overexpression of a proto-oncogene.

Loss of tumor suppressor genes and checkpoint proteins can also stimulate cancers. Tumor suppressors (such as the Rb protein; see Fig. 8.4) normally act as governors on cell growth. However, if a protein such as Rb is mutated or deleted, it cannot bind and suppress cell proliferation activators such as E2F, leaving the signal for cell growth to proceed uninhibited.

Mutation and overexpression of cell cycle genes, especially early G1 phase cyclins, can also lead to cancer. For example, mutation of cyclin D, such that it cannot be degraded as it would normally be, would allow the cell to continue to move through the cell cycle, and such mutations have been implicated in some cancers.

8.5.3   Stem cells and cancer

Many cancers seem resistant to chemotherapy and radiation. These treatments target fast-growing cells, damaging their DNA and killing them. But in many cases the cancer returns, even though the treatment seemed to have killed all of the fast-growing cancer cells. There is now growing evidence that the surviving cells that result in the cancer's return, and in fact in the original cancer itself, may be slower-growing aberrant stem cells. Tissue stem cells are programmed to continue growing, usually for the life of the individual, but only grow more rapidly in response to a need for tissue maintenance and repair. However, because of their growth potential, it is very possible that these stem cells are prime targets for initiation and maintenance of a cancer. Once mutated so that they lose their normal growth control and responsiveness to external inhibitors, the cancer stem cell could continue to grow and produce progeny that we identify as a tumor. This means that for effective cancer treatments, we need to consider not only the rapidly growing cancer cells, but also the slower-growing cancer stem cell that originates and maintains a cancer. Such strategies that target aberrant (not normal) stem cells are now being studied.

Suggested Reading

Clarke, M.F. and Fuller, M. (2006). Stem cells and cancer: Two faces of eve. *Cell.* 124: 1111–1115.

Kastan, M.B. and Bartek, J. (2004). Cell-cycle checkpoints and cancer. *Nature.* 432: 316–323.

Monitoring stem cell research: A report of the president's council on bioethics. Washington, D.C.: Government Printing Office; available at: *http://bioethics.gov/reports/stem-cell/index.html.* (in-depth discussion of science and ethics of stem cell research)

Murray, A.W. (2004). Recycling the cell cycle: Cyclins revisited. *Cell.* 116: 221–234.

Prentice, D.A. (2002). *Stem cells and cloning*, 1st ed., (Michael A. Palladino, series ed.), Pearson Education/Benjamin-Cummings, San Francisco. (background primer).

References

Prentice, D.A. (2006). The current science of regenerative medicine with stem cells. *J Investig Med.* 54: 33–37; available at: *journals.bcdecker.com/pubs/JIM/volume%2054,%202006/issue%2001,%20January/JIM_2006_05043/JIM_2006_05043.pdf*

Prentice, D.A. and Tarne, G. (2007). Treating diseases with adult stem cells. *Science.* 315: 328.

Rajagopalan, H. and Lengauer, C. (2004). Aneuploidy and cancer. *Nature.* 432: 338–341.

Chapter

9

Cellular Development

Michael W. King, Ph.D.

OBJECTIVES

- To present fertilization of the ovum by sperm
- To portray the processes by which the embryo is formed and develops
- To explain the role of programmed cell death in development

OUTLINE

9.1 Primordial Germ Cells   234
9.2 Fertilization   236
9.3 Gastrulation and the Establishment of the Germ Layers   237
9.4 Specification and Axis Formation   239
9.5 Limb Development: A Model of Pattern Complexity   246
9.6 Apoptosis in Development   250

Normally, all cells of a mammalian organism are descended from a single fertilized egg. They all have recognizable similarities and carry the complete set of genes of the organisms, as described in previous chapters. But during cellular development, on the way to form a new organism, cells change. How they change is described in this chapter. This knowledge provides the basis for the study of tissue engineering, regeneration, cloning, and stem cell technology as presented in the clinical boxes and the application box of this chapter.

Hidden in the beauty, grace, and symmetry of the animal form lies a complex and seemingly improbable series of genetic connections and interactions. The immense complexity begins with the simple union of two cells, the egg and the sperm. This union brings together the near identical, yet surprisingly different, genomes of the male and female of the species. Together these genomes mix and exchange their life code to direct the basic development of the organism. When one grasps the hierarchical nature of the mechanisms that must work in perfect symphony to generate an essentially flawless organism, it can appear that the end result is the rarity (perhaps the error), not the normal outcome of genetic union. The aim of this chapter will be to provide an overview of the complex processes that result in the formation of a new organism.

9.1   Primordial Germ Cells

9.1.1   Eggs

The egg, produced within the ovary, contains all of the material necessary for the early growth and development of the embryo (see Fig. 9.1). During **oogenesis**, the egg is called an **oocyte**. Maturation to an egg is stimulated by the hormone progesterone which renders the egg capable of being fertilized. Fully developed mammalian oocytes are blocked at prophase of the first meiotic division (see Sec. 10.1.2), and progesterone stimulates progression through the second meiotic division.

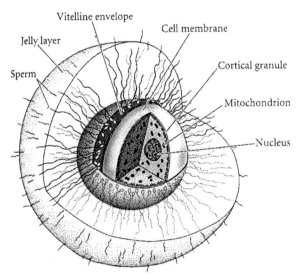

Figure 9.1 Cutaway diagram of a vertebrate egg showing relevant structures. For details, see Sec. 9.1.1. (Reproduced with permission from Gilbert, S.F., *Developmental Biology*, 7th ed., Sinauer Associates Inc., 2003.)

Because of the time required for the embryo to develop to a point where it can feed itself or obtain food from its mother, the egg must be a storehouse of nutrients. The storage of food in the egg is mainly protein termed yolk. The early embryo will also need to make its own proteins and this is accomplished by using maternally stored messenger RNAs (mRNAs) as well as ribosomes and transfer RNAs (tRNAs) (for explanation of RNAs, see Sec. 7.4.1).

As the early embryo divides, there are events taking place that drive particular inductive pathways (programs of gene expression) leading to the establishment of the **germ layers** (see Sec. 9.4) and ultimately to the full differentiation of the various cell types that make up the body plan of the organism. These early inductive events occur due to the presence, in the egg, of localized factors termed **morphogens**. In many species, embryonic development occurs outside the mother and thus the embryo needs protection from the environment and predators. The eggs of many species, such as amphibians, have pigmented membranes that protect against ultraviolet radiation. Some eggs contain distasteful molecules to repel predators.

There are many species whose eggs are coated with a gelatinous glycoprotein mesh termed the *jelly layer* or egg jelly (see Fig. 9.1). The glycoprotein meshwork serves to either attract or activate the sperm so that fertilization can occur.

9.1.2 Sperm

Sperm are the reproductive cells of the male (see Fig. 9.2). Unlike the egg, sperm are streamlined cells that contain only the protein components needed for movement and ultimate entry into the egg. Each sperm cell contains a minimal cytoplasmic constituent, a nucleus bearing a haploid complement of the male DNA, and a propulsion mechanism allowing them to move so that they can join with the egg.

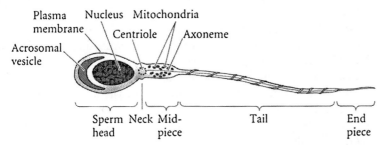

Figure 9.2 Cross-section of a mature mammalian sperm showing relevant structures. For details, see Sec. 9.1.2. (Reproduced with permission from Gilbert, S.F., *Developmental Biology*, 7th ed., Sinauer Associates Inc., 2003.)

In mature sperm nuclei, the DNA is highly compact. Lying at the head of the sperm cell is the *acrosomal vesicle,* which is a modified secretory vesicle (see Fig. 9.2). The **acrosome** is loaded with enzymes that digest complex carbohydrates and proteins on the outer surface of the egg. Digestion of these components of the egg thus allows for entry of the sperm.

The motility of most sperm is imparted by the "tail" which is a flagellum. Flagellae contain microtubules which are composed primarily of the protein tubulin but also contain other proteins necessary for flagellar function such as dynein. For morphology of microtubules, see Sec. 2.3.2 and Fig. 2.10. The energy necessary for flagellar movement is the adenosine triphosphate (ATP) produced in the mitochondria that reside in the central portion of the sperm. Mammalian sperm have the ability to swim when released by ejaculation but require a further maturation process, termed **capacitation** that occurs within the female reproductive tract.

9.2 Fertilization

Fertilization refers to the union of the sperm and the egg and is an exquisitely complex process. First, chemoattractants secreted by the egg induce the sperm to swim toward it. The enzymic contents of the sperm acrosomal vesicle are released by exocytosis either just prior to or after the sperm binds to the extracellular surface of the egg, which is called the **zona pellucida** or the vitelline envelope in mammals (see Fig. 9.1). The acrosomal enzymes begin to degrade the surface of the egg so that finally there can be fusion of the sperm and egg membranes. When sperm enters the egg, its **haploid** nucleus combines with the haploid nucleus of the egg to form a **diploid** nucleus. The early fertilized egg is now referred to as the **zygote.**

Because there would be disastrous consequences if multiple sperm were able to penetrate the egg, a process called **polyspermy**, rapid changes must take place upon the entry of the first sperm. An event termed "the fast block to polyspermy" is the result of a change in the electrical potential of the egg membrane and occurs within seconds of sperm entry. These electrochemical potential changes are caused by an influx of sodium ions. The altered membrane potential prevents other sperm from fusing with the egg membrane. Many animals also have a second mechanism to prevent multiple sperm entry into the egg called the "slow block to polyspermy" which occurs within a minute of the first sperm fusion. This second block is necessary because the fast block is a transient event. The slow block results from the fusion of cortical granules with the egg membrane. These cortical granules reside just under the membrane and release several enzymes whose function is to dissolve the proteins that connect the zona pellucida to the egg membrane. In addition, the granules release mucopolysaccharides that shift the osmotic gradient resulting in water accumulation in the space between the zona pellucida and the egg membrane.

As early cell division proceeds following fertilization, the overall size of the zygote does not change; the size of the individual cells continues to get smaller. This is termed the **blastula** stage. The next stage of embryonic development is called **gastrulation**. The cells of the zygote will begin to undergo a massive reorganization and the embryo will increase in size at the onset of gastrulation (see Sec. 9.3). Following gastrulation, the structures of the nervous system begin to develop, during the process called **neurulation**. The final steps of embryonic development include **patterning** and **organogenesis** leading to a fully formed embryo.

9.3   Gastrulation and the Establishment of the Germ Layers

Gastrulation, a process whereby complex cell migrations begin in the embryo, represents one of the most critical stages of early embryonic development. Although there are minor differences in the process of gastrulation when comparing amphibians, fish, birds, and mammals, the outcomes are the same: establishment of the three germ layers (see Sec. 9.4 and Fig. 9.3).

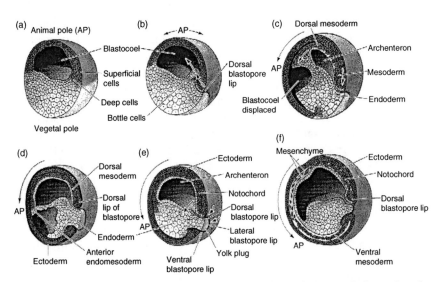

Figure 9.3 Meridional sections showing cell movements during frog gastrulation and establishment of the germ layers. For details, see Sec. 9.3. (Reproduced with permission from Gilbert, S.F., *Developmental Biology*, 7th ed., Sinauer Associates Inc., 2003.)

Gastrulation involves a complex series of cell movements with some of the most critical migrations involving cells on the outside of the embryo migrating to the inside. During the cell movements of gastrulation, the three embryonic cells layers, **ectoderm, mesoderm,** and **endoderm** establish. During the cell movements of gastrulation, outer cells destined to form the endodermally derived organs are moved inside, the embryo becomes surrounded by cells destined to form the ectoderm, and mesodermally destined cells are positioned within the embryo in their correct locations. Gastrulation has been most extensively studied in amphibians because these embryos are large and develop autonomously outside the female. The sea urchin embryo was one of the first organisms in which the cell migrations of gastrulation were captured by time-lapse cinematography showing cells moving from outside to inside in less than 60 min.

In frog embryos, gastrulation is initiated on the future **dorsal** side near the equator referred to as the **marginal zone**. The marginal zone cells begin their involution forming what is referred to as the lip of the blastopore (see Fig. 9.3). The **blastopore** is literally a hole in the side of the embryo filled with yolky cells originating from the vegetal half of the embryo. The orientation of the early amphibian embryo can be distinguished into animal (top) and vegetal (bottom) halves (see Fig. 9.3) by pigmentation where the animal hemisphere is brown to black and the vegetal hemisphere is cream colored. The **dorsal blastopore lip** is a major structural feature of amphibian gastrulation and in amphibian embryos is the site of the **Spemann organizer**. This region is so called because of pioneering work done by Hans Spemann and his collaborator Hilde Mangold in the early 1920s demonstrating that transplantation of the dorsal lip of the blastopore of a donor embryo to the future **ventral** side of a recipient embryo would result in a twinning of the embryo such that all of the anterior structures of the embryo are duplicated. Spemann was awarded the Nobel Prize in 1935 for this discovery.

During the early cleavage stages prior to gastrulation, a cell-free space termed the **blastocoel** forms in the animal hemisphere. The involuting cells from the marginal zone will migrate into and replace the roof of the blastocoel. The involuting marginal zone cells, which were originally fated as ectoderm, come into contact with factors secreted from the **vegetal pole** cells near the blastopore and their fate is changed to that of mesodermal derivatives. This is the earliest and most critical inductive events that occur during early embryonic development.

As the marginal zone cells involute, the cells of the animal hemisphere begin **epiboly**, a process whereby those **animal pole** cells eventually cover the embryo and converge at the blastopore. The yolk plug filling the blastopore is eventually internalized as the process of epiboly completely covers the embryo with cells originating in the animal hemisphere. At this point the embryo is covered with ectodermal cells, the endodermal cells have migrated to the interior of the embryo, and the **mesoderm** is situated between the ectoderm and the **endoderm** (see Fig. 9.3).

In birds, reptiles, and mammals the major structural feature of gastrulation is the **primitive streak**. The primitive streak arises from cells in the posterior marginal region of the early embryo. The cells of the streak undergo convergent extension leading to progression of the streak anteriorly. During the convergence of the cells making up the primitive streak, a groove, or depression, forms called the primitive groove. This formation is analogous to the blastopore of amphibian embryos. The future axes of the embryo are defined by the primitive streak. The streak moves from posterior to anterior and cells enter the streak from the dorsal side and migrate to the ventral side. At the anterior end of the primitive streak is a group of cells termed **Hensen node**. These cells are functionally equivalent to the Spemann organizer cells of the dorsal lip of the blastopore in amphibian embryos.

Cells that migrate through the anterior end of the streak will eventually form head mesoderm and **notochord**; cells migrating through the posterior end give rise to most of the endodermal and mesodermal tissues. The notochord is a flexible rod-like tissue down the back of the vertebrate embryo around which the backbone will be deposited. As the presumptive mesodermal and endodermal cells of the chick embryo are migrating inward, the ectodermal cells are migrating to surround the yolk through the process of epiboly. The process of epiboly in chick embryos is thus similar to that in amphibian embryos as both processes eventually surround the yolk with ectoderm.

It is from the initial differentiated germ layers that the vast diversity of the organism derives (see Fig. 9.4). From the ectoderm will arise tissues of the outer surfaces, the central nervous system, and the neural crest. The notochord, bone, blood, and muscles are just a few of the many derivatives of the mesoderm. The endoderm will lead to the digestive tract and associated specialized cells, the tubes of the respiratory system, and the esophagus.

9.4   Specification and Axis Formation

Initially, upon fertilization the polarity of the single cell embryo is radially symmetric and distinguishable primarily by the animal and vegetal hemisphere axes (see Sec. 9.3 for definition of these axes). As development proceeds, the vertebrate embryo will establish three primary axes: anterior-posterior (AP), dorsal-ventral (DV), and left-right (LR). Although outwardly the body plan of vertebrates appears bilaterally symmetric (i.e., the left leg is a mirror image of the right), it is in fact asymmetrical. Many organs exhibit specific left-right axes (e.g., the left and right chambers of the heart) and then there is the overall spatial orientation of organs; for example, the spleen is usually on the right side of the body and

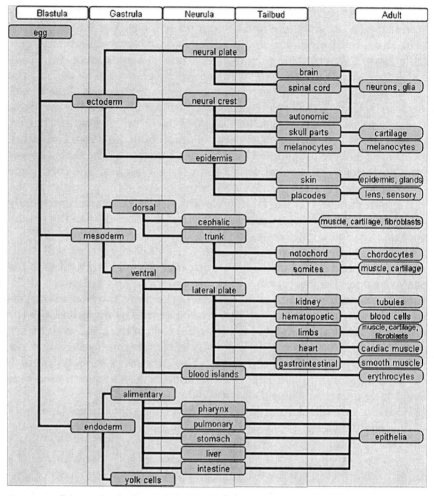

Figure 9.4   Schema for the formation of the basic vertebrate body plan from the three primordial germ layers. Extraembryonic tissues such as the yolk sac and the syncytiotrophoblast are excluded.

the liver on the left. The axes of the vertebrate embryo arise through a series of interactions that occur between neighboring cells and whose interactions are affected by both external and internal factors.

9.4.1   Dorsal-ventral (DV) axis

In the developing chick embryo, the establishment of the DV axis occurs when cleaving cells of the blastoderm create a barrier between the basic (pH 9.5) albumin (a protein of egg white) compartment in the egg and the acidic (pH 6.5) subgerminal cavity region of the embryo. A membrane potential difference results when sodium and water are transported from the albumin domain to the subgerminal cavity. The side of the embryo facing the basic and negative albumin domain becomes the dorsal side with the ventral side originating from the side facing the acidic fluid of the subgerminal cavity.

In the amphibian embryo, the future dorsal side of the embryo can be determined from the site of sperm entry. The side opposite sperm entry will be the dorsal side of the embryo. This DV axis is established as a consequence of cortical rotation following sperm entry. The rotation of the cortex of the embryo places signaling proteins (of the Wnt pathway) in the proximity of the cells of the future dorsal lip (see Fig. 9.3) of the blastopore, which in turn sets up a cascade of transcriptional events leading to dorsalization of the embryo.

The major protein regulating dorsalization of the amphibian embryo is *β-catenin*, which upon stabilization migrates to the nucleus and activates transcription of a set of genes (see Fig. 9.5). The cortical rotation that occurs following sperm entry moves granules containing the protein disheveled (Dsh) into the area of the future dorsal blastopore lip. Dsh inhibits a kinase called glycogen synthase kinase-3 (Gsk-3) preventing phosphorylation of $β$-catenin. The phosphorylated $β$-catenin on the future ventral side of the embryo is *degraded* while the nonphosphorylated form is *stabilized*. Therefore, only $β$-catenin target genes (genes activated as a consequence of the activity of $β$-catenin), such as *siamois*, are activated on the future dorsal side (see Fig. 9.5). Siamois in turn activates the transcription of the homeodomain-containing gene *goosecoid* which controls the expression of secreted growth factors and growth factor antagonists that mediate the inductive activities of the *Spemann organizer* on neighboring cells. (A homeodomain is a protein derived from a homeobox, a DNA sequence that is virtually identical in most organisms.)

9.4.2   Anterior-posterior (AP) axis

Formation of the anterior-posterior (AP) axis in vertebrates follows formation of the DV axis. During amphibian embryogenesis, it is the movement of the involuting mesoderm that establishes the AP axis. The mesoderm that migrates through the dorsal lip of the blastopore first gives rise to anterior structures while that migrating through the lateral and ventral lips of the blastopore give rise to posterior structures. As in establishment of the DV axis, the transcription factor goosecoid induces the expression of soluble factors that organize the AP axis.

As the involuting mesoderm passes through the dorsal lip of the blastopore and is positioned under the overlying ectodermal tissue, the ectoderm is induced to become neuroectoderm. The secreted factors that are responsible for this induction are members of the bone morphogenetic protein (BMP) family, primarily BMP4. In the absence of

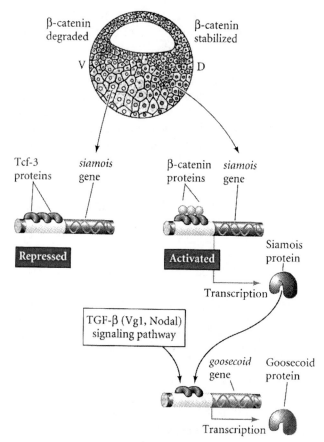

Figure 9.5   Events by which β-catenin action brings about the induction of the Spemann organizer of the amphibian embryo. For details, see Sec. 9.4.1 on dorsal-ventral axis formation. (Reproduced with permission from Gilbert, S.F., *Developmental Biology*, 7th ed., Sinauer Associates Inc., 2003.)

underlying mesoderm, the default state of the animal pole ectoderm is neural. The induction of epidermal fate is a consequence of the organizer cells inhibiting the action of BMPs. A major secreted protein involved in this mesoderm induction is noggin. Noggin binds to both BMP2 and BMP4 and prevents them from binding to their receptors. Other secreted BMP antagonists that have been identified to have inductive capabilities are chordin and members of the nodal family termed nodal-related proteins (e.g., Xnr3 from *Xenopus laevis*, the African clawed frog).

Along the overall AP axis, there exists regional specificity. The forebrain, midbrain, hindbrain, and spinal regions of the **neural tube** (which forms on the dorsal side of the embryo) must be properly organized along the AP axis. In amphibian embryos, the Wnt signaling pathway is primarily responsible for posteriorizing the neural axis. The most likely member of the *Xenopus* Wnt family responsible for this action is Xwnt8. During early development, a gradient of Wnt signaling is greatest in the posterior and is absent in the anterior part of the embryo. The role of Wnt signaling in the AP axis has been demonstrated by either overexpression of Wnt proteins such as Xwnt8 in the anterior region, thereby posteriorizing this area, or inhibition of Wnt signaling in the posterior region leading to anteriorization. Thus, a gradient of Wnt signaling is important in establishing the AP axis similarly to the role of a BMP gradient in establishing the DV axis.

9.4.3 Left-right (LR) axis

All vertebrates establish a left-right (LR) axis, and it has been shown that this axis arises as a consequence of the expression of nodal-related proteins in the lateral plate mesoderm on the left side of the embryo. LR axes are most apparent in the heart (i.e., left and right ventricles and atria) and gut (i.e., direction of coiling). In *Xenopus,* it is the Xnr1 protein that establishes the LR axis. Although it is known that Xnr1 is important in creating the LR axis, the cause of left side-specific expression of Xnr1 and the pathway by which it instructs the formation of the LR axis is incompletely understood. One gene that is activated by Xnr1 that is itself a key gene in establishing the LR axis is Pitx2. Expression of Pitx2 on the right side of an early frog embryo results in randomization of the heart and gut coiling.

The complexity of the genetic program for establishing the LR axis can best be appreciated from studies on early chick embryos (see Fig. 9.6).

When the primitive streak is near its maximum extension, expression of *sonic hedgehog (shh)* is extinguished on the right side of the embryo. Loss of shh expression comes about as a consequence of expression of *activin* and its receptor on the right side of the embryo. Activin signaling blocks expression of shh. While activin signaling is repressing shh expression, it is activating the expression of a member of the *fibroblast growth factor family, FGF8*. FGF8 in turn prevents expression of the *caronte (Car)* gene which normally functions to inhibit members of the *bone morphogenetic protein (BMP)* family. With no caronte protein present, BMP signaling blocks the expression of *nodal* and *lefty-2*. In the absence of nodal and lefty-2, the *snail family* of zinc-finger transcription factor genes is active. (A zinc finger is a DNA-binding protein configuration that resembles a finger; see also Sec. 6.2.1.) Snail function is downstream of nodal and upstream of *Pitx2*. Thus, as in LR axis formation in amphibian embryos, Pitx2 is crucial in establishing the LR axis in chick embryos.

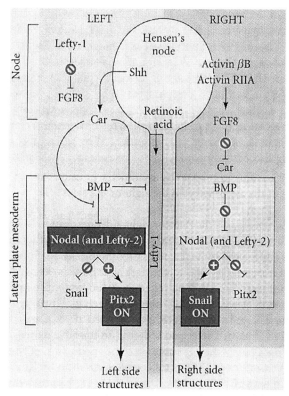

Figure 9.6   Gene expression patterns leading to establishment of the left and right axes of the chick embryo. For details, see Sec. 9.4.3. (Reproduced with permission from Gilbert, S.F., *Developmental Biology*, 7th ed., Sinauer Associates Inc., 2003.)

CLINICAL BOX 9.1
Regenerative Medicine

Tissue regeneration is a process that is essential to life. Tissue regeneration in multicellular organisms is accomplished by mitotically competent cells residing within tissues. In some cases these cells are fully differentiated. For example, hepatocytes of the vertebrate liver respond to partial hepatectomy by mitosis while maintaining all their differentiated functions. This mode of regeneration is called compensatory hyperplasia. However, the tissues of most multicellular organisms are regenerated by undifferentiated stem cells. In the majority of cases, these are reserve stem cells, suspended in an early state of differentiation as cells around them continue to differentiate during embryonic development. These adult stem cells may produce only a single type of tissue (unipotent) or possibly several tissue types (multipotent). Reserve stem cells have long been recognized as the source of renewal and regeneration for vertebrate tissues such as blood, immune cells, epithelia, bone, muscle, and olfactory bulb and nerve. Human embryonic stem cells (ESCs) exhibit pluripotency (potential to form nearly all tissues of the body) and unlimited growth potential. Regenerative biology and medicine encompasses research on the potential use of both ESCs and adult stem cells or their derivatives for transplantation to replace tissues damaged by injury or disease. Work is also progressing in the area of *in vivo* induction of stem cells in order to regenerate tissues.

Different species of vertebrates vary in their ability to regenerate. Amphibians such as frogs and salamanders have the greatest regenerative capacity. Amphibians use compensatory hyperplasia and reserve stem cells to regenerate tissues. What sets amphibians apart from other vertebrates is their ability to regenerate complex structures by dedifferentiation. Dedifferentiation is a unique method of producing stem cells, seen otherwise only in fish, which regenerate fins, and lizards, which regenerate tails.

Because the amphibians exhibit all of the known mechanisms of regeneration, they are an important experimental tool for expanding our understanding of regeneration and how to apply this knowledge to induce the regeneration of nonregenerating mammalian tissues and complex structures.

CLINICAL BOX 9.2
Stems Cells, Embryonic and Adult

As demonstrated in this chapter, one of the first critical events in embryonic development is the establishment of the three germ layers: ectoderm, mesoderm, and endoderm. The segregation of cells of the embryo into these three lineages lasts into adulthood such that mature cells, progenitor cells, and stem cells from these three germ layers maintain their early specification.

Stem cells are a specialized type of cell that remain in an undifferentiated state and have the unique property of self-renewal. They were introduced in Sec. 8.3. Stem cells are responsible for homeostatic cell replacement and regeneration of tissues. When stem cells respond to signals inducing tissue repair or regeneration they undergo differentiation to produce mature progeny, yet one daughter cell from the first cell division will remain undifferentiated (self-renewal). Stem cells are classified on their capacity to differentiate into different cells types. **Totipotent** stem cells can give rise to any embryonic cell type, **pluripotent** give rise to almost all cell types, and **multipotent** give rise to a more restricted set of cell types.

There exist two distinct pools of stem cells: embryonic and adult. **Embryonic stem cells** (ESCs) are derived from the pluripotent inner cell mass of the preimplantation blastocyst-stage embryo. Culture of blastocysts will give rise to an undifferentiated population of cells that can be grown indefinitely as clonal populations; thus, they can be used to create ES cell lines. The controversy with the use of ESCs of human origin is that the embryo is destroyed during the process of isolating the blastocyts. However, stem cells have been identified in numerous compartments of the adult organism.

Adult stem cells are not totipotent but range from multipotent to pluripotent. Human adult stem cells have been most effectively isolated and studied from the blood and the nervous system. Blood-derived stem cells are referred to as **hematopoietic stem cells** (HSCs). Neural stem cells can be found in the central and peripheral nervous systems. HSCs hold the most promise for adult stem cell technologies primarily because their numbers are larger than neural stem cells as well as being easier to isolate. Although HSCs serve the natural function of replenishing all of the cell types of mature blood, outside the context of the bone marrow they exhibit striking plasticity being able to contribute to nonhematopoietic lineages. Work is progressing on many fronts with numerous different sources of adult stem cells to determine how to manipulate their inherent plasticity.

The ability of stem cells to serve as the starting material for a wide array of differentiated derivatives is, of course, the reason why they may be so useful for purposes of cell-based therapies and regenerative biology. As yet, there have been limited but exciting examples where differentiated derivatives of stem cells were transplanted successfully in animal models of diseases or injuries. In mice, an insulin-producing cell line derived from ES cells resulted in normalized blood glucose levels when they were implanted into the spleens of drug-induced diabetic mice. Neuronal cells derived from human stem cells were able to improve motor and cognitive deficits in rats with stroke, and these cells are now being tested for use in human patients with basal ganglia stroke. Whereas the studies are as yet few in number, the results are consistent with the belief that stem cell-based therapies will provide an effective approach for ameliorating the effects of some devastating diseases and injuries.

9.5   Limb Development: A Model of Pattern Complexity

Development of the vertebrate limb provides an ideal model for studying the fundamental complexities of **morphogenesis**. Morphogenesis refers to the process(es) by which specific structures arise in specific locations during early embryonic development. Much of what we know about the development of the vertebrate limb has been obtained from morphological and molecular studies on chick limbs. Although there is a left and right side to the limbs, the principal axes are dorsal-ventral (DV), anterior-posterior (AP), and proximal-distal (PD). Using the human forelimb (the arm) as a model, the tips of the fingers are at the distal end of the limb and the shoulder is at the proximal end; the little finger represents the posterior side of the limb and the thumb the anterior side; the back of the hand (i.e., the knuckles and nails) is the dorsal axis and the palm is the ventral axis (Fig. 9.7).

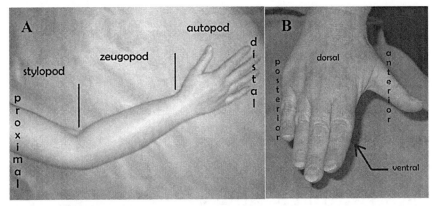

Figure 9.7   Orientation of the three primary axes of the limb: proximal-distal (A), dorsal-ventral, and anterior-posterior (B). As indicated, the back of the hand is the dorsal aspect, the digit tips are the distal aspect, and the anterior aspect is denoted by the thumb.

Initiation of tetrapod limb development requires the determination of a group of cells within the lateral plate mesoderm. These determined cells comprise what is referred to as the **limb field**. Limbs do not develop just anywhere along the body, and the cells that are committed to form limbs can be identified by transplantation experiments. In the simplest experiment, removal of the lateral plate mesoderm from the prospective limb field results in loss of limb outgrowth. Retinoic acid is critical for limb bud outgrowth since blocking its synthesis will prevent initiation of the limb fields. The specification of the limb fields arises as a consequence of correct retinoic acid-induced expression of Hox genes along the AV axis. Members of the fibroblast growth factor (FGF) family of growth factors are also involved in limb bud initiation. If FGF-soaked beads are implanted into the flank of a chick embryo between the forelimb and hindlimb lateral plate mesoderm, an extra limb is induced. Recent evidence has shown that the specific FGF family member FGF10 is expressed in the lateral plate mesoderm in the prospective limb-forming regions (see Fig. 9.8).

Early in limb bud outgrowth, forelimbs and hindlimbs look essentially the same. Many of the genes that have been identified as being crucial for limb development display the same patterns of expression in forelimbs and hindlimbs. How then do the structural differences between the limbs arise? Two genes that are members of the T-box (Tbx) family of genes, Tbx4 and Tbx5 display limb-specific expression. Tbx5 is expressed in the forelimb but not the hindlimb, and Tbx4 displays the reciprocal pattern of expression. In experiments involving FGF induction of ectopic limbs, it was determined that forelimb determination is in fact controlled by Tbx5 expression and hindlimbs by Tbx4 expression. When ectopic limbs are induced by placing FGF beads close to the hindlimb, the ectopic

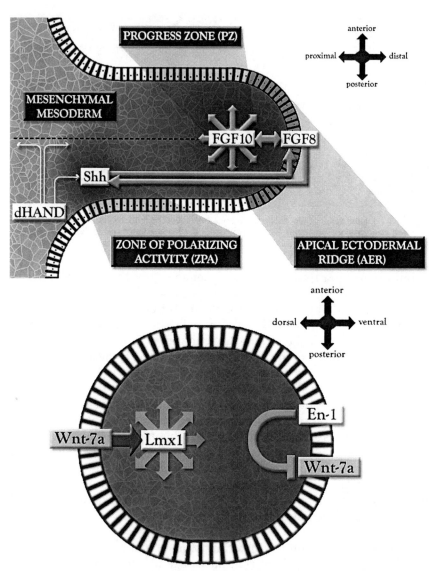

Figure 9.8   Factors involved in the early axis specification in the tetrapod limb. For details, see Sec. 9.5. (Figure produced by Barrett Fricke.)

limbs express Tbx4 and become hindlimbs, with the reciprocal result occurring when induction takes place near the forelimb.

Once initiated, the three axes of the limb must be properly patterned. Limb bud outgrowth is controlled by a specialized group of ectodermal cells running along the distal margin of the limb bud. These cells form a structure called the **apical ectodermal ridge** (AER), (see Fig. 9.8).

The AER is initiated as a consequence of the synthesis of FGF10 in the underlying mesenchymal cells. Removal of the AER early during limb bud outgrowth results in a dramatic reduction in limb development. The role of the AER is to continuously interact with the cells of the underlying **mesenchyme** as limb outgrowth proceeds, ensuring the proper PD orientation of the limb. The distal mesenchyme underlying the AER is referred to as the **progress zone** (PZ), (see Fig. 9.8). As indicated above, FGF10 is expressed in the PZ mesenchyme and induces formation of the AER. The AER in turn is induced to express *FGF8*. Secretion of FGF8 from the AER maintains the PZ mesenchyme in a constant mitotic state which allows them to continue to synthesize FGF10. Thus, there is a reciprocal interaction between the AER and PZ as limb outgrowth proceeds.

As limb outgrowth proceeds, there is a progressive change in the expression of Hox gene family members, the consequences of which are the determination of the **stylopod, zeugopod,** and **autopod** domains of the limb. Using the arm bones as a guide, the *stylopod* encompasses the humerus, the *zeugopod* the radius and ulna, and the *autopod* the bones of the hand (see Fig. 9.7). The critical role of Hox genes becomes apparent when their expression is blocked. Humans homozygous for a defective HOXD13 gene exhibit abnormalities in their hands and feet. (Homozygous refers to two identical copies of the gene on the two corresponding chromosomes.)

The AP axis of the limb is specified very early before the limb bud is identifiable. The initiation of the AP axis is the result of the action of a small group of mesodermal cells that reside near the posterior region of the early limb bud near the body wall. This area of cells is referred to as the **zone of polarizing activity** (ZPA), (see Fig. 9.8). If one transplants the ZPA tissue to the future anterior side of the limb, there will be a mirror image duplication of the posterior elements of the limb. Evidence has shown that the secreted protein *shh* is responsible for the activity of the ZPA. Implantation of shh-soaked beads into the anterior side of the limb results in the same consequences as transplantation of the ZPA. Expression of shh in the ZPA cells results from the action of FGF8 secreted from the AER cells. Restriction of shh expression to the posterior cells of the limb has been shown to be controlled by two transcription factors, HOXB8 and *dHAND* (see Fig. 9.8). When HOXB8 is ectopically expressed throughout the limb, shh expression occurs in both the posterior and anterior regions of the limb leading to formation of two ZPAs. The role of dHAND was demonstrated in mice lacking its expression. In these mice shh is not expressed in the limb and the result is severely truncated limbs.

The specification of the DV axis of the limb is determined by the ectoderm. In experiments on chick limb development it was shown that rotation of the limb ectoderm by 180°, with respect to the underlying mesenchyme, results in a reversal of the DV axis. A member of the Wnt family of secreted factors, *Wnt-7a* is expressed in the dorsal but not the ventral ectoderm (see Fig. 9.8). In mice harboring a Wnt-7a knockout, the limbs harbor ventral footpads on both surfaces. In the limb, Wnt-7a induces expression of the transcription factor *Lmx-1* in the dorsal mesenchyme. If Lmx1 expression is induced in the ventral mesenchyme, a dorsal phenotype will result in those cells. Humans harboring a loss-of-function mutation in Lmx-1 suffer from nail-patella syndrome, characterized by ventralization of the dorsal side of the limbs and manifests in digits with no nails and lack of kneecaps. Control from the ventral side is exerted by the homeobox gene *engrailed-1 (En-1)* which inhibits expression of Wnt-7a, thus ensuring ventral character to the ectoderm. The specificity of this interaction is demonstrated by the fact that loss of En-1 expression results in dorsalization of the ventral axis.

9.6   Apoptosis in Development

During embryogenesis and throughout life, cells continually die. In the adult organism, many of these cell death events are the consequences of damage or cellular injury. However, during development there is a coordinated series of physiological events leading to cell death. This type of programmed cell death is called **apoptosis**. The apoptotic pathway ensures that cells can be removed from the organism without inducing an inflammatory response. Most of the apoptosis occurring during development is mediated by a family of cysteine proteases termed **caspases** that cleave secreted proteins rendering them inactive or modifying their activity.

Activation of caspases occurs via three pathways, two that are internal and one that is external (see Fig. 9.9). The *external pathway* for activating apoptosis involves a family of extracellular proteins that bind to and activate **death receptors**. Death receptors are members of the tumor necrosis factor (TNF) receptor family of proteins. Ligand-mediated activation of death receptors results in activation of *caspase-8* and/or *caspase-10 (Casp-8,10)*. The intracellular pathways respond to cellular stress and lead to activation of caspases present in the mitochondria (*caspase-9*) or in the endoplasmic reticulum (*caspase-12*). Each of these three pathways ultimately impinges upon activation of *caspase-3* and/or *caspase-7 (pCasp-3,7)*, although active caspase-12 can itself lead to proteolytic modification of target proteins. For apoptotic signaling, see also Sec. 4.5 and Fig. 4.11.

The tetrapod limb serves as an ideal model for examining apoptotic events in development. During limb development, there is a requirement

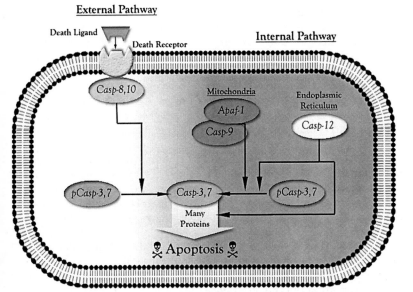

Figure 9.9 Roles of caspases in the major apoptotic pathways via extrinsic death receptors, mitochondria, and endoplasmic reticulum. For details, see Sec. 9.6. (Figure produced by Barrett Fricke.)

APPLICATION BOX 9.1
Bioartificial Tissues

Cell transplantation is the basis for the *in vitro* construction of bioartificial tissues. Bioartificial tissues are composed of a support matrix impregnated with cells that can then be implanted into the body to replace original tissue lost to injury or disease. The cells that can be used in bioartificial tissues can be of any type. They can be stem/progenitor cells, differentiated cells, or even genetically engineered cells. The support matrix must be of a type that will allow maintenance of the specialized function of the differentiated cells or promote the differentiation of stem/progenitor cells. The ideal support matrix is one that is biodegradable so that it is replaced by a natural matrix following implantation and the integration of the cells into the host. For example, in July 2005, Spiegel and Kessler published the results of a study in *Otology & Neurology* that described the repair of tympanic membrane perforation repair with acellular porcine submucosa. The authors concluded, "These results suggest that small intestine submucosa is a viable alternative to autologous and cadaveric grafts in tympanoplasty. A larger randomized study in humans is indicated to evaluate this material in clinical practice."

Collagen-based support matrices are the most common for both bioartificial tissues and the promotion of *in vivo* regeneration. This is because collagens are the most abundant proteins of the extracellular matrix of human tissues. Type I collagen is plentiful and easily extracted and purified and can be molded into a variety of shapes and importantly, collagen is biodegradable. Other materials that are frequently used in the construction of bioartificial tissue matrices are polylactic acid, polyglycolic acid, and polydioxanone. Each of these, like collagen, is biodegradable. An additional matrix material that is not biodegradable but highly useful is alginate.

Bioartifical skins are extremely useful for treating burn victims and excisional wounds. These are composed of a variety of matrices into which are seeded dermal fibroblasts. The cells in the matrix are eventually replaced by host fibroblasts and keratinocytes and the matrix is degraded. Whereas, these bioartifical skins are highly effective, the regenerated skin is not perfect lacking hair and glands.

In May 2005, Gore described the use of an acellular allograft dermis in the care of elderly burn patients. Traditional thickness skin grafts can't always be obtained in elderly burn patients without creating new full-thickness wounds at the skin graft donor site. The author concludes, "This initial experience suggests that use of AlloDerm may allow more elderly burn patients to undergo operative wound closure, thus improving functional outcome and reducing hospitalization."

for the coordinated activation of the apoptotic process such that refined sculpting of the structures of the limb will occur. Animals whose feet (as well as hands in humans) are not webbed lose the cells between the digits of the early limb via apoptosis. This programmed death is evident even if the interdigital tissues are transplanted to another area of the embryo or grown in culture. Apoptosis occurs in other regions of the developing limb and is required for proper architecture. In the zeugopodial region (**zeugopodium** is the lower part of the limb) of the limb, apoptosis leads to separation of the ulna and the radius and the tibia and fibula. Initiation of the apoptotic processes in the limb is the result of the activity of BMPs, most likely BMP4 and BMP7. Blocking BMP activity in the interdigital spaces prevents apoptosis.

Suggested Reading

Capdevila, J. and Izpisua Blemonte, J.C. (2001). Patterning mechanisms controlling vertebrate limb development. *Annu. Rev. Cell Dev. Biol.* 17:87–132.

Stocum, D.L. (1998). Regenerative biology and engineering: Strategies for tissue restoration. *Wound Repair Regen.* 6:276–290.

Vaux, D.L. and Korsmeyer, S.J. (1999). Cell death in development. *Cell.* 96:245–254.

Wagers, A.J. and Weissman, I.L. (2004). Plasticity of adult stem cells. *Cell.* 116:639–648.
Weissman, I.L. (2000). Stem cells: Units of development, units of regeneration, and units of evolution. *Cell.* 100:157–168.

References

Gore, D.C. (2005). Utility of acellular allograft dermis in the care of elderly burn patients. *J. Surg. Res.* 125(1):37–41.
Spiegel, J.H. and Kessler, J.L. (2005). Tympanic membrane perforation repair with acellular porcine submucosa. Middle ear and mastoid disease. *Otol. Neurotol.* 26(4):563–566.

10

From Cells to Organisms

Gabi Nindl Waite, Ph.D.

OBJECTIVES

- To describe the transition from unicellular to multicellular organisms
- To explain what makes cells different from one another
- To outline different cell types and tissues
- To introduce the interdisciplinary approach of systems biology

OUTLINE

10.1 From Unicellularity to Multicellularity   256
10.2 Cell Features   262
10.3 Determination and Differentiation   266
10.4 Morphogenesis   276
10.5 Systems Biology   282

It is fascinating how many common features exist in every cell, independent of whether it is a bacterial cell or a human neuronal cell. It is equally fascinating in how many ways the building blocks can be arranged so that the resulting arrangements lead to an immense variety of structures, functions, and phenotypes. This chapter aims at laying the groundwork for understanding how cells differentiate to build tissues, organs, and eventually one whole functional organism. It is the ultimate goal of engineers and scientists to develop an understandable model of the organism to be able to study the interdependence of the individual functional units.

This book has thus far explained the fundamental biology of a eukary-otic cell and the scientific findings that have resulted from molecular studies investigating eukaryotic cells. But molecular cell biology cannot provide the complete picture of how a eukaryotic cell functions, since most cells live as part of multicellular organisms through cooperation and specialization. For instance, an estimated 3 million cells cooperate in every square inch (~6.5 cm^2) of the human skin to provide cover and protection to the cells inside the body. During their life cycle of about 35 days, skin cells also specialize in that they become flatter and kera-tinized while moving from the basal layer of the epidermis toward the skin surface. Cell specialization can lead to amazing cellular abilities. Heart cells produce enough useful work in a human lifetime equivalent to that required to raise 30 tons from sea level to the top of Mount Everest. Nerve cells transport information along their axons at a speed faster than the average speed of race cars at the Indianapolis 500. Special hair cells in the inner ear use piezoelectric resonance to improve the sensitivity and frequency selectivity of the cochlea. This piezoelec-tric effect in hair cells is about 5 orders of magnitude greater than the piezoelectric effect of ceramic crystals (see Clinical Box 10.1). It is the goal of this final chapter to introduce the hierarchical cooperation of cells within multicellular organisms and their specialization for division of labor. We will first ask how multicellularity came about in evolution and then take a closer look at the processes that lead to cell differentiation and determination in order to introduce biology at the system level.

10.1   From Unicellularity to Multicellularity

During evolution, the cooperative interaction between cells as individ-ual functional entities working toward a common goal, opened the path for development of complex multicellular organisms. The transition from single cell to multicellular organisms most likely happened numer-ous times in independent ways, very early during the evolution of life. How early is still a topic for discussion, since the first single cell organ-isms did not contain hard body parts and consequently rarely survived as fossils. In a first step toward multicellular organisms, single cells might have stuck together to form a bigger mass, or multicellularity might have occurred as a consequence of cells failing to separate fol-lowing division. The initial selective advantage of such clusters of undif-ferentiated cells still remains unclear. Whatever the advantages might have been of staying together, it was the mass of eukaryotic cells that further progressed and gave rise to the variety of today's multicellular organisms. Before looking at this process, let us first look at the differ-ence between prokaryotic and eukaryotic cells.

CLINICAL BOX 10.1
Otoacoustic Emissions

Otoacoustic emissions are the sound that a mammalian inner ear produces and that a practitioner detects with sensitive microphones at the external ear to determine if there is sensorineural hearing loss (damage to auditory cells, nerve or brain structures, in contrast to damage to conduction structures). Yes, you read correctly. The inner ear is not only receiving and transmitting sounds to the brain, but also producing them as well. The sources of these tiny otoacoustic sounds are outer hair cells.

The cochlea contains two types of hair cells: inner and outer hair cells, as can be seen in Fig. 2.9. Inner hair cells send electrical signals toward the brain via the auditory nerve. Inner hair cells transform the mechanical disturbances caused by the sound waves into electrical information by opening membrane channels at the tip of their displaced stereocilia. The opening of the channels causes ion fluxes according to the electrochemical potential of the ions, which consequently changes the inner hair cell membrane potential (see Sec. 10.3.3). Membrane potential changes provide signals that can be transmitted from cell to cell until reaching the brain, where they are perceived as sound. Outer hair cells operate very similarly to inner hair cells in that they open membrane channels when their stereocilia are displaced, which causes a change in membrane potential. However, in contrast to inner hair cells, the change in membrane potential of outer hair cells leads to their vibration along the cells' long axes in synchrony with the frequency of the incoming sound or in a stable phase relation. This cellular movement, or electromotility, is transmitted by the ear's conducting structures to the outside, hence in opposite direction to the externally produced sound waves, and can be detected at the ear auricle as internally produced sound waves, or otoacoustic emissions.

In that way, outer hair cells can be compared to piezoelectrical crystals such as quartz or barium titanate: When subjected to a voltage, the shape of the crystal/hair cells changes; when subjected to mechanical stress, a voltage/membrane potential is induced. By comparison, the hair cell's piezoelectric coefficient ($\sim 20 \times 10^{-6}$ Coulombs/Newton) is extraordinary and about 5 orders of magnitude greater than that of the crystals ($2\text{-}4 \times 10^{-12}$ Coulombs/Newton for quartz and 140×10^{-12} Coulombs/Newton for barium titanate). Biologically, mammalians use the piezoelectricity of outer hair cells to extend their hearing range and to improve frequency discrimination which allowed for instance the development of speech. Clinically, the piezoelectricity of outer hair cells is detected as otoacoustic emissions. The clinical procedure has become the standard of care in the screening of cochlear function, at least for infants and uncooperative patients.

10.1.1 Prokaryotes and eukaryotes

Although a matter of debate, many scientists believe that life on Earth arose only once and all living cells have one last universal common ancestor (LUCA) which appeared about 3.5 billion years ago. The most widely accepted theory is still that the first life forms to arise from LUCA were **prokaryotes**, organisms that today are very numerous and live in almost every habitat. Based on cell structure and metabolism, prokaryotes comprise bacteria and **archaea**, the latter being formerly called archaebacteria. In a phylogenetic tree based on RNA data, archaea are an independent third branch of life, parallel to bacteria and eukaryotes. Prokaryotes are nearly always unicellular, but may form colonies, masses of undifferentiated cells, that stay together even though they are capable of living individually. The main characteristic to distinguish prokaryotes from eukaryotes is that prokaryotes are organisms without a membrane-bound nucleus ("pro"—Greek: before, "karyo"—Greek: nut or nucleus). Their DNA is contained mostly in a *nucleoid region*, but in some bacteria it is also dispersed throughout the cell in the form of small rings called *plasmids* (see Fig. 10.1).

A **eukaryote** is an organism made of one or more complex cells with each containing the typical set of organelles that were presented in

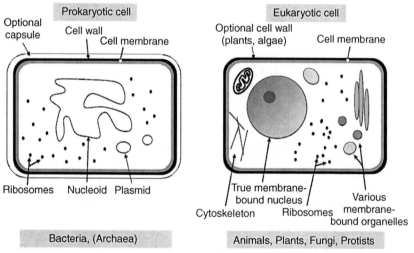

Figure 10.1 Idealized prokaryotic and eukaryotic cells. The prokaryotic DNA is contained in a nucleoid region and dispersed throughout the cell in plasmids, while the eukaryotic DNA is contained in a membrane-bound nucleus. Eukaryotes possess more types of organelles than prokaryotes and also membrane-bound organelles. Prokaryotes have a cell wall composed of peptidoglycans. Many types of eukaryotic cells also have cell walls, but none made of peptidoglycan. Please note that a eukaryotic cell is typically 10 times the size of a prokaryotic cell. Archaea look like a typical prokaryotic cell, but were recently phylogenetically separated from bacteria.

Chap. 2. Eukaryotic organelles are structurally more complex compared to prokaryotic organelles and several eukaryotic organelles are enclosed by lipid membranes that resemble the cells' plasma membranes. The feature that gave the eukaryotic domain (also called eukaryotic superkingdom or empire) its name is the fact that the genetic material is enclosed in a membrane-bound nucleus ("eu"—Greek: true). Eukaryotes comprise animals, plants, and fungi, which are mostly multicellular, and protists, which can be uni- or multicellular. Protists include animal-like protists (for instance protozoa), plant-like protists (for instance algae), and fungus-like protists (for instance molds). Figure 10.1 shows models of a typical eukaryotic cell and a typical prokaryotic cell that demonstrate the key differences between them.

The first eukaryotes appeared about 1 to 2 billion years ago and since then, eukaryotes have developed into multicellular species with increasing complexity. Many creative forces have been discussed that led to multicellularity and cell differentiation. The larger size of multicellular organisms may have been of advantage for predator evasion and for enhanced uptake and storage of essential nutrients. Cell differentiation may have followed the vertical orientation of the organisms, with bottom cells attaching to the ground and resisting waves and top cells competing effectively for light and nutrients. Cell differentiation required the development of the ability to exchange information and resources between the cells. Increasingly more complex interactions between cells furthered the division of labor and led to increasingly more complex interactions with the environment, which promoted the development of new species in a sort of "ecological engineering process." Numerous forms of ecological interactions, including mutualistic, symbiotic, and parasitic interactions, have been presented to explain natural selection and the advancement of species. Another fundamental force in the emergence of new species with increased fitness was most likely cooperative behavior, which includes activities that benefit other members of the same species despite a potential cost to the individual. The evolutionary process of new biological species filling many ecological niches is called **adaptive radiation**. At present, there are close to 100 million different multicellular species living in different habitats on Earth.

10.1.2 Sexual reproduction and meiosis

The adaptive radiation of species would not have happened if reproduction of the early multicellular organism occurred only by mitotic duplication of each of its individual cells. The organism would have remained little more than a colony of eukaryotic cells with a low level of specialization. The reason is that the undifferentiated cells that retained their original totipotent condition would have been better at reproducing compared to the specialized cells, with the consequence that the generalists would have continually outgrown the specialists. **Sexual reproduction** solved this problem by freshly recreating each new organism from one single cell while transmitting the command for cell differentiation from cell to cell. It must have been an advantage for certain eukaryotes to develop double strands of genes (diploid) for redundant information storage. Such a double set of genes also allowed for the repair of a damaged gene by the undamaged copy via genetic recombination. To reproduce, the diploid organisms then found a way to combine genes from two parents rather than simply cloning their cells so that sexual reproduction evolved into meiosis followed by fertilization.

Meiosis is the production of gametes (eggs and sperm) with a haploid (50% reduced) set of genetic material. The reduction of the genome happens during the first division of meiosis when the diploid genetic material is separated into two clusters, around each of which forms a haploid daughter cell. During the first meiotic division, genetic recombination occurs, similar to the gene exchange in case of damage. Paired homologous chromosomes exchange some portion of their DNA by crossover, which means that each DNA strand breaks apart and reconnects to the homologous strand. The haploid cells generated by the first meiotic division divide again by normal cell division with the end result of four haploid cells, the gametes. The second meiotic division is similar to mitosis, which was explained in Chap. 8, except that there is no S phase.

As gametes, four sperm are created in the male sex, and one egg and three polar bodies are created in the female sex. Crossover, genetic recombination, and any kind of mutation ensure that no two gametes formed as the result of sexual reproduction will be exactly the same. Figure 10.2 shows the diplontic gametic life cycle, of which humans are part. One of the haploid gametes from each sex combines during fertilization to create a diploid zygote of a new individual, which is genetically different from the male or female parent. The diploid zygote then divides by mitosis to create further diploid cells of the growing organism. Hence, both meiosis and fertilization allow for the mixing of the gene pool and the generation of offspring that differ genetically from the parents. The development of new organisms that are not merely copies of the parent organisms allowed selective adaptation to occur since there were always some descendants that could deal with changing environmental conditions that might have led otherwise to the extinction of the species.

Sexual reproduction is not only part of the gametic life cycle but also of other animal life cycles. These life cycles are mostly different from the gametic life cycle in that the living adult organisms are either haploid or alternate between haploid and diploid states. Additionally, in plant life

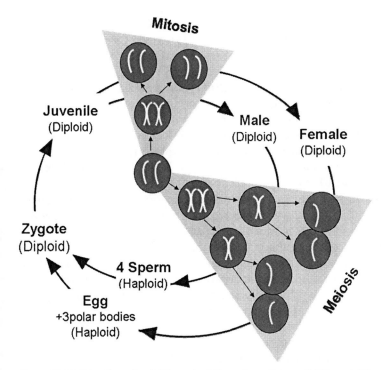

Figure 10.2 Gametic animal life cycle. Life cycle patterns of different life forms vary considerably, but they all result in offspring (juvenile) that resemble their parents (typically male and female). In organisms with sexual reproduction, the process of meiosis forms haploid gametes (sperm, egg) that have a 50% reduction in the amount of the genetic material. Gametes fuse during fertilization to form the diploid zygote with the full amount of genetic material. From the zygote, the organism grows and differentiates by mitotic cell division.

cycles there is an independent gametophyte stage between the meiosis process and the production of gametes. But in all cases, meiotic cell division serves to reduce the chromosome number from diploid to haploid, and fertilization initiates biological reproduction.

On the other hand, sexual reproduction is not necessary for an organism to survive and to grow. In fact, **asexual reproduction** without the involvement of another individual is a well-known strategy of many multicellular eukaryotic organisms. Similarly, prokaryotes don't use the complex processes of meiosis or mitosis for cell division. A prokaryotic cell basically grows, replicates its DNA, and then divides into two new cells in a process called **binary fission**. Recent genetic evidence seems to suggest that prokaryotic evolution relied mostly on **horizontal gene transfer** (transfer from one species to another) to create new species. (By contrast, vertical gene transfer occurs when genetic material is passed from the parents to the offspring). Today, DNA sharing and mutations are the main tools to generate bacterial diversity. The new molecular evidence has made some scientists revisit the models of the origin of life. They propose that prokaryotes followed eukaryotes in evolution, rather than the reverse as discussed in Sec. 10.1.1. In the new model, prokaryotes developed from eukaryotes by losing the cell nucleus and the related complex processes for reproduction and found a quicker way to change their genetic material to adapt to new food sources and environmental changes. Either way, the fact remains that during several billion years of evolution, prokaryotic organisms remained unicellular with some rare exceptions, while eukaryotes progressed toward multicellular complex life forms.

Although the development of sexual reproduction is mainly praised for the generation of offspring with genetic diversity, the alternation between haploid and diploid states was of equal evolutionary importance for the success of multicellular eukaryotes. We discussed the importance of haploid cells, which served as *gametes*, fusing to form new diploid cells. Since all diploid cells of the eukaryotic organism derived from a single cell by mitosis, all cells were principally identical and could be governed by one common organizational plan with increasingly more complexity. The capacity of eukaryotes for morphological differentiation, multicellularity, and large size allowed them to outpace prokaryotes and led to the modern world of complex organisms (Butterfield, 2000).

The oldest known eukaryotic multicellular organisms with sexual reproduction seem to be a type of red algae named *Bangiomorpha pubescens* that was found fossilized within a 1.2 billion-year-old rock (Butterfield, 2000). For comparison, the Earth is thought to be 4.6 billion years old, the great extinction of the dinosaurs was about 65 million years ago, and *Homo sapiens* arose in East Africa about 200,000 years ago. The red algae fossil indicates a minimum date for the beginning of the eukaryotic evolution. There have been records of older multicellular organisms but they are difficult to distinguish as an independent sexually reproductive group from groups that include colony-building eukaryotes.

10.2   Cell Features

Our detour into the evolution of sex and multicellularity reminds us of a principle that is important to recognize for answering the question: "What determines the features of a cell?" This principle is that every cell within our body originated from one common ancestor, the egg or oocyte, and hence has the same set of genes despite the fact that most cells are differentiated cells with special features. The heart cell that becomes an expert in repetitive contraction also contains the genes that make neurons transmit information, as well as the genes that instruct gastrointestinal cells to absorb nutrients. This principle applies not only to the cells in the human body, but also to the cells in a mouse, in a fly, or in moss; in other words to all cells within any living organism. Hence, let's first have a look at what specialized cells have in common and what is different between them.

10.2.1 Common cell features

All living cells use genes for transmission of hereditary information, which distinguishes the growth of living matter from the growth of nonorganic material such as a crystal. The genetic information is programmed in a chemical code (DNA), which is transcribed into an intermediary form (RNA) and subsequently translated into proteins. Nucleotides are the building blocks for the synthesis of DNA and RNA in all cells, and amino acids are the building blocks of all proteins (see Chap. 1). The universal use of DNA and RNA means that a human gene introduced into a plant can be read, interpreted, and copied by the plant, and in reverse, a human cell would be able to read and understand genes from a plant. An emerging biotechnology industry uses this principle to make plants and bacteria produce human substances such as insulin, vaccines, or hemoglobin-like molecules. Such gene transfer happens fairly commonly in nature between prokaryotic bacteria and is one reason for the development of antibiotic-resistant strains, but gene transfer is rare between eukaryotic cells.

Genes instruct cells how and when to build the proteins that allow the cells to create their structures and perform their specific functions. Until recently, it was thought to be a universal principle that one gene generates one protein. When the human genome sequencing project revealed that a human cell has about 25,000 genes, researchers needed to explain how these relatively few genes could be enough to account for a complex organism such as a human being. After all, it was previously estimated that humans have at least 100,000 genes due to the presence of about 100,000 different proteins. Obviously, one gene must result in the synthesis of several proteins, a process that is discussed in more detail in Sec. 10.2.2. The one-gene, one-protein model has been revised to say that a gene is the DNA sequence that is transcribed into RNA, but that there is no one-to-one correlation of RNA to protein (Ast, 2005). Some RNAs are never translated into proteins but instead have regulatory functions on gene expression. The remaining RNAs undergo various modifications before being translated into multiple proteins. Although the conundrum has not yet been fully resolved, it is obvious that the number of genes does not correlate with the sophistication of the species. After all, a corn cell has about 40,000 genes and the roundworm *Caenorhabditis elegans*, a model organism in the laboratory, has about 19,000 genes.

The proteins that are generated from the DNA code are used by all cells in a similar way. Proteins form transporters and channels for material that must cross the plasma membrane of cells (Chaps. 4 and 6). Proteins are used as signaling molecules for information transfer and information processing within cells (Chap. 4) and between cells (Chap. 6). Proteins are used as regulators of gene processing (Chap. 7) and as regulators of cell cycle progression (Chap. 8). And finally, proteins are used by all cells as enzymes (Chap. 3). Enzymes are necessary for the cell to survive since they act as catalysts and hence allow cellular reactions to occur in a timely manner. Enzymes act by lowering the energy that is required for a given reaction to occur, which introduces another common principle that applies to all cells. The principle states that maintaining a cell is only possible through the continuous input of free energy. This was explained in Chap. 1, together with the introduction of adenosine triphosphate (ATP) as the universal energy carrier of all cells. Every cell produces and consumes large amounts of ATP to drive other chemical reactions. Chapter 5 introduced these energy conversion processes and presented mitochondria and chloroplasts as places of ATP generation.

Billions of years ago, as now, the availability of energy was of great importance for a cell to be competitive and a major driving force for the development of eukaryotes from prokaryotes. The endosymbiontic theory proposes that originally mitochondria and chloroplasts were small prokaryotic cells that were able to produce large amounts of energy. This was so attractive to larger prokaryotic cells that they engulfed the smaller cells without digesting them. The double membranes around mitochondria and chloroplasts and the presence of DNA within these organelles are features that support the endosymbiontic theory. The benefit to the host and the symbiont was so great that, presently, there are almost no eukaryotic cells without mitochondria, and photosynthetic eukaryotes cannot survive without chloroplasts.

10.2.2   Features that make cells different

By now, we acknowledge that during the growth of the organism, cells don't lose genes in order for the cells to become different from one another. To understand why a heart cell differs from a kidney cell, or why one cell continues to divide while a different cell lost this ability, we have to look at **differential gene expression**. Different cells activate and express different genes at different times at different intensities, and this leads to an individual gene expression pattern for a particular cell type. In humans, one would expect this pattern to differ in about 2,000 to 4,000 genes between a heart cell and a kidney cell, or a skin cell and a neuronal cell. Cells that are more closely related such as B cells and

T cells (both immune cells) might differ in the expression of about 400 genes. These estimates result from DNA microarray analysis, a gene expression profiling technique that is presented in Application Box 7.2. Such analyses, as well as other genomic analyses, have also revealed that only a small fraction of all genes are expressed in any given cell at one time and that about 100 to 200 of these are **signature genes** that are commonly expressed by all cell types in an individual.

Genome analyses further revealed that the genome of the modern human is 99% identical with the genome of the chimpanzee and 99.9% identical with the genome of a Neanderthal. From evolutionary trees created on the basis of genomic comparisons, it is estimated that hominins (the ancestors of the human race) finally separated from the great apes only about 5.3 million years ago. This is very different from the case with bacteria, where a disease-causing variant of the gut bacteria *Escherichia coli (E. coli)* differs in its genome by 26% from the harmless variant, which belongs to the same species. A similar difference imagined in the human population would mean that about one-third of the genes would be different from individual to individual, instead of the 0.1% currently believed. This number might be corrected to 1% due to a recent discovery that many individuals have more than two copies of their genes and that this "multiple copy number" might be highly variable between individuals.

However the correct numbers might turn out, they are still relatively small. This leaves us with the questions of how modern humans have become so different from apes and how individuals can be so different from one another despite their gene similarities. The answer is very similar to the one that provided the explanation for the differences between cell types within an organism that operate with the same gene collection: It is not primarily the number of genes that determines the characteristics of an organism, but the differential expression of genes into mRNAs and proteins. Despite DNA similarities, the mRNA pattern of a human cell is quite different from the pattern of a comparable cell in apes and between one person and the next. The mRNA pattern is made unique by **alternative splicing**, which refers to the modifications that occur at the first DNA transcript (pre-mRNA) and then lead to the subsequent substantially larger number of mature mRNAs which are translated into proteins (for details, see Chap. 7). Although alternative splicing has been known for a long time, the extent of its importance for the development of diversity was only recently realized.

Complex gene regulation is the basis for the presence of distinctive cells and organisms. There is a complicated control system that dictates not only which genes are read in which cell, but also at what time. It is well known that cells can respond within minutes to changes in their environment with changes in their gene expression. For instance, one response of the pancreatic beta cell to high blood glucose levels is to activate genes for production of more insulin. This minute-to-minute control of differential gene expression happens mostly at the initiation of gene transcription as described in Chap. 7. In addition to fast and highly flexible gene control, eukaryotic cells developed long-term measures that prevent inadequate gene transcription at inappropriate times. These include packaging of eukaryotic DNA into chromosomes, silencing of DNA parts by methylation, and biochemical alteration of histone proteins.

Recent evidence shows that gene expression is controlled in a more global way than previously thought. For instance, it was believed that gene regulatory proteins bind to DNA sequences close to the sites where RNA polymerase begins to read the code. While that is still true for prokaryotes, it is now known that the DNA of eukaryotes bends and loops in such a way that regulatory proteins can influence the transcription of genes close by, at the same time that they influence genes that are thousands of nucleotides away. Gene regulatory proteins always influence the transcription of several genes, even in prokaryotes. For instance, the catabolite gene activator protein (also known as the cyclic AMP receptor protein), when bound to the DNA of E. coli, influences the transcription of several genes that code for enzymes involved in the metabolism of sugars, and hence globally regulates the bacterial energy metabolism. One can imagine the added level of complexity in eukaryotic gene regulation, where a multiplicity of regulatory proteins and transcription factors create **gene regulatory networks** which often include dynamic feedback loops, and where one protein in one regulatory complex might play a different role in another complex.

Gene expression can also be regulated after transcription. Changes can be made to mature mRNAs, and the transport of mRNAs from the nucleus to the ribosome can be regulated. Lastly, proteins can be modified in many ways after they are synthesized, a process that is called **posttranslational modification**. In summary, regulation at the DNA, RNA, and protein level allows for the fine-tuning of gene activity in response to multiple stimuli. It seems clear that there is a long road ahead to travel in unraveling the mystery of 25,000 genes. Although we have already come a long way toward understanding how cells coordinate the well-timed activation and inactivation of genes, there has been little success in inducing these changes artificially. Gaining such control over the cell's regulatory processes will open new therapeutic approaches for a wide variety of diseases.

10.3   Determination and Differentiation

By now, we have learned that the characteristic features of a cell are determined by differential gene expression. In the human body, differential gene expression leads to 210 to 220 distinct cell types. The process by which cells gradually become committed to develop into a specific cell type with a specific pattern of gene expression is called **cell determination**. It is followed by **cell differentiation** which results in a characteristic cell type such as a muscle cell or a nerve cell. The regulatory mechanisms that control the cell-specific pattern of gene expression are passed from cell to cell as a "developmental memory." The study that investigates the molecular nature of this memory is called **epigenetics**. Two mechanisms that are involved in the epigenetic preservation of gene activity patterns are DNA methylation and posttranslational modification of histone proteins. The addition of methyl groups to DNA is the most common chemical modification of DNA. It can alter gene expression without affecting the DNA sequence and it can be inherited. Similarly, it can be shown that the posttranslational modifications that occurred to histones in the parental cells are preserved through cell division and provide for the daughter cells a "molecular bookmark" about previously active locations of the genome. For transcription to actually occur at these memory marks at the end of mitosis, other molecules such as non-histone proteins and noncoding RNAs have to be involved. In general, the memory of cell transcription is closely associated with significant cell cycle events such as mitosis, nuclear organization, and cell replication which were presented in Chap. 8.

Now that we understand the heritable nature of cell determination, let's have a look at the process of becoming committed to a specific pattern of gene activity to eventually mature into a differentiated cell.

10.3.1 Cell lineage

The line of the cell divisions that occur to generate specific cell types is called **cell lineage**. The fate of a cell within a cell lineage is usually predetermined in lower organisms, but flexible in higher organisms. Roundworms are an extreme example of lower organisms in which every single cell has a predefined function. For instance, the complete cell lineage of *Caenorhabditis elegans*, a 1 mm-long roundworm that lives in soil, has been recorded from the zygote to the adult worm. The adult worm has exactly 959 somatic cells, (after growing 1090 cells and eliminating 131 by apoptosis), with 300 of them being neurons and 81 of them being muscle cells. In contrast, the fate of fetal cells from higher organisms is largely determined by the types of environmental signals that each cell receives. Each cell reacts in response to the chemicals produced in its immediate environment and responds by activating or inactivating certain genes, a principle that was described in Chap. 9 for the development of a limb (see Fig. 9.8). The information that a particular cell receives depends on its position within the embryo. This positional information can be traced back to the zygote, which contains cytoplasmic differences that give rise to different cell lineages. These positional differences within the fertilized egg, including variations in proteins and mRNAs, are called **cytoplasmic determinants**.

Early during embryogenesis, the signals are transient in nature. But when undifferentiated stem cells divide and gradually become committed to specific pattern of gene activity and hence to a specific phenotype, the process becomes irreversible. Once a mammalian cell has become specialized, it generally cannot turn into any other type of cell. Within a cell lineage, the determination of the cells is increasingly constrained as stepwise cellular differentiation occurs so that stem cells change from being totipotent to pluripotent to multipotent to, finally, being specialized cells (for details, see Chap. 8). Constraining differentiation is accomplished by constraining gene expression, by which initial regulatory genes constrain subsequent regulatory genes, ultimately restricting the offspring cells to the same cell types as the parental cells. Important mechanisms that are involved in these transitions are known as positive feedback cycles.

10.3.2 Size and shape of cells

It is well known that differentiated cells can have many sizes, forms, and shapes. For instance, Fig. 6.6 shows large Purkinje cells and small granule cells as two types of nerve cells with fundamentally different appearances. The size and shape of these neurons is determined by their functions in that the large dendritic tree of the Purkinje cell serves as a collection area for information input from many other neurons, while the small projections of the granule cell are sufficient to transmit information onto neighboring cells. This raises the question of whether a cell can infinitely increase its size if it serves its function. The answer is no. As a cell grows larger, the volume becomes too large relative to the surface area to allow sufficient movement of water, nutrients, oxygen, and waste across the cell membrane.

Figure 10.3a presents a cube as a simplified model of a cell or a single-cell organism to show that the changes of the total surface area to volume ratio become unfavorable for the cell, or organism, with increasing size. As the length (L) increases, the surface area to volume ratio (S/V) decreases. The result is that when the cell or organism gets too big, it will starve from lack of incoming nutrients or poison itself with wastes that cannot be excreted rapidly enough. The maximal size of a spherical cell with tolerable S/V ratio is close to 10 to 12 μm (Fig. 10.3a). One example is lymphocytic white blood cells, which typically vary in size from 6 μm up to 15 μm.

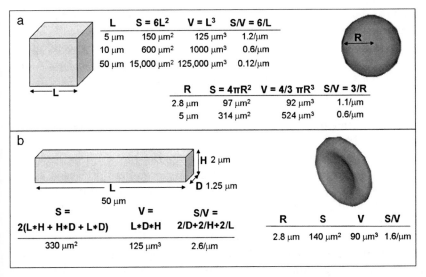

Figure 10.3 Surface area (S) to volume (V) ratio of a cell. (a) Changes of the total surface area to volume (S/V) ratio become unfavorable for a spherical cell or organism with increasing size (L: length of a cube, R: radius of a sphere). (b) A cuboid with the same volume as the cube has an increased surface area, resulting in a more favorable S/V ratio. By having a biconcave shape, a red blood cell increases its surface area by about 44% and becomes a cell with a more favorable S/V ratio compared to an isovolumetric spherical cell.

There are several ways that cells overcome the problem of being limited by diffusion and transport capacity across their membranes. Somatic cells often overcome the limitations by taking on a shape with a high S/V ratio. Many cells are elongated instead of being spherical and/or develop cytoplasmic branches. That way, the surface area is increased with a proportionally smaller increase in cell volume. Figure 10.3*b* shows that a cuboid (rectangular parallelepiped) with the same volume as the cube (125 μm^3) has an increased surface area (330 μm^2 compared to 150 μm^2), which indicates that the S/V ratio of elongated cells is more favorable than the ratio of spherical cells. The previously mentioned Purkinje cells are examples of cells with enormously increased total surface area. Another example are bone cells, which extend long thin cellular processes deep into the extracellular matrix to overcome the problems associated with a low S/V ratio while at the same time being able to deposit large amounts of extracellular bone material.

Many other cells have developed characteristic shapes that increase their surface area without compromising their function. A good example is the red blood cell. The circular biconcave shape (see Fig. 10.3*b*) increases its surface area by about 44% compared to a spherical cell with the same volume, but still allows the red blood cell to fit through small vessels and to endure shear stress encountered when leaving the vessel and being squeezed into the surrounding tissue. While some cells adjust their overall shape, like red blood cells, other cells develop special intracellular structures that help ensure vital exchange of nutrients and waste with the environment. An extreme example is the green alga *Caulerpa*, a marine single-celled organism that can be up to 1 m in length. *Caulerpa* developed fluid-filled sacs that squeeze its cytoplasm toward the cell wall into a 5 μm-thick layer, and therefore has a S/V ratio comparable to other, smaller cells.

While *Caulerpa* is the largest single-celled organism, the egg of an ostrich is the largest cell that a living higher organism produces. It can weigh 1.5 kg (3.3 lb)! Birds, together with reptiles and mammals, belong to the taxonomic unit of amniotes, which owe their name to their evolutionary invention of an amniotic egg. An amniotic egg is massive in size and has yolk content, which resulted in the evolutionary advantage of providing bigger offspring. To keep the large egg cell alive, a unique set of supporting extraembryonic structures developed—the amniotic fluid for water supply, the allantois for waste collection, and the chorion for gas exchange in combination with the allantois. The yolk sac allows storage of nutrients that provide for the rapid series of cell divisions after fertilization.

Another possibility to compensate for the changes in the S/V ratio that are associated with increases in cell size is the variety of membrane transport processes that have developed and which are presented in the following section.

10.3.3 Membrane transport

Cells use four basic types of transport processes across membranes as shown in Fig. 10.4:

1. *Simple diffusion* that requires concentration gradients,

2. *Facilitated diffusion* that requires concentration gradients and carrier proteins,

3. *Active transport* that requires carrier proteins and ATP, and

4. *Pinocytosis* or *phagocytosis* that involves vesicle formation and requires ATP.

Diffusion is the tendency of molecules to spontaneously move from a region where they are more concentrated to a region where they are less concentrated (down the concentration gradient). The rate at which the molecules move through a defined area depends on the concentration gradient of the molecules. Cells frequently use simple diffusion for transport of molecules, not only through membranes, since it has the enormous advantage of not requiring energy. The diffusion of water

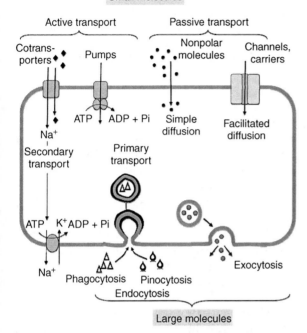

Figure 10.4   Membrane transport. Small molecules can be passively transported by simple or facilitated diffusion, which does not require energy, or they can be actively transported by primary or secondary transport, which requires energy. Large molecules are expelled from cells by exocytosis, and ingested into cells by endocytosis, either by phagocytosis in case of particles, or by pinocytosis in case of solutes. For details, see Sec. 10.3.3.

down its concentration gradient is called osmosis. But diffusion has two major disadvantages for cell functioning. First, it is slow since it has a square root dependence on time indicating that moving 10 times farther will take 100 times longer. This does not affect the transport across a membrane so much as it does the transport across a cell or the transport between cells. Second, diffusion through a membrane barrier is influenced by the size, the charge, and the lipid solubility of the molecules to be transported. This limits transport across membranes since most biologically relevant substances except for small hydrophobic molecules cannot readily pass through lipid bilayer membranes.

For diffusion of large and hydrophilic molecules across membranes, a broad range of carrier proteins have been developed to facilitate diffusion. The carriers bind molecules and transport them across membranes down the concentration gradients and hence still don't require energy. An example is the GLUT-1 carrier which facilitates glucose transport across the human red blood cell membrane. Although **facilitated diffusion** often substantially speeds up diffusion, it is much slower than transport through channels because the binding of the molecules and the conformational changes of the carrier during transport take time. The transport of molecules through channels was presented in Chap. 6.

Transport against concentration gradients requires energy-driven pumps (ATPases) and is called **active transport**. **Primary active transport** uses energy directly to transport molecules across a membrane (see Fig. 10.4). **Secondary active transport** takes advantage of an ion gradient which was established using energy-driven ATPases (for details, see Sec. 10.3.4). There are three classes of ATPases, all of which are present in most cells with few exceptions. Mitochondria possess ATPases that are used to synthesize ATP (F-type, see Fig. 5.10). Lysosomes and other vesicles contain ATPases that lower the pH within the vesicles (V-type). Plasma membranes and membranes of organelles such as endoplasmic reticulum contain ATPases that are critical to the maintenance of the characteristic ion gradients found in nearly all cells (P-type). They include Na^+-K^+-ATPases and Ca^{2+}-ATPases.

When molecules are too large to fit through channels, or cannot be transported by carriers, they are brought into the cell by **endocytosis** or expelled from the cell by **exocytosis** (see Fig. 10.4). Large particles are incorporated by *phagocytosis*; large amounts of solutes are ingested by *pinocytosis*. These elaborate and highly regulated processes are described in Chap. 2.

10.3.4 Membrane potential

In every living cell, ions are asymmetrically distributed across the cell membrane with higher K^+, lower Na^+, lower Cl^-, and lower Ca^{2+} concentrations intracellularly compared to extracellularly (see Fig. 10.5). The unequal distribution of ions also causes a separation of charge across the cell membrane with relative excess negative charge on the inside, and relative excess positive charge on the outside. Consequently, ion gradients possess not only chemical potential energy due to the concentration gradient, but also electrical potential energy due to the unequal distribution of charges. The fluxes of ions across membranes according to their electrochemical potential, which depend on the permeability of the membrane to these ions, are balanced so that a stable **membrane potential** is created. The asymmetric distribution of Na^+, K^+, and Cl^- across membranes and the presence of impermeant mostly negatively charged macromolecules inside cells are the major contributing factors to the membrane potential. In resting cells, the energy

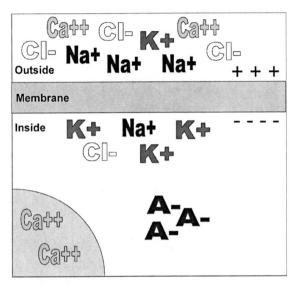

Figure 10.5 Separation of charges across cell membranes. There are more Na^+, Cl^-, and Ca^{2+} ions outside the cell than inside. There are more K^+ and organic anions (A^-) inside the cell. Ca^{++} is stored inside the cell in organelles. The relative excess of positive charges outside and the relative excess of negative charges inside the membrane of a cell give rise to an electrical potential difference of about 70 mV in excitable cells such as neurons, sensory cells, and muscle cells. This potential difference is called resting membrane potential. For details, see Sec. 10.3.4.

of ATPases is used to counterbalance any disturbances of the membrane potential by transporting ions against their concentration and electrical gradients.

The membrane potential is roughly 10 mV in nonexcitable tissue and 60 to 90 mV in excitable cells, but a wide variety of potentials in between these values can be found in various cells. Excitable cells are special cells including neurons, sensory cells, and muscle cells that generate meaningful electrical signals by selectively changing their membrane permeability to Na^+, K^+, or Cl^- via opening and closing of the related channels. When fluxes of these ions lead to meaningful signals, that is, action potentials or generator potentials, the cells are said to be "excited." When the cells are not excited, their membrane potential is generally called **resting membrane potential**. Changes in membrane potentials are signals that are meaningful to all cells, not only to excitable cells, and all cells react with cell-specific responses.

Quantitatively, the resting membrane potential (V_m) of a cell is given by the Goldman-Hodgkin-Katz equation 10.1:

$$V_m = \frac{RT}{F} \ln \frac{P_K [K^+]_o + P_{Na}[Na^+]_o + P_{Cl}[Cl^-]_i}{P_K [K^+]_i + P_{Na}[Na^+]_i + P_{Cl}[Cl^-]_o} \qquad (10.1)$$

where V_m = Resting membrane potential
R = Universal gas constant (2 cal/mol/K)
T = Temperature in K (at 37°C = 310 K)
F = Faraday constant (23 cal/mV/mol)
P = Permeability of charged particles
o = Outside of cell
i = Inside of cell

Much can be said about this equation. In our context, it is important to recognize that the membrane potential is a weighted average of the individual equilibrium potentials of the biologically important ions Na^+, K^+, and Cl^-. The "weight" stems from the membrane permeability (P). Since in many eukaryotic cells the membrane at rest is about 50 times more permeable to K^+ than Na^+, K^+ is the ion that contributes most to the potential at rest in these cells. As soon as the membrane permeability changes, for example, by opening Na^+ channels, Na^+ is the main ion to determine the overall magnitude of the potential.

The asymmetric distributions of ions other than Na^+, K^+, and Cl^- are quantitatively unimportant for determining the magnitude of the membrane potential in most cell types, but they are equally important for the proper functioning of the cell. For instance, the asymmetric distribution of Ca^{2+}, intracellularly 10,000 times less Ca^{2+} compared with the extracellular space or the space within storage organelles, is a universal mechanism for cell signaling (see Chap. 4). Every cell recognizes a change in the intracellular free Ca^{2+} concentration and responds to it, which makes Ca^{2+} the most common cell messenger that directly or indirectly regulates nearly all cell processes.

After changes in membrane potentials and ion gradients have occurred, either passively by membrane leakage or actively by opening channels, the resting cell state has to be reconstituted by primary active transport of ions. This active transport allows the membrane to maintain a constant resting membrane potential, constant concentration gradients, and a steady-state condition even though most of the ions are out of equilibrium. The fact that 75% to 85% of all ATP in the nervous system, the muscles, and the kidneys is used to drive Na^+-K^+-ATPases and Ca^{2+}-ATPases underscores the importance of membrane transport for the proper functioning of a cell.

10.3.5 Cell polarity

Cellular asymmetry is a characteristic of almost all eukaryotic cells and it is absolutely critical for the functioning of epithelia. To explain the latter statement, let us use the example of an enterocyte, the epithelial cell that lines the intestinal tract and absorbs nutrients through microvilli (see Fig. 2.8). If we imagine the idealized cell in Fig. 10.4 to be an enterocyte, the upper membrane part containing the carriers, pumps, and channels would be the apical membrane, which faces the intestinal lumen. The rest of the cell membrane would be the basolateral membrane, which is close to blood vessels that transport the absorbed nutrients away. In such an enterocyte, Na^+-K^+-ATPases would need to be restricted to the basolateral membrane, while Na^+ carriers would need to be limited to the apical membrane as shown in Fig. 10.4.

Basolateral Na^+-K^+-ATPases transport Na^+ out of the cell, in exchange for K^+, using an ATP-dependent process that results in a low sodium concentration inside the cell. Consequently, extracellular Na^+ wants to enter the cell along its electrochemical gradient (see Sec. 10.3.3); it is, however, restricted from doing so by the Na^+ impermeable lipid membranes. Na^+ carriers in the apical membrane allow Na^+ from the gastrointestinal lumen to enter the cell. Once inside the cell, the Na^+ is transported across the basolateral membrane into the extracellular space, from where it enters nearby blood vessels. Cell polarity, characterized by the unequal distribution of Na^+ carriers and pumps, is necessary to maintain this Na^+ transport across the enterocyte, from the gastrointestinal lumen to the plasma.

In enterocytes, Na^+ transport proteins often allow the passage of other molecules such as amino acids or glucose, together with the passage of Na^+, and are then called **cotransporters** (see Fig. 10.4). Since the coupled transport does not directly use ATP hydrolysis, but instead uses an ATP-dependent Na^+ gradient to move molecules against their concentration or voltage gradient, this type of transport is called secondary active transport as already introduced in Sec. 10.3.3. Different transporters exist for different substances that can be transported together with Na^+. In addition to cotransporters, there are also **countertransporters** that transport substances across cell membranes in opposite direction of the Na^+ transport. Again, loss of the cell polarity would abolish these transport mechanisms used by enterocytes for absorption of nutrients. Enterocyte transport provides humans an intestinal absorptive capacity of 3.6 kg (7.9 lb) of glucose per day. This is an impressive safety factor since it is about 12 to 18 times more than a person of normal weight ingests daily. While the capabilities of enterocytes are impressive, they cause us to face the problem of obesity when overindulging since everything that's ingested is absorbed, and could make one question the use of the term "safety factor."

Although epithelial cells are good examples to present cell polarity, note that asymmetry in cell shape, distributions of biomolecules, and cell functions are a feature of all eukaryotic cells in one form or another. Several mechanisms play a role in how cells maintain this polarity. Actin filaments and microtubules (see Sec. 2.3.2), which are polarized themselves (having + and − ends), establish and maintain a certain cell morphology as well as regulate intracellular transport between morphologically different areas. The cytoskeleton additionally regulates the lateral distribution of integral membrane proteins. The cytoskeleton also maintains **lipid rafts**, which are specialized membrane domains enriched in cholesterol and certain lipids. Scientists increasingly recognize the importance of lipids in regulating cell polarity and cell functions. For instance, it is well known that glycosphingolipids (glycolipids derived from sphingosine) provide a platform for clustering of signal transducers and can act as second messengers to control signal transduction events.

In many cells, the filaments of the cytoskeleton inside cells are linked to the junctions between cells and to the **extracellular matrix** that fills the gap between cells. This material, directly outside of cells, is presented next on our journey from cells to organisms governed by morphogenesis.

10.4   Morphogenesis

Morphogenesis is the process that gives a developing organism a specific three-dimensional form. In the developing organism, cells tend to find their positions before they specialize, rather than specializing before they find their specific positions. Morphogenesis includes the hierarchical organization of cells into tissues, organs, and organ systems to provide the external shape of the organism as explained in more detail in the following sections.

10.4.1   Cell junctions

Cell junctions are the links that physically or functionally connect neighboring animal cells and are hence an important step toward the development of tissues. There are three types of cell junctions: occluding junctions, anchoring junctions, and communication junctions (see Table 10.1). Plant cells do not need structural reinforcement between cells since they are encapsulated by a rigid cell wall, but they do interact among each other via communication junctions.

Occluding junctions are more commonly called **tight junctions**, or *zona occludens* when reaching around a cell like a belt. They are crucial elements that seal cells together so that they act as a whole functional unit instead of individual entities. Tight junctions prevent, or occlude, the movement of membrane proteins between the apical and basolateral cell membranes and hence are important structural elements to maintain cell polarity (see Sec. 10.3.4). Tight junctions also occlude, or better regulate,

TABLE 10.1   Cell Junctions

	Alternate Names and Subcategories	Characteristic	Function
Occluding junctions	Tight junctions	Membrane proteins such as claudins and occludins form seal between adjacent cells	Seal cells together Regulate paracellular transport
Anchoring junctions	1. Adherens junctions (zonula adherens or belt desmosome)	Cadherin proteins linked to actin	
	2. Desmosome junctions (macula adherens or spot desmosome)	Cadherin proteins linked to keratin intermediate filaments	Hold cells in fixed position
	3. Hemidesmosomes	Integrins link keratin filaments to laminin of basal lamina	
Communication junctions	1. Gap junctions 2. Plasmodesmata	Connexons form channels across plasma membranes of adjoining animal cells Connect plant cells to neighboring cells	Regulate passage for small chemical or electrical signals

the inter- or paracellular passage of small molecules and water. Although tight junctions allow epithelia to act as barriers, the name "tight" is somewhat misleading. For instance, tight junctions in the gastrointestinal **epithelium** are semipermeable, having the highest permeability in the duodenum and the lowest permeability in the rectum, consistent with the physiological requirements for absorption of nutrients and water. Nevertheless, tight junctions create compartments such as in the gastrointestinal epithelium where they successfully separate the luminal contents from body tissue and even create a potential difference of about 3 mV in the duodenum and 40 mV in the rectum that could be measured if electrodes were to be placed on both sides of the epithelium.

There are three types of cell **anchoring junctions**: adherens junctions, desmosomes, and hemidesmosomes (see Table 10.1), which function to hold cells together when tissues are stretched. In all types of cell connections, anchoring proteins extend through the plasma membrane to link cytoskeletal proteins from one cell to the proteins of the next cell and to the extracellular matrix.

The last type of junction, communication junctions, allows exchange of small molecules and ions between cells. In animal cells, **gap junctions** are formed when two identical plasma membrane channels, one on each of the neighboring cells, join. Gap junctions create a **syncytium**, which is the term to describe cytoplasmic continuity between cells. Information exchange via gap junctions is crucial for many tissues and organs to function properly. For instance, gap junctions connect heart muscle cells and allow the heart to contract as one functional unit. Similarly, most smooth muscle cells are connected by gap junctions.

Plant cells are also able, like animal cells, to communicate with one another. In plant cells, the plasma membranes and cytoplasm of neighboring cells are connected by thin strands, called plasmodesmata, which reach through pores in the cell wall. A **plasmodesma** allows the passive transport of small molecules as well as the active, energy-requiring transport of larger molecules. Due to the presence of plasmodesmata, plants can also be viewed as a syncytial cytoplasmic mass with many cell nuclei instead of a multicellular organism.

10.4.2   Extracellular matrix

Extracellular matrix is any part of a tissue directly outside of animal cells. It is produced and secreted by the cells and contains fibrillar proteins such as collagens, which give strength, and elastins, which give resilience. Extracellular matrix further contains nonfibrillar proteins such as fibronectins, which bind to membrane receptors such as integrins and to other components in the matrix such as collagen, fibrin, and heparin. A soluble form of fibronectin exists in blood plasma. The last family of common proteins in extracellular matrix is proteoglycans, which provide the ground substance that fills intracellular gaps, but most likely also contributes to intracellular signaling. Proteoglycans, also called mucopolysaccharides, are composed of a protein core to which are attached long chains of repeating disaccharide units termed glycosaminoglycans (GAGs). Intercellular substance can be abundant, as in the matrix of bone tissue or the plasma of blood, and it can be minimal, as in epithelial tissue in which epithelial cells are aligned side by side with minimal filling material between them. The intercellular matrix may contain special substances such as hydroxyapatite $[Ca_{10}(PO_4)_6(OH)_2]$ crystals of bone.

A special extracellular component that anchors epithelial cells to each other at their basal ends is the **basement membrane**. The membrane is composed of two cell-free layers, the basal lamina and the reticular lamina, and has very important functions in various organs. For instance, it acts as a molecular sieve in the kidney (Farquhar, 2006), and lends stability to muscle cells. The importance of the latter becomes obvious when it is lacking in congenital muscular dystrophy, where a defective protein of the basement membrane causes progressive, ultimately lethal muscle weakness.

10.4.3   Tissues

Combinations of extracellular matrix and functionally united cells are called **tissues**. Tissue engineering is the discipline that aims at repairing or replacing tissues with combinations of cells, engineering materials, and chemical factors (Langer and Vacanti, 1993). Application Box 10.1 describes an approach for the engineering of a blood vessel. Let us consider the four main tissue types of advanced animal tissue and the three main tissue types of advanced plants:

APPLICATION BOX 10.1
Vascular Tissue Engineering

Vascular tissue engineering is an interdisciplinary field that uses engineering principles as well as biological knowledge to develop effective prosthetic vessels. The major clinical problem that is addressed by vascular tissue engineering is that suitable autologous (from the same individual) vessels are not available in up to 40% of patients who need them as bypass vessels or replacement vessels (Yow et al., 2006). Although synthetic vascular prostheses have been used successfully, they exhibit a low patency (refers to the openness of a tube) rate in small diameter vascular grafts (i.e., they clog due to blood clots). Hence the need for a tissue-engineered replacement vessel.

The general approach to vascular tissue engineering is to attempt to construct a graft which avoids thrombogenicity (blood clot responses) and immunogenicity (immune responses) by using autologous cells to repopulate some type of tubular, biocompatible scaffold. These scaffolds may be biological matrices or synthetic polymers, but in either case they must allow for and encourage the growth of autologous smooth muscle cells and endothelial cells in a configuration which allows the vessel to exhibit material characteristics similar to those of native blood vessels. Biological scaffolds include (1) allogenic (from the same species) scaffolds or xenogenic (from different species) scaffolds. They may also include (2) decellularized nonvascular conduits and (3) collagen gels or matrices.

In 1998, L'Heureux et al. used smooth muscle cells and fibroblasts, separately cultured, to create a tissue-engineered blood vessel. The generated sheet of smooth muscle was wrapped around a tube to produce the tunica media, or middle layer of the blood vessel. A fibroblast sheet was wrapped around the smooth muscle layer to produce the tunica adventitia, or outer coat of the blood vessel. After maturation, the tube was removed and endothelial cells were seeded onto the lumen to produce the endothelium, the vessel layer facing the blood. At this stage, it was demonstrated that all three cell types were viable. In 2006, L'Heureux et al., now at Cytograft Tissue Engineering, Inc., California, reported that these tissue-engineered blood vessels had mechanical properties that compared favorably with human arteries and surpassed values reported for other *in vitro* vascular conduits. The burst pressures of the engineered vessels were in excess of 3,000 mm Hg compared to a burst pressure of approximately 2,000 mm Hg in normal human arteries. However, it took approximately 28 weeks to produce a blood vessel. At the time of writing of this book, the efficacy of the vessels is still being tested in animal studies. One critical issue will be decreasing the time required to produce an autologous tissue-engineered vessel, but the future looks very promising for this technology.

Animal Tissues   **Epithelia** line all body surfaces, cavities, and tubes in animals. Epithelia are found for instance in the skin, the intestine, the salivary glands, and the liver. Epithelia provide protection, regulate exchange with the environment, and form glands. Epithelia are divided into three major categories according to the shape of the cells. Squamous epithelium contains flattened cells, cuboidal epithelium has cube-shaped cells, and columnar epithelium consists of elongated cells. Further classification occurs according to the number of cell layers. For instance, simple epithelia are formed by one cell layer only, while stratified epithelia are composed of two or more cell layers.

Connective tissue provides the framework for all other tissues of the animal. It is formed by cells that produce extracellular matrix. Connective tissues exist in various forms and shapes, and the reader is referred to histology books for their classification. Most types of connective tissue contain collagen, which makes it the most abundant protein of the human body. Blood is also considered a form of connective tissue since it is composed of cells and the extracellular matrix called plasma.

Muscle tissue is composed of highly specialized elongated animal cells that contain the proteins actin and myosin. Actin and myosin form filaments that slide past one another when activated to cause cell contraction. Contractions of skeletal muscle cells contribute to coordinated voluntary movements and to breathing. Contraction of cardiac muscle cells are the basis for automaticity and the rhythmic beating of the heart. Contractions of smooth muscle cells lead to slow involuntary movements of the gastrointestinal wall and contribute to the regulation of blood pressure and blow flow when contracting in the arterial walls.

Nervous tissue contains neurons that generate and conduct electrical signals in the body and glial cells that provide support to neurons. There are three types of neurons: (1) sensory or afferent neurons that transmit information from sensory organs to the central nervous system, (2) motoneurons or efferent neurons that transmit information from the central nervous system to muscles and glands, and (3) interneurons that associate sensory neurons with motoneurons. The Purkinje cells and the granule cells of the cerebellum that are shown in Fig. 6.6 and discussed in Sec. 10.3.2 are motoneurons and interneurons, respectively. Glial cells influence neuronal development, provide stability and nutrients to mature neurons, and regulate the amount and composition of the fluid around neurons. Overall, nervous tissue allows an organism to sense and integrate internal and external stimuli to provide commands for appropriate body responses.

Plant Tissues   **Dermal tissue** forms the outer surface of the leaves and of the young plant body. It is responsible for interaction with the environment such as gas exchange and prevention of excessive water loss. The dermal tissue of roots is involved with water and ion uptake.

Ground tissue is the tissue of the leaf which manufactures nutrients by photosynthesis, the tissue of the stem that provides stability, and the tissue of the root that stores nutrients.

Vascular tissue provides the conduction structures of plants called xylem and phloem. Xylem mainly transports fluid and nutrients from the soil throughout the plant. Phloem transports sugars, amino acids, and other small molecules from the leaf down to the roots and up to the flowers, fruits, and seeds.

10.4.4 Organs, organ systems, and organisms

Multicellular organisms are hierarchically organized using structurally and functionally distinct elements that increase in size and complexity from the bottom of the hierarchy to its top (see Fig. 10.6). The hierarchical level above tissues is **organs**. Organs are generally made of several tissues, although one tissue predominates and determines the

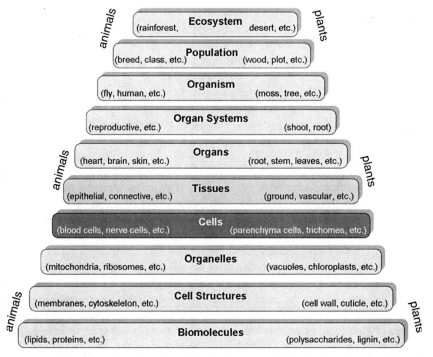

Figure 10.6 Organization of complex biological systems. Cells, the functional entities of life, are composed of organelles and other cell structures, which are built from biomolecules. Cells structurally and functionally unite in tissues, tissues in organs, and organs in organ systems, which are the functional entities of an organism. Multiple organisms make up a population, composed of individuals that have similar attributes or are confined to a similar space. Populations are part of ecosystems. Examples from the animal kingdom are given on the left, and from the plant kingdom on the right. Most organelles, cell structures, and biomolecules of animals are present in plants as well.

principal function of the organ. For instance, the liver primarily consists of glandular tissue which is a subcategory of epithelial tissue, but connective, muscle, and nervous tissues are present as well. Examples of plant organs are the root, the stem, the leaves, the buds, and the fruits.

Groups of organs that contribute to a particular set of functions are called **organ systems**. For example, the liver is part of the digestive organ system, which further includes other organs and tissues such as mouth, stomach, intestine, liver, and pancreas. The human organism is said to have 11 organ systems: circulatory, digestive, endocrine, excretory (urinary), immune (lymphatic), integumentary, muscular, nervous, reproductive, respiratory, and skeletal. A higher plant organism such as a tree typically has two organ systems, one above the ground (the shoot system) and one below the ground (the root system).

A **population** is a group of individuals that share one or more characteristics and are part of an *ecosystem*. It might seem like a stretch in a cell and molecular biology book to mention functional entities of biology that are higher in the hierarchy than organisms. But in fact, it might be the tasks of future bioengineers and other scientists to uncover similarities in the principles that govern biological systems and to apply this knowledge for improvement of societal well-being. Hence, we finish the chapter and book with an outlook toward systems biology.

10.5   Systems Biology

The recent progress in molecular biology, especially in genome sequencing and high-throughput measurements, has provided an overwhelming amount of data (1) on the structural components and molecular machineries of the cell and its organelles and (2) on the complexity of intra- and intercellular communication. It was the purpose of this book to provide an insight into the fascinating world of a cell. But just as a book does not automatically create knowledge in an individual reader, data are not automatically meaningful when collected. Databases on gene regulatory and biochemical networks have provided an amazing breakthrough in our understanding of cell components. Nevertheless, the meaning of the networks can only be unraveled when studying their dynamics, control, and design. One approach, **systems biology**, attempts to develop models that integrate individual components at all levels of biological organization, from genes to cells to higher hierarchical elements. The complexity of biological systems on the one hand, and the complexity of advanced computer technologies and engineering tools on the other hand have been a barrier to dialog. However, whenever the barrier is overcome, major advances can be made at the interface of disciplines. For instance, a commonality between complex biological systems and advanced engineering systems is the combination of their robustness and fragility at the same time. Just as a Boeing 777 is a robust fully fly-by-wire system with obvious weaknesses, an organism is a highly robust, autonomically controlled system that allows for exploratory weaknesses. In his book, *To Engineer Is Human*, Henry Petroski writes that in engineering design, "All successes lead to failures, and all failures lead to success." This principle is explained further in the following section using the example of homeostasis.

10.5.1 Homeostasis

The proper functioning of biological systems: cells, tissues, organs, organ systems, and ecosystems relies on maintenance of a state that allows acceptable efficiency. The maintenance of such an internal balance, **homeostasis**, is a common property of highly complex open systems (for a discussion on complexity, see also Application Box 10.2). Homeostasis of an organism is the ability to react to every change in the environment through a series of modifications of equal size and opposite direction called **negative feedback**. A response that amplifies a change in the environment is called **positive feedback**. Although positive feedback further destabilizes the system, it is used for fast dramatic changes that bring a certain condition to another level which then can be fine-tuned by negative feedback. The goal of these modifications is to maintain the balance of physical and chemical parameters in the internal environment.

APPLICATION BOX 10.2
The Complexity of an Aircraft and a Worm

Interdisciplinary research between engineers, mathematicians, and biologists requires dialog and the understanding of each other's specific language. To begin the dialog, common terms used in all the three disciplines must be defined. Let us make an attempt to define the term complexity and ask the stimulating question whether we consider an airplane or a worm to be more complex?

The definition of complexity might relate the degree of complication of a system to the number and intricacy of interfaces. In a similar way that the complexity of a life form is increased by a higher number and sophistication of sensory organs (interfaces with the environment), the number and sophistication of input/output devices raise the complexity of an engineering system. A recently updated model of the 1-mm worm *Caenorhabditis elegans* estimates the external connection cost to consist of 200 sensory and motor neurons wired to 20 sensory organs, 95 body wall muscles, and 1 representative muscle for the vulva and anus, respectively (Chen et al., 2006). In a modern aircraft, the number of interfaces would need to include the number of gauges, meters, radios, control surfaces, and interactions between people and the aircraft machine, the computer programs, and other complex tools.

A more complete definition of complexity would include the number and intricacy of conditional branches and the degree of nesting. Conditional branches are some of the most powerful tools in any programming language because they allow for decision making. Nesting is important because it allows us to work out problems in multiple dimensions using "do-while" or "if-then" type statements. The wiring diagram of our worm (about 97% complete) lists 6,393 chemical synapses, 890 electrical junctions, and 1,410 neuromuscular junctions. The estimation of the modern aircraft would include all the cables, switches, and other electronic components.

Our estimation thus far has focused mainly on numbers. But our estimation of complexity would also need to include an estimation of the intricacy of interfaces, branch points, and regulation loops. This more sophisticated type of analysis would first need to take into account the number of branch points and number of loops involved in every single process in which DNA produces some protein. It would need to classify and weigh the control circuits. Biological circuits are of a different type compared to man-made ones in that they operate with indeterminate parameters. Computer programs are not very efficient in performing tasks without well-defined range of inputs, so that the most sophisticated robot still cannot understand a joke. To take it a step further, the complexity of the worm genome could be estimated by the amount of information it encodes about the world in which it has evolved.

Although we cannot find a simple answer to our question, how an airplane and a worm compare in terms of complexity, without a much more detailed research project, our intent is to demonstrate advantages that might arise from interdisciplinary communication. Advanced computer models will give insight into the regulation of biological systems as well as improve engineering control systems.

The internal environment within a body is the extracellular fluid, which is regulated by blood and other body fluids. Every cell relies on the automatic control that keeps parameters such as nutrients (glucose, oxygen, and so forth), waste (urea, carbon dioxide, and so forth), building materials (carbon, hydrogen, calcium, and so forth), and temperature within tolerable limits despite changes in the external environment.

A multicellular organism expends much effort to maintain homeostasis. Every cell contributes to this task. As long as normal conditions are maintained in the internal environment, the cells live and function properly. This means that every cell also benefits. The combination of contribution and benefit is the basis for the automaticity of organisms and their robustness. When some cells or cell systems lose their ability to contribute their share, all cells of the organism suffer. Moderate dysfunction leads to sickness, while extreme dysfunction leads to destruction of the system. Engineering design uses the same principles. The robustness of a nuclear power plant depends on the competence of its containment isolation valves (which shut down the plant after a loss-of-coolant accident) to fail in the safest mode when they fail. On the other hand, it benefits the valve if the plant is designed with robustness so that the valve will rarely encounter the stressful conditions of a loss-of-coolant accident.

To avoid destruction, a complex system, be it a cell, a human, or a state, must also respond to modifications of the environment. While biological systems adapt by changing their genotype and phenotype, engineering systems rely on redundancy and insensitivity to noise. As in biological systems, negligence leads to disaster, exemplified by the Kansas City Hyatt Regency Skywalk collapse in 1981, which could have been avoided by a redundant support system that would continue to support the skywalk after some component failed or an alternative load support pathway which could have supported additional loads after the first component failure. The necessity for simultaneous maintenance and modification of a complex system, or the simultaneous presence of robustness and fragility, is an integral feature of the wonder that we call life. Insight into it can shed light on the functioning of biological and advanced engineering systems.

Suggested Reading

Alberts, B., Johnson, A., Lewis, J., Raff, M., Roberts, K. and Walter, P. (2002). Control of Gene Expression. In: *Molecular Biology of the Cell*, (B. Alberts, A. Johnson, J. Lewis, M. Raff, K. Roberts, P. Walter, eds.). Garland Science, New York, NY, pp. 375–465.

Blaustein, M.P., Kao, J. P.Y. and Matteson, D.R. (2004). *Cellular Physiology*. Elsevier, Philadelphia, Pennsylvania.

Kierszenbaum, A.L. (2002). *Histology and Cell Biology: An Introduction of Pathology*. Mosby, Missouri.

References

Ast, G. (2005). The alternative genome. *Sci. Am.* 292(4):58–65.

Butterfield, N.J. (2000). *Bangiomorpha pubescens* n. gen., n. sp.: Implications for the evolution of sex, multicellularity and the Mesoproterozoic/Neoproterozoic radiation of eukaryotes. *Paleobiology.* 26(3): 386–404.

Chen, B.L., Hall, D.H. and Chklovskii, D.B. (2006). Wiring optimization can relate neuronal structure and function. *Proc. Natl. Acad. Sci. USA.* 103: 4723–4728.

Csete, M.E. and Doyle, J.C. (2002). Reverse engineering of biological complexity. *Science.* 295: 1664–1669.

Langer, R. and Vacanti, J.P. (1993). Tissue engineering. *Science.* 260: 920–926.

L'Heureux, N., Dusserre, N., Koenig, G., Victor, B., Keire, P., Wight, T.N., Chronos, N.A.F., Kyles, A.E., Gregory, C.R., Hoyt, G., Robbins, R.C. and McAllister, T.N. (2006). Human tissue-engineered blood vessels for adult arterial revascularization. *Nat. Med.* 12(3): 361–365.

Petroski, H. (1992). *To Engineer Is Human: The Role of Failure in Successful Design*. Vintage Books, New York.

Yow, K.H., Ingram, J., Korossis, S.A., Ingham, E. and Homer-Vanniasinkam, S. (2006). Tissue engineering of vascular conduits. *Br. J. Surg.* 93(6): 652–661.

Glossary

Chapter 1. Biomolecules

α-helix: Common secondary protein structure composed of spirally arranged amino acids stabilized by hydrogen bonds.

Anhydride bond: Bond created by subtracting the molecules of water.

Amphiphilic: Molecules that have both hydrophilic and hydrophobic regions; also called amphipathic.

Buffer: Substance in solution that maintains pH within narrow limits when acids or bases are added by absorbing or releasing hydrogen ions.

Chiral molecule: Molecule that is the mirror image of another molecule.

Covalent bond: Biologically important bond between atoms that share one or more electrons and which has a higher bond energy compared to a hydrogen bond.

ΔG: Change in Gibbs free energy, which is defined as the energy available to do work.

ΔG°: Standard Gibbs free energy change at 1 atm pressure and 1 M concentration.

ΔG°′: Standard Gibbs free energy change at 1 atm pressure, 1 M concentration, and pH 7.

Dipole: Molecule with positive and negative charges on opposite sides.

Disulfide bond: Covalent bond between SH groups of two cysteine residues, which stabilizes folded structures.

Enantiomer: Molecule mirroring another molecule and having identical chemical and physical properties except for the ability to rotate plane-polarized light in the opposite direction.

Endergonic: Characteristic of a system that absorbs energy from the surroundings.

Enthalpy (H): In biology, measure of the heat content of molecules, primarily derived from molecular collisions and from forming and breaking of bonds.

Entropy (S): Measure of randomness or disorder in a system.

Exergonic: Characteristic of a system that releases energy to the surroundings.

Gibbs free energy (G): Thermodynamic quantity of the amount of work that can be extracted from some process. See also ΔG.

Glycosidic bond: Special covalent bond that links the carbon of one monosaccharide to the hydroxyl group of a second molecule such as another saccharide.

Homeostasis: The ability of a functional entity to automatically maintain a balanced internal environment.

Hydrogen bond: Biologically very important weak bond between a hydrogen atom and another molecule.

Internal energy (E): In biology, measure of the covalent forces that stabilize atoms in the molecule, which is equal to enthalpy (H) at constant temperature.

Liposome: Synthetic microscopic sphere of amphiphilic molecules in water, typically phospholipids plus cholesterol, used to transport biological molecules to target cells.

Micelle: Spherical aggregates of amphiphilic molecules in water, with the hydrophobic, nonpolar portions in the interior and the hydrophilic, polar portions exposed to the polar aqueous environment.

Nucleic acid: Biologically significant heterocyclic organic compound derived from purine or pyridine and present in all living cells and viruses as deoxyribonucleic acid (DNA) and ribonucleic acid (RNA).

Nucleoside: Nucleic base (e.g., adenine) coupled to ribose (e.g., adenosine) or 2-deoxyribose (e.g., deoxyadenosine).

Nucleotide: Basic consitutent of DNA and RNA comprised of a nucleoside (nucleic base plus sugar) and a phosphate group.

Peptide bond: Special covalent bond that links amino acids into polymeric structures such as polypeptides.

pH: Measure of the acidity or alkalinity of a solution based on the concentration of hydrogen ions, using a scale from 0 to 14 (7 being neutral, <7 being acid, and >7 being alkaline).

Phospho anhydride bond: Interphosphate linkage of nucleotides such as ATP, which are formed in a reaction that involves the removal of water and which release high energy when broken (hydrolyzed).

Van der Waals forces: Forces between molecules that arise from the polarization of molecules into dipoles.

Chapter 2. Cell Morphology

Amphipathic: Molecules that have both hydrophilic and hydrophobic regions; also called amphiphilic.

Cell wall: Rigid outmost layer of a plant cell, and certain prokaryotic cells, which is not present in animal cells.

Centriole: One of two cylindrical structures comprised of nine triplet microtubules and involved in spindle formation during mitosis.

Centrosome: Cytoplasmic region near the cell nucleus, which contains centrioles and serves as microtubule-organizing center.

Chloroplast: Plastid containing DNA, chlorophyll, and other pigments, and being present in cells that perform photosynthesis.

Cilium: Fine cellular folding of the cell membrane extruding from the apical surface of a cell or a unicellular organism that is capable of motion.

Cis: Literally, the "near side of something"; used here for the side of the Golgi apparatus facing the ER.

Crista: One of the folds of the inner mitochondrial membrane extending toward the mitochondrial cytoplasm.

Cytoplasm: Gelatinous or watery fluid of the cell that is contained by the cell membrane.

Cytoskeleton: Any of the various filamentous elements within the cytoplasm of eukaryotic cells, which maintain the shape of a cell and are involved in cellular movement.

Differential permeability: Feature of cell membranes to selectively prevent or regulate passage of some molecules (e.g., large and charged molecules) while allowing passage of others (e.g., small and lipid-soluble molecules).

Endocytosis: Process of bringing something into a cell from the outside by engulfing it with the cell membrane.

Endoplasmic reticulum (ER): Organelle, composed of tubules, that modifies proteins, generates macromolecules, and transfers substances through the cell.

Endosome: Cellular compartment within the cell formed after endocytosis.

Euchromatin: Form of chromatin, the condensed mass of genetic material within the nucleus, that is rich in genes and genetically active.

Eukaryote: Unicellular or multicellular organism comprised of cells that contain a distinct membrane-bound nucleus.

Exocytosis: Process of secreting or excreting the contents of vesicle-enclosed substances from the cell by fusion of the vesicular membrane with the cell membrane.

Flagellum: Long thin attachment of sperm cells and some bacteria and protozoa that is used for locomotion.

Fluid mosaic model: Model of cell membrane taking into account the mobility of proteins within the double layers of phospholipids.

Golgi apparatus: Functionally polarized organelle, which is composed of a stack of flattened cisternae with different membranous composition, and which processes and sorts proteins within vesicles.

Hair cell: Sensory cell with fine apical projections resembling hairs, present in the auditory and vestibular part of the inner ear, and transmitting information on sound and movement to the brain.

Heterochromatin: Tightly coiled form of chromatin, the condensed mass of genetic material within the nucleus, that is genetically inactive.

Intermediate filament: Member of one of the three prominent cytoskeletal filaments with a size between that of microfilaments and microtubules.

Lysosome: Membrane-bound organelle that contains enzymes for degrading and recycling of molecules and that forms endosomes when combined with endocytosed vesicles.

Microfilament: Smallest member of the three prominent types of cytoskeletal filaments; composed of actin.

Microtubule: Largest member of the three prominent cytoskeletal filaments; composed of tubulin molecules which form hollow cylinders.

Microvillus: Cell projection containing a core bundle of actin filaments.

Mitochondrion: Organelle with smooth outer membrane and highly folded inner membrane, being the principal energy source of the cell and having other specialized roles, for instance, related to cell immunity and cell death.

Nissl substance: Rough endoplasmic reticulum of nerve cells.

Nucleolus: Round body within the nucleus containing RNA and being involved in protein synthesis.

Nucleus: Membrane-bound organelle containing the cell's hereditary material.

Organelle: Structural and functional entity of a cell, analogous to an organ being the discrete entity of an organism.

Peroxisome: Organelle containing enzymes for oxidative reactions and forming endosomes when combined with endocytosed vesicles.

Phagocytosis: Cellular process of engulfing particulate material.

Pinocytosis: Cellular process of engulfing fluid droplets.

Plasma membrane: Semipermeable membrane around cells separating the external and internal environments and regulating the exchange between them.

Prokaryote: Unicellular organism that has no membrane-bound nucleus, membrane-bound organelles, and chromosomes.

Ribosome: Small organelle composed of RNA and protein responsible for translation, the assembly of polypeptides encoded by mRNA.

Rough ER (RER): Cell element composed of endoplasmic reticulum and ribosomes, which manufactures, modifies, and transports proteins.

Sarco-: Word stem relating to striated skeletal muscle.

Semifluid: Characteristic of a substance with flow properties between those of solids and liquids such as the cellular cytoplasm.

Smooth ER (SER): Endoplasmic reticulum that is not associated with ribosomes, and that takes part in synthesis of various biomolecules, in metabolism of carbohydrates, in detoxification of cells, in calcium storage, and in transport of nutrients.

Stereocilium: Specialized microvillus resembling a cilium that projects from the surface of hair cells.

Trans: Literally "across something"; used here for the side of the Golgi apparatus directed toward the plasma membrane.

Chapter 3.   Enzyme Kinetics

Active site (enzyme): Part of the enzyme where the reaction chemistry takes place.

Allosteric: Enzymes that change their response to substrate binding due to regulators that bind at sites other than the active site.

Bi Bi system: Type of multisubstrate enzyme system characterized by two substrates and two products.

Briggs-Haldane relationship: Model of relationship of substrate-velocity data that does not require the chemical step to be rate-limiting.

Catalyst: Substance that changes the rate of a chemical reaction without being permanently altered.

Coenzyme: Small molecule supporting an enzymatic reaction.

Competitive inhibition: Reaction where the inhibitor (e.g., a poison) binds to the same site on the enzyme as the substrate.

Cooperative enzyme: Enzyme with multiple active sites that cooperate with one another to influence the rates of reaction at each active site.

Cooperativity: Interaction by which the binding of a substrate to one substrate site influences the binding at a second site.

Dissociation constant: Constant for the dissociation of a complex into its components; describes how tightly a ligand (e.g., substrate) binds to a protein (e.g., enzyme).

Eadie-Hofstee/ Hanes-Woolf plot: Linear graphical representation (velocity vs. velocity/substrate concentration) of the Michaelis-Menton equation to determine the Michaelis constant (K_M) and the maximal initial velocity of an enzyme reaction (V_{max}).

End-product inhibition: Process by which a product of a metabolic pathway inhibits an enzyme in that same pathway that led to the product formation.

Enzyme (E): Biological macromolecule (generally a protein) that catalyzes a chemical reaction.

Enzyme inhibitor: Substance that reduces velocity of an enzyme reaction.

Equilibrium constant: Factor relating the concentration of reactants and products in a chemical reaction at equilibrium.

ES complex: The complex formed when the substrate S is bound to the enzyme E.

Feedback inhibition: See "End-product inhibition."

First-order reaction: Reaction in which the rate of disappearance of a chemical substance is directly proportional to the concentration of that substance.

First-order velocity expression: Reaction in which the rate is proportional to the concentration of one reactant.

Hill coefficient: Measure of cooperativity in a binding process.

Hill plot: Graph to determine the number of cooperative binding sites on an enzyme (e.g., the Hill coefficient).

Initial rate condition: Initial condition of enzymatic activity much before equilibrium conditions are approached.

Isoenzyme: Enzyme that is chemically different from another but catalyzes the same reaction.

k_{-1}: Breakdown rate of the enzyme/substrate complex (ES) to enzyme (E) and substrate (S).

k_1: Rate constant for the formation of the enzyme/substrate complex (ES) from enzyme (E) and substrate (S).

k_{cat}: Rate constant for the formation of product (P) and enzyme (E) from the enzyme/substrate complex (ES).

Lineweaver and Burk plot: Double-reciprocal plot (reciprocal of velocity vs. reciprocal of substrate concentration) of the Michaelis-Menton equation to linearize substrate-velocity data for determination of the Michaelis constant (K_M) and the maximal initial velocity of an enzyme reaction (V_{max}).

Michaelis constant (K_M): Concentration of substrate at which the rate of the enzyme reaction proceeds at half the maximal velocity.

Michaelis-Menton equation: Relationship to describe the rate of an enzyme in a single-substrate, noncooperative enzyme reaction, at which the concentration of substrate (S) is higher than the concentration of enzyme (E) and the concentration of the enzyme/substrate complex (ES) is constant.

Noncompetitive inhibition: Reaction where a reversible inhibitor binds to a site different from the substrate binding site and decreases the rate of reaction.

Product (P): Chemical resulting from the enzymatic conversion of a substrate (S).

Protein kinase: Enzyme catalyzing the phosphorylation of proteins.

Protein phosphatase: Enzyme catalyzing the dephosphorylation of proteins.

Specificity constant: Constant expressed in terms of k_{cat}/Michaelis constant (K_M), which describes the efficiency of an enzyme by incorporating the rate constants for all steps of the reaction.

Steady-state condition: Condition in which some specified characteristic remains steady over long periods.

Substrate (S): Chemical to be converted by enzyme (E) into product (P).

Ternary complex: The complex formed by the simultaneous binding of two substrates to an enzyme.

Turnover number: Maximum number of product molecules produced per enzyme per unit time.

Uncompetitive inhibition: Enzyme inhibition that is based on an inhibitor that binds only to the enzyme/substrate complex (ES) and decreases the rate of the reaction.

V_{max}: Maximum initial velocity of an enzyme reaction following Michaelis-Menton kinetics.

Zero-order reaction: Reaction in which the rate of disappearance of a chemical substance is independent of its concentration.

Chapter 4.   Cellular Signal Transduction

Adaptor protein: Accessory protein that mediates specific protein-protein interactions that drive the formation of protein complexes in a signal transduction pathway.

Adenylyl cyclase (AC): Key transmembrane enzyme in the AC-cAMP signaling pathway, which is activated by G-protein and converts adenosine triphosphate (ATP) to cyclic adenosine monophosphate (cAMP).

Affinity: Avidity with which a chemical entity binds a ligand.

Agonist: Molecule that selectively binds to a specific receptor and triggers a response in the cell.

Antagonist: Molecule that binds to a specific receptor but does not activate the receptor and blocks it from activation by agonists.

Apoptosis: Programmed cell death where cells disintegrate into membrane-bound particles that are eliminated by phagocytosis.

Apoptosome: Large multiprotein structure formed in the process of apoptosis in response to release of cytochrome c from the mitochondria.

Autocrine: Signaling process in which a cell responds to its own chemical messenger.

Calcium ATPase: Enzyme that transports calcium from the cytoplasm to the extracellular space (plasma membrane form of enzyme) or into the endoplasmic reticulum (ER form of enzyme) with the expenditure of ATP.

Calmodulin: A calcium-binding protein that is a key component of the calcium second-messenger system.

Caspase: Member of enzyme family with a crucial cysteine residue that cleaves other proteins after an aspartic acid residue.

Coactivator: Molecule that increases gene expression by associating directly or indirectly with a DNA-binding activator (e.g., transcription factor).

Corepressor: Molecule that decreases gene expression by associating directly or indirectly with a DNA-binding repressor.

Cyclic adenosine monophosphate (cAMP, cyclic AMP): Molecule derived from ATP that serves as a second messenger in signal transduction.

Death receptor: Cell surface receptor such as the Fas receptor that belongs to the tumor necrosis factor superfamily and triggers apoptosis upon ligand binding.

Diacylglycerol: A second messenger molecule consisting of glycerol bound to two fatty acids which is generated by phospholipase C acting on a phospholipid.

Dissociation constant (K_d): See Chap. 3.

Endocrine: A signaling process in which a cell produces a chemical that regulates the activities of distant cells.

G-protein: Guanine-nucleotide binding protein. Member of protein family involved in second messenger cascades that are activated by exchange of guanosine diphosphate (GDP) for guanosine triphosphate (GTP).

General transcription factor (GTF): Important proteins involved in the transcription of class II (protein coding) genes to mRNA.

GTPase: Member of enzyme family that can bind and hydrolyze guanosine triphosphate (GTP) to guanosine diphosphate (GDP), an important "switch mechanism" in signal transduction.

Histone: See Chap. 7.

Homeostasis: See Chap. 1.

Inositol triphosphate (IP$_3$): Important role as a second messenger in a cell based on the cyclic polyalcohol inositol.

Juxtacrine: Signaling process in which a cell-bound or matrix-bound chemical messenger acts on a nearby cell.

Ligand: Molecule that binds to another chemical entity to form a larger complex.

Nucleosome: See Chap. 7.

Paracrine: Signaling process in which a cell produces a chemical that regulates nearby cells.

Partial agonist: Molecule that binds to specific receptors but produces only some of the actions of a full agonist.

Phospholipase C (PLC): Enzyme that hydrolyzes phospholipids and leads to increases of cytosolic calcium in the PLC signaling pathway.

Phosphorylation: The addition of a phosphate (PO$_4$) group to a protein or other molecule.

Promoter: See Chap. 7.

Protein kinase: Enzyme that transfers a phosphate group (PO$_4$) from ATP to specific target molecules.

Protein kinase A (PKA): Enzyme that catalyzes the activity of intracellular proteins in the Ac-cAMP signaling pathway when activated by cyclic adenosine monophosphate (cAMP).

Protein kinase C (PKC): Enzyme family that catalyzes phosphorylation of tyrosine residues in certain proteins and plays an important role in the PLC signaling pathway when activated by diacylglycerol.

Protein tyrosine phosphatase (PTP): Enzyme that removes a phosphate group (PO$_4$) from phosphorylated tyrosine residues on proteins.

Receptor: See Chap. 6.

Response element: DNA sequence recognized by a specific molecule (e.g., transcription factor) that allows regulation of gene expression.

RNA polymerase: Enzyme responsible for making RNA from a DNA template.

Selective estrogen receptor modulators (SERMs): Molecule that exerts only a subset of the actions of full estrogens.

Sensitivity (cell): Ability of a cell to respond to a stimulus.

Signal transduction: Process by which a cell converts one kind of signal or stimulus into another.

Signaling pathway: Cellular circuits composed of proteins and other molecules that sense and transmit biological input signals to generate cellular responses.

Transactivation function: Stimulation of transcription mediated by a transcription factor which binds to DNA and activates proteins (i.e., coactivators).

Tyrosine kinase (TK): Enzyme that phosphorylates tyrosine residues and plays important role in activating receptors that are involved in cell growth.

Chapter 5. Energy Conversion

Absorption spectrum: Shows the fraction of the electromagnetic radiation (e.g., visible light) absorbed by a molecule over a range of frequencies.

Action spectrum: Rate of a physiological activity plotted against the wavelength of light.

Adjuvant: Agent that stimulates the immune system and increases the response to a vaccine.

Aerobic: Requires oxygen.

Allosteric: See Chap. 3.

Anabolism: Metabolic process that builds larger molecules from smaller ones.

Anaerobic: Does not require oxygen.

Apoptosis: See Chap. 4.

ATP synthase: Enzyme through which protons flow in the mitochondrial membrane and trigger phosphorylation of adenosine diphosphate (ADP) to adenosine triphosphate (ATP).

Bifunctional (enzyme): An enzyme that has two separate, sometimes opposing activities.

C3 plant: Plant in which the first stable products after fixation of CO_2 are three carbon molecules.

C4 pathway: See Hatch-Slack pathway.

C4 plant: Plant in which the first stable products after fixation of CO_2 are four carbon molecules.

Calvin cycle: Mechanism to fix carbon in C3 plants, which are more than 95% of all green plants.

CAM plant: Crassulacean acid metabolism plant mainly belonging to succulents (i.e., crassulaceae) in which carbon fixation leads to formation of an acid.

Carbon fixation: Process in which carbon dioxide is converted into organic compounds.

Catabolism: Metabolic process that breaks down complex molecules into simpler ones.

Cellular Respiration: Process in which chemical bonds of energy-rich molecules such as glucose are converted into other forms of energy.

Chemiosmosis: Phosphorylation of adenosine diphosphate (ADP) to adenosine triphosphate (ATP) when hydrogen ions move along an electrochemical gradient across the inner mitochondrial membrane.

Chlorophyll: Green photosynthetic pigment found in most plants, algae, and cyanobacteria.

Coenzyme: See Chap. 3.

Cytochrome: Generally a membrane-bound protein that contains a heme group and carries out electron transport or catalyzes a redox reaction.

Dehydrogenase: Enzyme that oxidizes a substrate by transferring one or more protons and a pair of electrons to an acceptor.

ΔG, $\Delta G°$, and $\Delta G°'$: See Chap. 1.

Electron transport system: Series of redox reactions during the aerobic production of adenosine triphosphate (ATP) which are associated with four protein complexes in the inner mitochondrial membrane and transfer electrons from nicotinamide adenine dinucleotide (NADH) and the reduced form of flavin adenine dinucleotide (FADH$_2$) to oxygen.

Endergonic: See Chap. 1.

Energy transduction: Converting energy from one form to another.

Exergonic: See Chap. 1.

Fermentation: Anaerobic process in which pyruvate is metabolized to lactate, or to ethanol and carbon dioxide.

Free radical: A highly reactive atomic or molecular species with unpaired electrons.

Gluconeogenesis: Generation of new glucose from organic molecules like pyruvate, lactate, glycerol, or amino acids.

Glycolysis: Series of biochemical reactions by which a molecule of glucose is oxidized to two molecules of pyruvate.

Hatch-Slack pathway (C4 pathway): Mechanism to fix carbon in C4 plants such as sugarcane, corn, and sorghum.

Hydrolysis: The process in which a molecule is split by reacting with a molecule of water.

Kinase: See Protein kinase Chap. 4.

Nicotinamide adenine dinucleotide (NAD$^+$): Important cofactor in redox reactions in cells that accepts two electrons (oxidizing agent) and a proton to form its reduced form NADH.

Nicotinamide adenine dinucleotide phosphate ($NADP^+$): Phosphorylated form of NAD^+ that accepts two electrons and a proton to form its reduced form NADPH, an important reducing agent.

Oxidase: Enzyme that catalyzes a reaction involving molecular oxygen (O_2) as the electron acceptor.

Oxidative decarboxylation: Transition process between the end product of glycolysis, pyruvate, and the start product of the citric acid cycle, acetyl CoA.

Oxidative phosphorylation: The terminal process of cellular respiration in which energy from electrons is used indirectly to form ATP from inorganic phosphate and ADP.

Photon: Elementary particle responsible for electromagnetic interactions and light.

Photophosphorylation: Production of ATP using the energy of sunlight.

Photorespiration: Alternate pathway for Rubisco (the main enzyme of the Calvin cycle) in which the enzyme uses oxygen to produce a glycolate and a glycerate.

Photosystem: Large pigment-protein complex that converts light energy into chemical energy.

Proton motive force: The potential energy in a proton (H^+) electrochemical gradient that can be used for the synthesis of ATP by oxidative phosphorylation.

Q cycle: Chain of redox reaction of the mitochondrial energetic metabolism, in which electrons are transferred from coenzyme Q to Complex III.

Reaction center: Complex photosynthetic molecule including chlorophyll which initiates photophosphorylation.

Redox potential (reduction potential): Tendency of a chemical species to acquire electrons and thereby be reduced.

Standard free energy: See $\Delta G^{\circ\prime}$ in Chap. 1.

Standard redox potential: Redox potential measured under standard conditions of 25°C, a 1 M concentration for each ion participating in the reaction, a partial pressure of 1 atm for each gas that is part of the reaction, and metals in their pure state.

Stoma (plural stomata): Opening or pore in a leaf that is used for gas exchange.

Substrate-level phosphorylation: Chemical reaction in which adenosine triphosphate (ATP) is formed by the direct transfer of a phosphate group from a reactive intermediate to adenosine diphosphate (ADP).

Superoxide: The anion O_2^{-1}, which has one unpaired electron and is considered to be a free radical.

Synthase: Enzyme that catalyzes a synthetic process.

Thylakoid: Membrane-bound compartment found within chloroplasts that are arranged in stacks called grana.

Transgenic (plant): Plant whose genetic material has been altered using recombinant DNA technology.

Translocation: The process of transferring a substance to a new location.

Tricarboxylic acid cycle (citric acid cycle, Krebs cycle): Series of chemical reactions that convert carbohydrates, fats, and proteins into carbon dioxide and water, and produce high-energy phosphate compounds which serve as the main source of cellular energy.

Chapter 6.   Cellular Communication

Affinity: See Chap. 4.

Agonist: See Chap. 4.

Antagonist: See Chap. 4.

Axonal growth cone: The mobile tip of a growing axon.

Biogenic amines: Member of a group of amines derived from amino acid decarboxylation and possessing physiological properties.

Chaperone: A protein which facilitates the folding of other unfolded or partially folded proteins.

Desensitization: Decreased response as a result of a constant amount and frequency of stimulation.

Downregulation: The cellular process of decreasing the synthesis of specific proteins following a specific stimulus.

Excitotoxicity: Damage or death of a nerve cell due to overstimulation by excitatory stimuli.

Heterodimer: A compound formed by two different molecules or subunits.

Homodimer: A compound formed by two identical molecules or subunits.

Internalization: The cellular process of incorporating membrane-associated proteins into the cytoplasm.

Miniature end-plate potential: Cell membrane depolarization measuring approximately 1 mV.

Motoneuron: Nerve cell that innervates muscles.

Negative antagonist: Substance binding to the inactive form of a receptor.

Neuromuscular junction: The site of interface between a nerve cell and a muscle cell.

Neurotransmitter: Substance released by a nerve cell that triggers a specific physiological response in other nerve cells or other types of cells.

Neurotrophin: A neuropeptide with the ability to regulate the growth, differentiation, and survival of nerve cells.

Neutral antagonist: Substance binding to the active or inactive form of a receptor and preventing the transformation of the receptor to its active state.

Perikaryon: The cell body of a cell excluding the nucleus; applied particularly to nerve cells.

Promoter: See Chap. 7.

Quantum: Amount of neurotransmitter necessary to elicit a miniature end-plate potential.

Receptor: Molecular complex on or within a cell that mediates a specific physiological response whenever it is bound by a specific substance.

Synapse: The point of interaction between two nerve cells.

Transporter: Specialized proteins which bind to and carry cellular products from one location to another.

Zinc finger region: A zinc-binding site on a protein that regulates transcription by binding to specific regions of a gene.

Chapter 7.   Cellular Genetics

A-form of DNA: Right-handed form of double helical DNA (similar to the common B-form of DNA) present under conditions of reduced hydration.

Alkylating agent: Chemotherapeutic agent reacting with DNA and disrupting DNA structure.

Aminoacyl-tRNA synthetase: Enzyme linking transfer RNAs (tRNAs) to their corresponding amino acid during protein synthesis.

Annealing (DNA): Reformation of double-stranded DNA after its denaturation.

Anticodon: Sequence of three nucleotides at the end of a transfer RNA (tRNA) that base-pairs with a codon messenger RNA (mRNA) during protein translation.

A-site (ribosome): Ribosomal binding site for a transfer RNA (tRNA) bound to an amino acid.

B-form of DNA: Right-handed form of double helical DNA (most common one of the three biologically active DNAs) present under physiological conditions.

Chromatin: Genetic material composed of DNA and protein that is located in the cell's nucleus and condenses to chromosomes during cell division.

Chromosome: Characteristically shaped DNA strand and associated proteins carrying genetic information.

Codon: Set of three adjacent nucleotides in DNA or RNA that code for an amino acid or signal the end of a polypeptide (stop codon).

Denaturation (DNA): Separation of DNA strands.

Deoxynucleoside triphosphate (dNTP): See Chap. 1.

Deoxyribonucleic acid (DNA): Nucleic acid present in the cell nucleus in three forms (A, B, and Z) and carrying genetic information.

Diploid: Having two of the basic sets of chromosomes in the nucleus.

Double-stranded RNA (dsRNA): RNA of some viruses with two complementary strands.

Elongation factor (EF): Nonribosomally associated protein required for addition of amino acids during translation.

Enhancer: Regulatory sequence on the same DNA/RNA strand that can elevate transcription from an adjacent promoter.

Exon: Gene-coding regions that remains in the RNA following posttranscriptional processing.

Gene: Segment of nonrepetitive DNA that contributes to cellular functioning.

Genome: Total complement of DNA of an organism.

Genomic imprinting: Refers to dependence of genes on whether or not they are inherited maternally or paternally.

Genomics: Comparative analysis of all genes that are transcriptionally active between any two types of cells or tissues under any given set of conditions.

Helicase: Enzyme unwinding duplex DNA during replication.

Histone: Major class of DNA-binding proteins involved in maintaining the compacted structure of chromatin.

Hybridization (DNA): Pairing of complementary DNA strands to form a double-stranded molecule.

Intitiation factor (IF): Nonribosomally associated protein required for initiation of translation.

Intron: Nucleic acid sequence in RNA that is removed during posttranscriptional processing.

Ligase (DNA): Enzyme closing the nicks in DNA strands.

Mass spectrometry: Important analytical technique in proteomics to determine the mass of intact proteins.

Melting temperature (Tm): Temperature midpoint during thermal denaturation at which 50% of the DNA exists as single strands.

Messenger RNA (mRNA): Genetic coding template used by the translational machinery to determine the order of amino acids incorporated into an elongating polypeptide.

Nonrepetitive DNA: Section of DNA with protein-coding exons since most genes are only present in one copy.

Nuclear matrix: Nonhistone protein scaffold that stabilizes the chromatin structure.

Nucleic acid: See Chap. 1.

Nucleoside: See Chap. 1.

Nucleosome: Subunit of chromosomes composed of DNA coiled around proteins.

Nucleotide: See Chap. 1.

Okazaki fragment: The lagging (or discontinuous) strand of newly synthesized 100 to 200 DNA bases with attached RNA primer.

Polymerase chain reaction (PCR): *In vitro* technique involving multiple rounds of DNA synthesis with a bacterial polymerase enzyme that is resistant to heat denaturation.

Polymerase: Member of enzyme family involved in replication and repair by linking deoxynucleoside triphosphates (dNTPs) into a DNA or RNA strand.

Preinitiation complex: Complex of enzymes involved in initiation of translation.

Primase: Enzyme synthesizing a short stretch of RNA as a primer for DNA polymerase.

Processivity: Ability of an enzyme to remain associated with the DNA template strand while being catalytically active.

Promoter: DNA sequence that promotes the ability of RNA polymerases to recognize the nucleotide at which initiation of transcription begins.

Proteomics: Global analysis of all proteins present in a given cell or tissue at a given time or condition.

P-site (ribosome): Ribosomal binding site for a transfer RNA (tRNA) bound to the peptide being synthesized.

Purine: Aromatic compound that forms the nucleobases adenine (A) and guanine (G).

Pyrimidine: Aromatic compound that forms the nucleobases cytosine (C), thymine (T), and uracil (U).

Releasing factor (RF): Nonribosomally associated protein required for termination of translation.

Repetitive DNA: DNA made of copies with identical or similar nucleotide sequence.

Replication fork: The site of the unwound DNA template strands.

Replication: Process of doubling the DNA content of cells prior to normal cell division.

Replisome: DNA replication structure at the replication fork including DNA polymerase dimers and other proteins.

Ribosomal RNA (rRNA): Nucleotide strand that forms ribosomes together with ribosomal proteins.

Semiconservative (DNA): Characteristic of DNA indicating that newly synthesized strands remain associated with their respective parental strands.

Small nuclear ribonucleoprotein particles (snRNPs): RNA-protein complexes that regulate nuclear mRNA splicing.

Spliceosome: Complex of RNA and small nuclear ribonucleoprotein particles (snRNPs) that removes introns from unprocessed mRNA.

Splicing: Modification of genetic information after transcription, in which introns are removed and exons are joined.

Topoisomerase: Enzyme relieving torsional stresses in duplexes of DNA during replication.

Transcription: Mechanism by which a template strand of DNA is utilized by specific RNA polymerases to generate one of the three different classifications of RNA.

Transfer RNA (tRNA): Small nucleotide strand recognizing the encoded sequences of mRNAs to allow correct insertion of amino acids into the elongating polypeptide chain.

Translation: Synthesis of a polypeptide with an amino acid sequence in accordance with the codon sequence of a messenger RNA (mRNA).

Watson-Crick base-pairing: Rule that in complementary strands of DNA, adenine (A) will only base-pair with thymine (T), and cytosine (C) with guanine (G). In RNA, T is replaced by uracil (U).

Z-DNA: One of the three biologically active double helical DNA structures in which the double helix turns in the left-handed direction producing a zigzag appearance.

Chapter 8.   Cell Division and Growth

Adult stem cell: Undifferentiated cell from any postnatal tissue source that can renew itself and differentiate into a number of cell types.

Anaphase promoting complex: Protein complex that is activated once all chromosome kinetochores have attached to spindle microtubules to initiate anaphase.

Anaphase: Third phase of mitosis (or meiosis) in which the two sister chromatids of each chromosome are separated and move toward opposite poles of the spindle.

Benign tumor: Cancer cells that uncontrollably proliferate without showing invasiveness and migration.

Centromere: Chromosome region that joins two sister chromatids and to which spindle fiber attaches during mitosis.

Centrosome: See Chap. 2.

Checkpoint (cell cycle): Molecular surveillance system that assesses the cell's advancement through the cell cycle for potential mistakes.

Chromatid: One of two identical condensed chromatin strands that are linked together at the centromere and into which a chromosome splits during mitosis.

Cyclin: Critical protein involved in the regulation of the cell cycle via activation and targeting of cyclin-dependent kinases.

Cyclin-dependent kinase (cdk): Enzyme that, when activated by a cyclin, regulates cell cycle processes via phosphorylation of substrates.

Cytokinesis: Process at the end of mitosis that partitions the cellular contents into the two daughter cells.

Dysplasia: Process of a cell developing abnormally in size, shape, or organization.

Embryonic stem cell: Undifferentiated cell taken from the inner cell mass of 5 to 7-day-old embryos and that can renew itself and differentiate into many different cell types.

Fluorescence-activated cell sorter (FACS): Instrument that sorts individual cells in a mixture of cells according to their abilities to scatter and/or emit light (fluorescence).

G0 phase: Resting phase of the cell cycle.

G1 phase: Cell cycle phase during interphase when cells respond to various signals for growth or differentiation and prepare for DNA synthesis.

G2 phase: Cell cycle phase during interphase when cells prepare for chromosome segregation and cell division.

Hematopoietic stem cell: Cell able to produce all types of blood cells.

Hyperplasia: Increase in cell number resulting in an increase in tissue or organ mass.

Interphase: Cell cycle phase between cell division that is subdivided into G1 phase, S phase, and G2 phase.

Kinetochore: Protein complex located at the centromeres of chromosomes to capture spindle fiber microtubules and help transport chromosomes.

M phase: Mitotic cell cycle phase when cells separate their chromosomes, ending with cell division into two new daughter cells.

Malignant tumor: Cancer cells that show invasiveness and metastasis.

Metaphase: Second phase of mitosis (or meiosis) in which the chromosomes move toward the equator of the cell and align themselves in the equatorial plane.

Metaplasia: Process of a differentiated cell type changing into another.

Metastasis: Process by which cancer cells spread to a place other than their place of origin.

Mitosis: Process of division of a somatic cell into two cells and that is subdivided into prophase, metaphase, anaphase, and telophase.

Multipotent: Having the ability to form multiple cell types.

Nondisjunction: Misallocation of chromosomes to daughter cells.

Oncogene: Gene that transforms a normal cell into a cancerous cell.

Oncogenesis: Development of a tumor.

Pluripotent: Having the ability to form most or all postnatal tissues.

Prophase: Early phase during mitosis (or meiosis) in which the chromosomes condense and become visible as paired filaments and the spindle apparatus forms at opposite poles of the cell.

Proto-oncogene: Normal gene that can become an oncogene which transforms a normal cell into a cancerous cell.

Quiescence (cell cycle): Period in which a cell is not growing or dividing.

Restriction point: Time in the late G1 cell cycle phase which is necessary to pass for normal completion of the cell cycle.

S phase: Cell cycle phase during interphase when DNA is replicated.

Senescence: Process of aging.

Somatic cell nuclear transfer (SCNT): Scientific term for cloning in which an egg without a nucleus is combined with the nucleus of a somatic cell to create an embryo.

Spindle fiber: Microtubules that grow out of centrosomes and provide the mechanism for chromosomal movement.

Telomerase: Enzyme involved in the synthesis of telomeres.

Telomere: Repetitive DNA sequence at the ends of chromosomes that are required for replication and stability and shorten with each replication in somatic cells.

Telophase: Final phase of mitosis (or meiosis) in which the nuclear membranes reform, chromosomes decondense, spindle fibers disappear, and the cell divides.

Totipotent: Having the ability to form all postnatal tissues as well as extraembryonic tissues.

Transdifferentiation: Converting non-stem cells and differentiated stem cells into another type of cell.

Ubiquitin: Common, highly conserved, molecule that binds to proteins, thereby labeling them for proteolytic degradation.

Unipotent: Having the ability to form only one differentiated cell type.

Chapter 9.   Cellular Development

Acrosome: Structure in the sperm head filled with enzymes that digest complex carbohydrates and proteins on the outer surface of the egg.

Adult stem cell: See Chap. 8.

Animal pole: Top half of the egg/embryo; the darker pigmented half of an amphibian egg/embryo.

Apical ectodermal ridge: Embryonic structure formed by special ectodermal cells that initiate limb development.

Apoptosis: See Chap. 4.

Autopod: Distal limb segment; corresponds to the bones of the hand or foot.

Blastocoel: The fluid-filled cavity inside a blastula-stage embryo.

Blastopore: Hole in the side of the early embryo (blastula stage) filled with yolky cells originating from the vegetal half of the embryo.

Blastula: Early embryonic stage following fertilization and prior to gastrula.

Capacitation: The process by which sperm are altered (usually during their passage through the female reproductive tract) that gives them the capacity to penetrate and fertilize the ovum.

Caspases: See Chap. 4.

Death receptor: See Chap. 4.

Diploid: See Chap. 7.

Dorsal blastopore lip: Area of cells that are migrating inward toward the animal pole during early gastrulation and will become the dorsal mesoderm.

Dorsal: Near the back or upper side of a structure.

Ectoderm: The outermost of the three primary germ layers of the embryo, from which the skin, nerve tissue, and sensory organs develop.

Embryonic stem cell: See Chap. 8.

Endoderm: The internal layer of cells of the gastrula, which will develop into the alimentary canal (gut) and digestive glands of the adult.

Epiboly: Process during gastrulation whereby animal pole cells eventually cover the embryo and converge at the blastopore.

Fertilization: Process that refers to the union of the sperm and the egg.

Gastrulation: The process of movements and infoldings of embryonic cells destined to become endoderm in early animal embryos, immediately following blastula (or blastoderm) stage, generating the blastopore.

Germ layer: The main divisions of tissue types in multicellular organisms.

Haploid: A single set of chromosomes (half the full set of genetic material).

Hematopoietic stem cell: See Chap. 8.

Hensen node: Embryonic cells of birds, reptiles, and mammals that are functionally equivalent to Spemann organizer cells of amphibian.

Limb field: Group of cells within the lateral plate mesoderm that initiate the development of the tetrapod limb.

Marginal zone: Area of frog embryos where gastrulation is initiated.

Mesenchyme: Connective tissue including bone, fat, cartilage, and blood vessels that are derived from embryonic mesoderm.

Mesoderm: The middle layer of cells in an embryo, from which the muscular, skeletal, vascular, and connective tissues develop.

Morphogen: Factors of embryonic tissue that stimulate the development of form in an organism such as the establishment of germ layers in the egg.

Morphogenesis: Refers to the process(es) by which specific structures arise in specific locations during early embryonic development.

Multipotent: See Chap. 8.

Neural tube: A cylindrical structure that runs through the midline of the embryo and expands in the head to form the brain and in the trunk to form the spinal cord.

Neurulation: Embryonic process during which the structures of the nervous system begin to develop.

Notochord: Flexible rod-like tissue down the back of the vertebrate embryo around which the backbone will be deposited.

Oocyte: Term for the egg during oogenesis.

Oogenesis: The formation and growth of the egg or ovum in an animal ovary.

Organogenesis: The process of formation of specific organs in a plant or animal involving morphogenesis and differentiation.

Patterning: The process by which the various structures of the body are defined.

Pluripotent: See Chap. 8.

Polyspermy: Penetration of the egg by multiple sperms.

Primitive streak: Major structural feature of gastrulation in birds, reptiles, and mammals.

Progress zone: Embryonic structure formed by mesenchymal cells.

Spemann organizer: Region of cells of the early gastrula stage embryo that secrete growth factors, which in turn direct a neural fate on cells migrating through the blastopore.

Stylopod: Limb segment closest to the hip or shoulder; corresponds to humerus of arm and femur of leg.

Totipotent: See Chap. 8.

Vegetal pole: The yolky, lightly pigmented half of an amphibian embryo opposite the animal pole.

Ventral: Near the front or bottom side of a structure.

Zeugopod: Radius.

Zeugopodium: Lower part of the limb corresponding to the radius and ulna of the forelimb and the tibia and fibula of the hindlimb.

Zona pellucida: Vitelline envelope (related to egg yellow) of mammalian egg which binds the sperm.

Zone of polarizing activity: Small block of mesodermal tissue near the posterior junction of the limb bud near the body wall; regulates the anterior-posterior polarity of the developing limb.

Zygote: Early fertilized egg.

Chapter 10.   From Cells to Organisms

Active transport: Movement of molecules against their concentration gradient or electrical potential using a carrier protein and energy derived from ATP.

Adaptive radiation: Development of a variety of species from a single ancestor.

Alternative splicing: Modifications of eukaryotic pre-mRNA, resulting in one gene producing several different mRNAs and proteins.

Anchoring junction: Type of connection between adjacent animal cells that physically joins them together.

Archaea: Group of primitive organisms that morphologically resemble bacteria but comprise a unique domain of life based on genetic differences.

Asexual reproduction: Process in which an organism produces a copy of itself without the involvement of another individual.

Basement membrane: Extracellular thin membrane underlying epithelial cells.

Binary fission: Method by which bacteria reproduce asexually.

Cell Determination: Commitment of a cell to a cell lineage.

Cell Differentiation: Process by which a nonspecialized cell commits to becoming a specialized cell.

Cell lineage: Description of the history of cell divisions that occur to generate a specific cell type.

Cell Polarity: Asymmetric form or distribution of molecules of a cell.

Connective tissue: One of four types of animal tissues, providing the framework for all other tissues.

Cotransporter: Integral membrane protein that regulates transport of two different types of substances in the same direction across cell membranes.

Countertransporter: Integral membrane protein that regulates transport of two different types of substances in opposite direction across cell membranes.

Cytoplasmic determinant: Substance in the cytoplasm of the egg that regulates gene expression of the embryo.

Dermal tissue: One of three types of plant tissues, forming the outer surface of the leaves and the young plant body.

Differential gene expression: Selective translation of the genomic information into functional products.

Diffusion: Spontaneous movement of molecules from a high concentration to a low concentration.

Endocytosis: See Chap. 2.

Epigenetics: Study of the processes that alter gene function without altering DNA sequences.

Epithelium (plural epithelia): One of four types of animal tissues, lining all body surfaces, cavities, and tubes.

Eukaryote: See Chap. 2.

Exocytosis: See Chap. 2.

Extracellular matrix: Any part of a tissue produced and secreted by cells into the intercellular space.

Facilitated diffusion: Movement of molecules down their concentration gradient using a carrier protein though no energy.

Gap junction: Type of connection between adjacent animal cells that allows communication.

Gene regulatory network: Complex arrangement of control elements that govern the rate and timing of gene transcription.

Ground tissue: One of three types of plant tissues, being present in leaves, roots, and stems.

Homeostasis: See Chap. 1.

Horizontal gene transfer: Transmission and incorporation of DNA from one species to another.

Lipid raft: Cell membrane microdomain, enriched in cholesterol and certain lipids.

Meiosis: A two-step cell division process in sexually reproducing organisms that results in the production of gametes (egg and sperm) with a haploid set of genetic material.

Membrane potential: Potential difference across a cell membrane, with the nearby area inside the cell being negative relative to the fluid outside.

Morphogenesis: See Chap. 9.

Muscle tissue: One of four types of animal tissues, being contractile through elongated cells containing actin and myosin proteins.

Negative feedback: Response in such a way as to reverse the direction of change.

Nervous tissue: One of four types of animal tissues, containing neurons that generate and conduct electrical signals in the body and glial cells that provide support to neurons.

Organ system: Groups of organs that contribute to a particular set of functions.

Organ: Part of organism capable of performing a special function.

Plasmodesma: Connection between adjacent plant cells that allows communication.

Population: Group of individuals that share one or more characteristics.

Positive feedback: Response that increases or amplifies the variable to which it is responding.

Posttranslational modification: Modification of a protein after its synthesis from mRNA.

Primary active transport: Movement of a chemical substance across a membrane against its chemical or electrical gradient while using energy derived from ATP.

Prokaryote: See Chap. 2.

Resting membrane potential: Potential difference across a membrane of an excitable cell at rest.

Secondary active transport: Movement of molecules against their concentration gradient, taking advantage of a previously established concentration or electrical gradient that required energy consumption.

Signature gene: Genes commonly expressed by all cell types.

Sexual reproduction: Process in which two haploid gametes fuse to form a diploid zygote with a different genome than that of either parent.

Syncytium: Tissue composed of cells with cytoplasmic continuity.

Systems biology: Scientific discipline studying the interactions between the components of complex biological systems.

Tight junction: Type of connection that joins adjacent animal cells together and regulates intercellular passage of molecules.

Tissue: Combination of extracellular matrix and functionally united cells.

Vascular tissue: One of three types of plant tissues, providing the conduction structures of plants called xylem and phloem.

Index

1, 3-bis- phosphoglycerates, 110
2-deoxyribose, 20
2-phosphoglycerate, 110
3-phosphoglycerate, 110, 136
3 to 5 exonuclease activity, 190
5 to 3 exonuclease activity, 190
6-carbon intermediate, 136
7-transmembrane receptors, 93

Absorption spectrum, 128
α-carbon, 15
Acetals, 21
Acetaminophen, 72
Acetic acid (Ac), 162
Acetylcholine (ACh), 158, 160–165,
    175
Acetylcholinesterase (AchE), 162
Acetyl coenzyme A (acetyl-CoA), 108,
    115, 116
Acrosomal vesicle, 236
Acrosome, 236
Actin, 50
Actin filaments, 49, 276
Action spectrum, 128, 131*f*
Activators of G-protein signaling (AGS)
    proteins, 95
Active site, of enzymes, 67
Active transport, membrane transport
    and, 272
Activin, 243
Adaptive radiation of species, 259
Adaptor proteins, 100
Adenine, 25, 178
Adenosine, 25
Adenosine diphosphate (ADP), 25, 107
Adenosine monophosphate (AMP), 25
Adenosine triphosphate (ATP), 25, 44,
    79, 106
  amino acid activation and, 199
  cell features and, 264

Adenylyl cyclase (AC), 95, 96
Adenylyl cyclase (AC)-cAMP pathway,
    94–97
Adherens junctions, 277
Adjuvants, 130
Adrenergic neurons, 164
Adult stem cells, 225, 227–229, 245
Aerobic respiration, 106, 114–126
Afferent neurons, 280
Affinity, 87, 151
A-form of DNA, 179
Aging, free radicals and, 123
Agonists, 89, 98, 148, 151
Agrin, 175
Albuterol, 98
Alcohol dehydrogenase, 67
Alcoholic fermentation, 111
Alcohol sensitivity, 68
Aldehyde dehydrogenase, 68
Aldehydic polyols, 20
Aldohexoses, 20
Aldopentoses, 20
Aldoses, 20
Alkylating agents, 192
Allosteric inhibitors, 79
Allosteric regulators, 78, 108
Alpha helix, 18
Alternative splicing, 265
Aluminum hydroxide, 130
Amide bonds, 15
Amino acids (AA), 14, 158, 164
  20 common, 15
  activation of, 199
  peptide bonds and, 14
Aminoacyl-adenylate intermediate, 199
Aminoacyl-tRNA synthetases, 199
Amniotic egg, 270
AMPA receptor, 148
Amphipathic character of phospholipids,
    40

Amphiphilic molecules, 33
Amplification, multistep pathways and, 95
Amyotrophic lateral sclerosis,
    excitotoxicity and, 167
Anabolic pathways, 106
Anaerobic respiration, 106, 107–114
    fermentation and, 164
    regulation of, 112–114
Anaphase, 216, 218
Anaphase promoting complex (APC), 219
Anchoring junctions, 277
Aneural pretzels, 174
Anhydride bonds, 25
Animal pole, 238
Animal proteins, 14
Animal starch, 23
Animal tissue, 280
Annealing, 181
Antagonists, 89, 98
    ionotropic receptors and, 148
    negative/neutral, 151
Antenna pigments, 129, 131f
Anterior-posterior (AP) axis, 239,
    241–243, 246, 249
Anthracyclines, 192
Antiapoptotic proteins, 103
Anticodon, 200
Antimetabolites, 193
Antioxidants, 123
Apical ectodermal ridge (AER), 248
Apoptosis, 87, 250–252
    lysosomes and, 44
    signaling in, 101–103
Apoptosis-activating-factor-1 (Apaf-1), 103
Apoptosomes, 103
Apoptotic bodies, 101
Application boxes
    C_4 pathways in C_3 plants, 141
    cell transplantation, 251
    diabetes, cellular communication
        and, 169
    DNA replication, anticancer drugs
        and, 192
    estrogen, 92
    free radicals, 123
    genomics, 196
    muscle machine, 50
    nondisjunction, 222
    otoacoustic emissions, 257
    phospholipid nanoparticles, drug
        delivery and, 33–35
    polymerase chain reaction, 189
    proteomics, 205

Application boxes (Cont.):
    reduction-oxidation, 120
    regenerative medicine, 244
    repairing damaged tissue with stem
        cells, challenges of, 227
    self-assembly, biology/technology and, 41
    tissue engineering, 279
Archaea, 258
Aromatic amino acid decarboxylase
    (AADC), 164
Artificial labeling of cells, 167
Asexual reproduction, 261
A-site, 203
Aspartate, 164
Aspartate transcarbamoylase, 78
Aspirin, 72
Ast universal common ancestor (LUCA),
    258
Atenolol, 98
ATPases, 272
ATP synthase, 124
ATP synthesis, 122–125, 135
Atypical protein kinase C (aPKC), 101
AUG codon, 200, 203
Autocleavage, 102
Autocrine communication, 169
Autocrine messengers, 86
Autopod, 249
Axis formation, 239–246
Axonal growth cone, 156
Axonal terminals, classical
    neurotransmitters and, 164

Bad, 103
Balcavage, Walter X., 1–35
Bangiomorpha pubescens, 262
β-arrestins, 88
Basement membrane, extracellular
    matrix and, 278
Basilar membrane, 50
Bax, 103
β-carbon, 15
β-catenin, 241
Bcl-2 family of proteins, 103
Benign tumors, 230
B-form of DNA, 179
Bi Bi systems, 67
Bid, 103
Bifunctional enzymes, 112
Binary fission, 261
Bioartificial tissue, 251
Biogenic amines, 158, 164
Biological sciences, 2

Biology, self-assembly and, 41
Biomechanics, muscle machine and, 50
Biomolecules, 1–35
Bionics, muscle machine and, 50
Blastocoel, 238
Blastopore, 238
Blastula stage, 237
Blood transfusions, agglutination
   and, 43
Blood vessels, tissue-engineered, 279
Body temperature, 9
Bone cells, 269
Bone morphogenetic protein (BMP), 241,
   243, 252
Bone repair protocols, adult stem cells
   and, 229
Bound ribosomes, 49
Brain ischemic attack, adult stem cells
   and, 228
Briggs-Haldane relationship, 62
Brown adipose cells, 44
Buffer, 13
Bundle sheath, 139
Burns, 252

C_3 plants, 136–138
   C_4 pathways in, 141
   photorespiration and, 139
C_4 pathways, 140, 141
C_4 plants, 139–142
Caenorhabditis elegans, 267, 283
Cajal, Ramon Y., 157
Calcium ATPases, 99
Calmodulin, 98
Calories per mole (cal/mol), 7
Calvin cycle, 136–138
CAM plants, 139, 142
cAMP response element (CRE), 96
cAMP response-element binding protein
   (CREB), 96
Camptothecins, 192
Cancer, 230–232
   anticancer drugs, DNA replication, 192
   peroxisomes and, 46
   stem cells and, 231
Cannabinoid (CB) receptors, 98
Capacitation, 236
Carbohydrate polymers, 20–24
Carbohydrate synthesis, 127, 136–142
Carbon dioxide, photorespiration and, 139
Carbon fixation, 136
Carbonic acid buffer, 13
Carbon nanotubes, 41

Carboxylation/decarboxylation pathways,
   140
Carboxylic acid, 15
Caronte (Car) gene, 243
Cartilage repair protocols, adult stem
   cells and, 229
Caspase 3, 103
Caspase 8, 102
Caspases, 102, 250
Catabolic pathways, 106
Catalysts, enzymes as, 58
Catecholamines, 150, 164, 167
Catechol-O-methyl transferase (COMT),
   167
Caulerpa, 270
Cell aging, 229
Cell cycle, 210–215
   control of, 212–215, 220–222
   phases of, 210–212
Cell cycle genes, mutation/overexpression
   of, 231
Cell cycle proteins, 216
Cell cycle regulatory proteins, 230
Cell determination, 266–276
Cell differentiation, 266–276
Cell division, 209–232
Cell growth, 209–232
   cancer and, 230–232
   cell cycle and, 210–215
   senescence and, 229
Cell junctions, 276–278
Cell labeling in vivo, 165, 166
Cell lineage, 267
Cell membrane, 38–42, 146
Cell morphology, 37–55, 54
Cell polarity, 274–276
Cells
   DNA composition in, 178–181
   features of, 262–266
   membrane transport and, 270–272
   size/shape of, 268–270
Cell secretion, 168–175
Cell sensitivity, 88
Cell surface proteins, 43
Cell transplantation, 251
Cellular communication, 145–176
   cell secretion and, 168–175
   neurotransmitters and, 86, 158–168
   receptors and, 147–158
Cellular development, 233–254
   apoptosis and, 250–252
   fertilization and, 236
   limb development and, 246–250

Cellular genetics. *See* Genetics of cells
Cellular respiration, 106
  aerobic, 114–126
  anaerobic, 107–114
Cellular signaling, 86. *See also* Signal
  transduction
Cellulose, 22
Cell walls, 54
Centrioles, 53
Centrosomes, 53, 218
CF_0 base, 135
CF_1 heads, 135
Chaperones, 154
Checkpoint proteins, loss of, 231
Checkpoints, cell cycle and, 220–222
Chemical energy, converted from light
  energy, 126, 130
Chemical transmission, 157, 160–162
Chemiosmosis, 122–125
Chemotherapy, alkylating agents
  and, 192
Chiral molecules, 15
Chlorambucil, 192
Chloride, high water concentration and, 11
Chlorophyll, 126, 128–131
Chloroplasts, 54, 127
Cholesterol biosynthesis, enzyme
  inhibitors and, 73
Choline (Ch^+), 32, 161, 162
Choline acetyltransferase (ChAt), 160
Cholinergic receptors, 161
Choline transporter, 162
Chordin, 242
Chromatids, 216, 218
Chromatin structure, 181–184
Chromatin proteins, 91
Chromatin-remodeling complexes, 91
Chromosomal rearrangement, 231
Chromosomes, 53, 181, 183
Chymotrypsin, 81
Cilia, 52
Cis site of Golgi apparatus, 48, 171
Citrates, 116
Classical neurotransmitters, 149, 158,
  164–168
Clinical boxes
  alcohol sensitivity, 68
  cholesterol biosynthesis inhibitors, 73
  digestion, 12
  in vivo labeling of cells, 166
  marijuana, cannabinoid receptors
    and, 98
  organ transplantation, 43

Clinical boxes (*Cont.*):
  partial estrogen agonists, 89
  phytol tail, vaccines and, 130
Closed thermodynamic systems, 3
Coactivator proteins, 90–93
Coactivators, 89
Cochlea, 257
Coding DNA, 182
Codon, 200
Coenzymes, 69, 107
Cohesin, 219
Collagen, 20, 252
Commercial gene chips, 196
Communication junctions, 277
Competitive inhibitors, 74
Complexity, 283
Concentration state, 7
Conditional branches, complexity and, 284
Connective tissue, 280
Cooperative enzymes, 78–80
Cooperativity, 78, 80
COPII-coated vesicles, 171
Copper atoms, 121
CoQ (coenzyme Q), 121
Corepressor proteins, 90, 93
Corticosteroid receptors, 153–156
Cortisol, 153–155
Cotransporters, 275
Countertransporters, 275
Covalent bonds, 5, 14
Covalent regulation, of enzyme activity,
  81
Crassulacean acid metabolism (CAM),
  139, 142
Cristae, 44, 114
Crohn disease, adult stem cells and, 229
Crystalline ice, 10
Cuboids, 269
Custom arrays, 196
Cyclic adenosine monophosphate (cAMP),
  95, 96, 155
Cyclic photophosphorylation, 135
Cyclin-cdk complexes, 215, 230
Cyclin-cdk pairings, 214
Cyclin-dependent kinases (cdks), 213
Cyclins, 214
Cyclophosphamide, 192
Cytidine triphosphate (CTP), 79
Cytochrome bc_1, 121
Cytochrome c (cyt c), 103
Cytochrome oxidase, 122
Cytochromes, 120
Cytokinesis, 214, 216, 221

Cytoplasm (cytosol), 42
Cytoplasmic determinants, 268
Cytosine, 25, 178
Cytoskeleton (cellular skeleton), 49–54
  filaments of, 49–51
  peripheral proteins and, 42
Cytosolic calcium, 99
Cytoxan, 192

Death receptors, 102, 250
Decanoate, 30
Decarbazine, 192
Dehydrogenase, 68
  GAP, 138
  NADH, 121
  NADPH-dependent malate, 141
  pyruvate, 115
D-enantiomer, 21
Denaturation, 181
Dendrites, 157
Deoxyadenosine, 25
Deoxynucleoside triphosphate (dNTP), 185
Deoxyribonucleic acid. *See* DNA
Deoxyribonucleosides, 25
Deoxyuridine, 25
Dermal tissue, 280
Desensitization, 152
Desmosomes, 277
Determination, 266–276
Developmental memory, 267
Dextrorotatory form, 15
Diabetes
  adult stem cells and, 228
  cellular communication and, 169
Diabetic retinopathy, adult stem cells and, 228
Diacylglycerol (DAG), 98
Differential gene expression, 264, 265, 266
Differential permeability, 40
Differentiation, 266–276
Diffusion, membrane transport and, 270
Dihydrofolate reductase (DHFR), 193
Dihydroxyacetone phosphate, 110
Dihydroxyphenylalanine (DOPA), 164
Diploid cells, 181, 260, 262
Diploid nucleus, 236
Dipolar water molecule, 9, 10
Dipoles, 10, 19
Disaccharides, 22, 137
Disorder, 4
Dissociation constant (K_d), 62, 88
Disulfide bonds, 19

DNA (deoxyribonucleic acid), 20, 27, 178–182
  cell features and, 263
  chromatin structure and, 181–184
  methylation of, 191, 267
  postreplicative modifications and, 191–193
  protein production and, 54
  structure of, 180*f*
  synthesis/repair of, 184–193
  transcription to RNA, 193–199
DNA-binding proteins, 182
DNA helices, 178–181
DNA ligases, 188
DNA replication, chemotherapy and, 192
DNA strands, 185–187
Docking mechanism, SNAREs and, 172
Domains, of α-helices, 18
Dopamine, 164
Dopamine-β-hydroxylase (DBH), 164
Dopaminergic neurons, 164, 166*f*
Dopamine transporter, 167
Dorsal blastopore lip, 238
Dorsal side, 238
Dorsal-ventral (DV) axis, 239, 240, 246, 249
Double reciprocal plot (Lineweaver-Burk equation), 64
Double-stranded RNAs (dsRNAs), 206
Downregulation, 152
Down syndrome, nondisjunction and, 222
Drug delivery, phospholipid nanoparticles and, 33–35
Duong, Taihung, 145–176
Dynein, 52, 219
Dysplasia, 230

E2F factors, 214
Eadie-Hofstee/Hanes-Woolf plot, 65
Ecosystems, 282
Ectoderm, 238, 249
Effector caspases, 102
Efferent neurons, 174, 280
Efficiency, enzymatic reactions and, 62
Egg, 234, 262, 270
Egg jelly, 235
Electrical communication, 170
Electron flow, through photosystems, 133–135
Electron transport, 106, 118–122
Electrospray ionization (ESI), 205
Elongation, 195
Elongation factors (EFs), 203

Elongation peptide chain, 203
Embryo, 235, 237–239
Embryogenesis, 268
Embryonic stem cells (ESCs), 225, 244, 245
Enantiomers, 15
Endergonic pathways, 106
Endergonic reactions, 6
Endocrine communication, 168
Endocrine messengers, 87
Endocytosis, 44, 272
Endoderm, 238
Endoplasmic reticulum (ER), 48
Endoplasmic reticulum Golgi
    intermediate compartment (ERGIC),
    170
Endosomes
    lysosomes and, 44
    peroxisomes and, 46
End-product inhibition, 72
Energy, 2–9
    cell features and, 264
    electron transport and, 118–122
    first law of thermodynamic and, 3
    internal energy (E) and, 5
    quantitative relationship between
        different forms of, 5–7
Energy conversion, 4, 105–143
    aerobic respiration and, 114–126
    anaerobic respiration and, 107–114
    photosynthesis and, 126–136
Energy transduction, 126
Enhancers, 195
Enterocytes, cell polarity and, 274
Enthalpy (H), relationship to entropy (S)
    and free energy (E), 5–7
Entropy (S)
    as driving force in chemical reactions,
        7–9
    relationship to enthalpy (H) and free
        energy (E), 5–7
    second law of thermodynamic and, 4
Enzyme-catalyzed reactions, 58*f*, 59,
    62, 72
Enzyme inhibitors, 72–78
Enzymes (E)
    bifunctional, 112
    cell features and, 264
    concentration of, 59
    cooperative, 78–80
    covalent regulation, of, 81
    efficiency and, 62
    kinetics of, 57–83
    multisubstrate, 66–72

Epac, 96
Ephaptic communication, 170
Epiboly, 238
Epigenetics, 267
Epinephrine, 94–97, 164
Epithelia, 280
Epithelial cells, cell polarity and, 274
Equations
    equilibrium, 5
    free energy, 8
    Gibbs, 5
    glucose oxidation, 6
    Lineweaver-Burk, 64
    Michaelis-Menton, 59–63
Equilibrium constant, 60
Equilibrium equation, 5
Escherichia coli (*E. coli*), 184
ES complex, 59, 63, 67
Estrogen, 87, 92
Estrogen receptor (ER), 88, 92
Estrogen receptor alpha (ER-α), 89
Ethanolamine, 32
Ethanol intolerance, 68
Etoposide, 192
Euchromatin, 54
Eukaryotes, 55, 181, 258
Eukaryotic cells, 258*f*
Eukaryotic initiation factors (eIFs), 201
Eukaryotic releasing factors (eRFs), 204
Eukaryotic RNAs, 195
Evolution, from unicellularity to
    multicellularity, 256–262
Excisional wounds, 252
Excitotoxicity, 167
Executioner caspases, 102
Exergonic pathway, 106
Exergonic reactions, 6
Exocytosis, 45, 173, 272
    Golgi apparatus and, 47
    smooth endoplasmic reticulum (SER),
        48
Exons, 182
Extracellular matrix, 276, 278
Extrinsic pathways, 103

Facilitated diffusion, 272
FAD coenzyme, electron transport and,
    118–122
FADH$_2$, electron transport and, 118, 121
Fas-associated death domains (FADD),
    102
Fas ligand, 87
Fas receptors, 87

Fats, 30
Feedback, 283
Feedback inhibition, 79
Fermentation, 111
Ferredoxin (F_d), 133
Ferredoxin-NADP reductase (FdNR), 133
Fertilization, 236
Fe-S proteins, 120
Fetal cells, 267
Fibroblast growth factor (FGF), 243, 247
First-order reaction, enzymatic reactions
    and, 62
Flagella, 52
Flag peptide, 166
Flavoproteins, 120
Fluid mosaic model of the cell, 41
Fluorescence-activated cell sorter (FACS),
    212
Free energy (E)
    Gibbs free energy (G) and, 4
    relationship to entropy (S) and
        enthalpy (H), 5–7
    units of, 7
Free radicals, 123
Free ribosomes, 49
Fructose 1,6-bisphosphatase (FBPase), 112
Fructose 1,6-bisphosphate (FBP), 110
Fructose 2,6-bisphosphatase (F2,6Bpase),
    112
Fructose 2,6- bisphosphate (F2,6-P), 112
Fructose-6-phosphate (F6P), 108

G0 phase, of cell cycle, 211
G1 phase, of cell cycle, 210, 212
G2 phase, of cell cycle, 210, 212
Gametes, 260, 261f, 262
γ-aminobutyric acid (GABA), 164, 175
Gap junctions, 170, 277
Gastric juice, 12
Gastric mucosa, 12
Gastric ulcer disease, 12
Gastrointestinal epithelium, tight
    junctions and, 277
Gastrulation, 237–239
Gene amplification, 231
Gene arrays, 196
General transcription factors (GTFs), 90
Gene regulation, 265
Gene regulatory networks, 266
Genes, 182, 260–268
    differential gene expression and, 264, 265
    signature, 265
Genetic damage, 220

Genetics of cells, 177–208
    DNA structure and, 178–181
    DNA transcription to RNA, 193–199
    RNA translation to protein, 199–207
Genome imprinting, 191
Genomes, 181, 234
Genome sequencing project, 263, 265
Genomics, 196
Germ layers, 235, 237–239
Gibbs equation, 5
Gibbs free energy (G), 4
Glial cells, 280
Glossary, 287–309
Glucocorticoid receptor, 153, 154
Gluconeogenesis, 108, 111
Glucose, 20
Glucose 6-phosphatase, 112
Glucose-6-phosphate, 108
Glucose oxidation, 4, 6, 125
Glucose transport, 271
GLUT-1 carrier, 271
Glutamate, 148, 164
Glycans, 166
Glyceraldehyde 3-phosphate (GAP),
    110, 137
Glycerol phospholipids, 32
Glycogen, 21, 22, 23
Glycogen phosphorylase, 82
Glycogen synthase, 82
Glycolysis, 107–110
Glycosaminoglycans (GAGs), 278
Glycosides, 22
Glycosidic bonds, 22
Goldman-Hodgkin-Katz equation, 273
Golgi apparatus, 46, 47, 170, 171
Goosecoid, 241
GPCR signaling, 88
G-protein-coupled receptors (GPCRs),
    93–99, 149–152
    down-regulation of, 88
    neurotransmitter interaction and, 161
Graft vs. host disease, 43
Granule cells, 157, 268, 280
GRK-phosphorylated receptor, 88
Ground tissue, 281
Growth factor receptor-bound protein 2
    (Grb2), 100
GTPase-activating proteins (GAPs), 95
Guanine, 25, 178
Guanine nucleotide exchange factors
    (GEFs), 95, 205
Guanosine diphosphate (GDP), 94, 203
Guanosine triphosphate (GTP), 94, 112

Hair cells, 50, 147, 257
Hanes-Woolf/Eadie-Hofstee plot, 65
Haploid cells, 260, 262
Haploid nucleus, 236
Hatch-Slack pathways, 140
H buffer, 13
Heart muscle cells, gap junctions
    and, 277
Heat shock protein 70 (hsp70), 154
Heat shock protein 90 (hsp90), 153, 155
Helicases, 186
Helicobacter pylori (*H. pylori*), 12
Hematopoietic stem cells (HSCs),
    224, 246
Heme-controlled inhibitor (HCI), 205
Hemidesmosomes, 277
Hemoglobin, 18–20
Hensen node, 239
Heterochromatin, 54
Heterodimers, 156
Heterologous ligands, 88
Heterotrimeric G-proteins, 93, 95
Hexanoate, 30
Hexokinase (HK), 108, 113
High-energy phosphate bonds, 25
Hill coefficient, 80
Hill plots, 80
Histocompatibility, 43
Histones, 90, 182
    histone-modification complexes and, 91
    nucleosomes formation and, 182–184
    phosphorylation of, 215, 217
    posttranslational modification of, 267
Homeodomain, 241
Homeostasis, 86, 283–285
Homeostatic state, 13
Homodimers, 156
Homologous ligands, 88
Homozygous, 249
Horizontal gene transfer, 261
Hormones
    nuclear receptors and, 153
    replacement therapy and, 89
Hox gene, 247, 249
Hughes, James P.
    energy conversion and, 105–143
    signal transduction and, 85–104
Human cells, vs. plant cells, 54
Human genome sequencing project,
    263, 265
Human leukocyte antigens, 43
Hurley, Thomas D., 57–83
Hybridization, 181, 196
Hydrogen, 11

Hydrogen bonds (H bonds), 10, 178
Hydrogen ion concentrations, pH of, 11
Hydrolysis, 106
Hydrophobic interface, 31
Hyperplasia, 230

Ibuprofen, 72
Ice, 10
Ifosfamide, 192
Immune rejection, of transplanted
    embryonic stem cells, 226, 227
Importin, 155
Inhibitors, 75–77
Initial rate conditions, enzymatic
    reactions and, 59
Initiation factors (IF), 201
Initiator caspases, 102
Inorganic ions, 146
Inositol, 32
Inositol trisphosphate (IP$_3$), 98
Insertional mutagenesis, 231
Insulin, apoptosis and, 103
Insulin-like growth factor 1 (IGF1), 86
Insulin receptor, 100
Insulin-receptor substrate (IRS), 100
Insulin signaling pathways, 100
Integral proteins, 41
Integration, multistep pathways and, 95
Intercellular communication,
    transmembrane proteins and, 42
Interdisciplinary research, dialog
    and, 283
Interferons, 206
Interhelix bonds, 19
Intermediate filaments, 51
Intermembrane space, 121–125
Internal energy (E), 5
Internalization, 152
Interneurons, 280
Interphase of cell cycle, 210
Intrinsic pathways, 103
Introns, 182
In vivo labeling of cells, 165, 166
Ion channels, 146
Ionotropic receptors, 147–149, 161
Iron-sulfur clusters, 133
Iron-sulfur proteins (Fe-S), 120
Isoenzymes, 68

Jelly layer, 235
Juxtacrine communication, 169
Juxtacrine messengers, 87
Kainate receptor, 148
Keratin, 51

Ketoses, 20
Kinases, 108
Kinesins, 51, 219
Kinetic expressions, for Bi-reactant
    systems, 70–72
Kinetics
    enzyme, 57–83
    presteady state, 59
    steady-state, 58–72
Kinetochore, 216, 218
King, Michael W.
    cellular development and, 233–254
    cellular genetics and, 177–208
Krabbe disease, adult stem cells
    and, 229

Lactate fermentation, 111
Lactose, 22
Lagging DNA strands, 187
Leading DNA strands, 187
Left-right (LR) axis, 239, 243–246
Lefty-2, 243
Leukemias, alkylating agents and, 192
Leukeran, 192
Levorotatory form, 15
Ligand-activated receptors, 91f
Ligand binding
    GPCR and, 151
    protein kinase-associated receptors
        and, 152
Ligand-gated channels, 146, 147
Ligands, 87, 88
Ligases, 188
Light-dependent reactions, 127, 135
Light energy, conversion to chemical
    energy, 126, 130
Light-harvesting complexes (LHCs), 128
Light-independent reactions, 127, 138
Limb development, 246–250
Limb fields, 247
Lineweaver-Burk equation, 64
Lipid rafts, 276
Lipids, 28–35
Liposomes, drug delivery and, 34
Liquid chromatography (LC), 205
Lmx-1 transcription factor, 250
Lumen, 128, 135
Last universal common ancestor (LUCA),
    258
Lysosomes, 44–46

Macular degeneration, adult stem cells
    and, 228
Major grooves, DNA and, 179

Malate, 141
Malignant tumors, 230
Maltose, 22
Mammalian cells, vs. plant cells, 54
Mangold, Hilde, 238
MAP/ERK (MEK) kinase, 101
Marginal zone, 238
Marijuana, cannabinoid receptors and, 98
Mass spectrometry (MS), proteomics
    and, 205
Matrix-assisted laser desorption
    ionization (MALDI), 205
Matulane, 192
Mechanically gated channels, 146, 147
Mechanoreceptors, 147
Mediator complexes, 91
Meiosis, 259–262
Melting temperature (T_m), 181
Membrane-bound organelles, 42–48
Membrane potential, 272–274
Membrane proteins, 40f, 41
Membrane receptors, 147–152
    neurotransmitter interaction with, 161
    signal transduction via, 93
Membrane transport, 270–272
Mesenchymal cells, 249
Mesenchyme, 249
Mesoderm, 238
Mesophyll cells, 140
Messenger RNAs (mRNAs), 48, 193, 235
Messengers, 86, 87, 95
Metabolic disorders, adult stem cells
    and, 229
Metabolic reactions, energy of, 8
Metabolism, 106
Metabotropic glutamate receptors, 149
Metabotropic receptors. See G-protein-
    coupled receptors
Metaphase, 216
Metaplasia, 230
Methylation of DNA, 191, 267
Micelles, 33, 34
Michaelis constant, 63
Michaelis-Menton equation, 59–63
    Eadie-Hofstee/Hanes-Woolf plot
        and, 65
    experimental data analysis and, 64
Microfilaments, 49
Microtubules, 51, 236, 276
Microvilli, 50
Mineralocorticoid receptor, 153, 154
Miniature end-plate potentials, 162
Minor grooves, DNA and, 179
Mitchell, Peter D., 122

Mitochondria, 42–44, 114
Mitogen-activated protein (MAP) kinases, 101
Mitosis, 210, 215–222
    mechanics/control of, 216–220
    stages of, 215
Mitotic apparatus, 218
Mitotic chromosomes, 183
Mixed-type noncompetitive inhibition, 76
Molecular biotechnology, 1
Molecular water, 9
Monoamine oxidase (MAO), 167
Monosaccharides, 20–22
Morphogenesis, 246, 276–282
Morphogens, 235
Motoneurons, 174, 280
Motor protein, 51
M phase, of cell cycle, 210, 212
Mucopolysaccharides, 278
Multicellularity, 256–262, 281, 284
Multi-phosphorylated
    phosphatidylinositols, 101
Multiple sclerosis, adult stem cells
    and, 228
Multipotent stem cells, 225, 245
Multisubstrate enzymes, 66–72
Muscarinic receptors, 161
Muscle cells, 44, 174
Muscle machine, 50
Muscle tissue, 280
Myosin, 50, 220

NAD coenzyme, 68, 118–122
NADH, electron transport and, 118, 121
Nail-patella syndrome (NPS), 250
Nanoparticles, drug delivery and, 33–35
Nanotechnology, self-assembly and, 41
Natulan, 192
Necrosis, 101
Negative antagonists, 151
Negative feedback, 283
Neosar, 192
Nerve cells, 156, 157, 268
Nervous tissue, 280
Nesting, complexity and, 284
N-ethylmaleimide sensitive factor (NSF), 173
Neural stem cells, 228, 246
Neural tube, 242
Neurodegenerative diseases, embryonic
    stem cells and, 226
Neuromuscular junction, 160, 174
Neurons, 280

Neuropeptides, 149, 158
Neuropeptide transmitters, 168
Neurosteroids, 158
Neurotransmitter-receptor interactions, 157
Neurotransmitters, 86, 148, 157, 158–168
    active release into synaptic cleft, 161
    chemical transmission process and, 160
    interaction with receptors on
        postsynaptic and presynaptic
        membrane, 161
    storage of, 161
    synthesis of, 160
    termination of, 161, 167
Neurotrophins, protein kinase-associated
    receptors and, 152
Neurulation, 237
Neutral antagonists, 151
Nicotinamide adenine dinucleotide
    (NAD^+), 107, 111
Nicotinamide adenine dinucleotide
    phosphate ($NADP^+$), 126
Nicotinic cholinergic receptors, 148, 174
Nicotinic receptors, 161
Nissl substance, 55
NMDA receptor, 148
Nodal-related proteins, 242, 243
Noncoding DNA, 182
Noncompetitive inhibitors, 75–77
Nondisjunction, 221, 222
Nonhistone proteins, 182
Nonmembrane-bound organelles, 48–53
Nonrepetitive DNA, 182
Nontransmembrane integral proteins, 41
Noradrenergic neurons, 164
Norepinephrine, 94–97, 164
Norepinephrine transporter, 167
Notochord, 239
Nuclear laminar proteins,
    phosphorylation of, 215, 217
Nuclear localization signals (NLS), 154
Nuclear matrix, 183
Nuclear pore complex, 155
Nuclear receptors, 90–93, 153–158
Nucleic acid, 24–28, 54, 185
Nucleoid region, 258
Nucleolus, 54
Nucleosides, 24–28
Nucleosomes, 90, 182–184
Nucleotide-gated, ion-gated channels, 147
Nucleotides, 24–28, 178, 263
Nucleotide sequence of DNA strands, 181
Nucleus, 38, 53

Occluding junctions, 276
Octanoate, 30
Oils, 30
Okazaki fragments, 187
Oligosaccharide chains, 166
Oligosaccharides, 22–24
Oncogenes, 230
Oncogenesis, 230
Oocyte, 234, 262, 270
Oogenesis, 234
Open thermodynamic systems, 3
Optical isomers, 15
Ordered-sequential Bi Bi reaction
    mechanism, 67, 70
Organelles, 38
    membrane-bound, 42–48
    nonmembrane-bound, 48–53
Organisms, 255–286
    cell features and, 262–266
    unicellularity to multicellularity and,
        256–262
Organogenesis, 237
Organs, 281
Organ systems, 282
Orphan receptors, 153
Osmosis, 271
Ostrich eggs, 270
Otoacoustic emissions, 257
Oxaloacetates, 116, 140
Oxidase, 122
Oxidation, pyruvate, 115
Oxidative decarboxylation, 115
Oxidative phosphorylation, 110, 115

P700 reaction center, 133
Paracrine communication, 169
Paracrine messengers, 86
Parietal cells, 12
Parkinson disease
    adult stem cells and, 228
    embryonic stem cells and, 226
Partial agonists, 89, 98
Partial estrogen agonists, 89
Patterning, 237
Peptide bonds, 14, 15
Peptides, 49, 203
Perikaryon, 157
Peripheral proteins, 41, 42
Peroxisomes, 46–48
pH, 11–14
Phagocytosis, 45, 272
Phenylalanine, 164
Phenylalanine hydroxylase, 164

Phenylethanolamine N-methyltransferase
    (PNMT), 164
Pheophytin (Ph), 133
Phloem, 281
Phosphatidic acid, 31
Phosphatidylcholine, 32, 33
Phosphatidylethanolamine, 32
Phosphatidyl glycerol, 31
Phosphatidylinositide 3-kinase [PI(3)K],
    101
Phosphatidylinositol (PI), 32, 99f, 101
Phosphatidylserine, 32
Phospho anhydride bonds, 25
Phosphodiester bonds, 178
Phosphoenolpyruvate (PEP), 110, 139
Phosphoenolpyruvate carboxykinase
    (PCK), 112
Phosphofructokinase-1 (PFK1), 110,
    112
Phosphofructokinase-2 (PFK2), 112
Phosphoglycolate (PG), 138
Phospholipase C (PLC) pathway, 98
Phospholipid bilayer, 40
Phospholipid nanoparticles, drug delivery
    and, 33–35
Phospholipids, 28, 31–35
    amphipathic character of, 40
    dispersion in water, 33
    mitochondria and, 43
Phosphoric acid buffer, 13
Phosphorylated tyrosines, 100
Phosphorylation, 81, 97
    of histone proteins, 215, 217
    of multiple tyrosine (Y) residues, 100
    of myosin, 220
    of nuclear laminar proteins, 215, 217
Photolysis, 132, 135
Photons, 126
Photorespiration, 138, 142
Photosynthesis, 54, 126–136
Photosynthetic pigments, 128–131
Photosystem I/photosystem II, 126, 131
    electron flow through, 133–135
    photolysis and, 135
Photosystems, 128
Phylloquinone (A$_1$), 133
Phytol tail, 128, 129f, 130
Ping-pong sequential Bi Bi reaction
    mechanism, 69, 71
Pinocytosis, 45, 272
Pitx2 gene, 243
Plant cells, vs. human cells, 54
Plant cell walls, 21

Plant tissue, 280
Plasma membrane, 38–42, 146
Plasmids, 258
Plasmodesma, 278
Plastocyanin (PC), 133
Plastoquinol (PQH$_2$), 133, 135
Plastoquinone A (QA), 133
Plastoquinone B (QB), 133
Pleckstrin homology (PH), 101
Pluripotent stem cells, 225, 245
Polymerase chain reaction (PCR),
    189, 196
Polymerases, 184
Polymers of carbohydrates, 20–24
Polymorphic variant aldehyde
    dehydrogenase, 68
Polypeptide chains, cooperative enzymes
    and, 78
Polypeptides, 15–17
Polysaccharides, 22–24
Polyspermy, 236
Polyubiquination, 215
Populations, 282
Porins, 128
Porphyrin ring, 128, 129f
Positive feedback, 283
Positive feedback cycles, 268
Postreplicative modifications, DNA and,
    191–193
Postsynaptic nerve cells, 157
Posttranslational modification, 266, 267
Prazosin, 98
Preinitiation complex, 202
Prentice, David A., 209–232
Presteady state kinetics, 59
Presynaptic nerve cells, 157, 159
Primary active transport, membrane
    transport and, 272
Primase, 185, 187
Primitive streak, 239
Primordial germ cells, 234–236
Proapoptotic proteins, 103
Procarbazine, 192
Procaspase 8, 102
Procaspase 9, 103
Processivity, 187
Processivity accessory proteins, 187
Products (P), 58, 59
Progress zone (PZ), 249
Prokaryotes, 55, 181, 258
Prokaryotic cells, 258f, 261
Proline, 18
Promoters, 90, 153, 195

Prophase, 216
Prosthetic vessels, 279
Proteases, 81
Protein folding, 19
Protein kinase A (PKA), 96, 113
Protein-kinase-associated receptors,
    99–101, 152
Protein kinase B (Akt), 101
Protein kinase C (PKC), 98
Protein kinases, 81, 99, 206
Protein phosphatases, 81
Proteins, 18–20
    animal, 14
    cell features and, 263
    cell membrane and, 40f, 41
    DNA/RNA and, 54
    DNA replication and, 185
    posttranslational modification
        and, 266
    proteomics and, 205
    rough endoplasmic reticulum and, 38
    three-dimensional structure of, 18
    translation from RNA, 199–207
Proteoglycans, 278
Proteolysis, 102
Proteolytic cleavage of proenzymes, 81
Proteomics, 205
Protists, 259
Proton motive force (pmf), 123
Proto-oncogenes, 231
Proximal-distal (PD) axis, 246
P-site, 203
Purine, 25, 178
Purkinje cells, 157, 268, 280
Pyrimidine, 25, 178
Pyrophosphate, 185
Pyruvate, 108, 115, 142
Pyruvate carboxylase (PC), 112, 114
Pyruvate dehydrogenase (PDH), 115
Pyruvate kinase (PK), 110, 113
Pyruvate orthophosphate dikinase
    (PPDK), 142

Q Cycle, 122
Quantum, 162
Quiescence, 211
Quisqualate receptor, 148

Raf, 101
Raloxifene (Evista), 89
Random Bi Bi reaction mechanism, 70
Ras, 101
Reaction centers, 129, 133

Reactions, 8
Receiving nerve cells, 157
Receptor proteins, 166
Receptors
    7-transmembrane, 93
    affinity of for ligands, 87
    binding, 87–90
    cellular, homeostasis and, 86
    estrogen, 88
    Fas, 87
    G-protein-coupled, 93–99
    insulin, 100
    membrane, 93, 147–152
    nuclear, 90–93, 153–158
    protein-kinase-associated, 99–101
    spare, 88
Receptor up-regulation/receptor
    down-regulation, 88
Red algae fossil, 262
Red blood cells, size/shape of, 269
Redox potential, 131
Reduction-oxidation (redox), 118, 120
Regenerative medicine, 244–246
Regulators of G-protein signaling (RGS)
    proteins, 95
Releasing factors (RFs), 204
Renaturation, 181
Repetitive DNA, 182
Replication fork, 186, 188
Replication of DNA, 181, 184
Replisomes, 187
Reserve stem cells, 244
Respiration
    aerobic, 114–126
    anaerobic, 112–114
Response elements (RE), 91f, 92
Resting membrane potential, 273
Restriction point, 211
Reticuloendothelial system (RES), 35
Retinal degeneration, adult stem cells
    and, 228
R groups, of amino acids, 15–18
Rhodopsin-adrenergic receptor, 149
Ribonucleic acid. See RNA
Ribonucleosides, 25
Ribose, 20
Ribosomal RNA (rRNA), 193
Ribosomes, 48, 235
    cell secretion and, 170
    nucleolus and, 54
Ribulose 1,5-bisphosphate (RuBP), 136
Ribulose bisphosphate carboxylase
    (rubisco), 136

RNA (ribonucleic acid), 20, 27
    cell features and, 263
    classes of, 193
    protein production and, 54
    synthesis of, 194
    transcription from DNA, 193–199
    translation to protein, 199–207
RNA polymerase II, 90
Robotics, muscle machine and, 50
Rough endoplasmic reticulum (RER), 38,
    48, 170

"Sarco-" prefix, 55
Saturated triglycerides, 30
Scaffolds, 279
Schwann cells, 175
Scleromyxedema, adult stem cells and,
    228
Secondary active transport, membrane
    transport and, 272
Second messengers, 95
Secretin-vasoactive intestinal peptide
    receptor, 149
Securin, 219
Selective estrogen receptor modulators
    (SERMs), 89, 92
Self-assembly, biology/technology and, 41
Semiconservative nature of DNA
    replication, 185
Semifluids, 42
Senescence, 229
Sensitivity, of cells to ligands, 88
Sensory neurons, 280
Separase, 219
Serine, 32
Serpins, 81
Sexual reproduction, 259–262
Siamois, 241
Sigmoidal saturation binding curves, 79
Signaling pathways, 93–99
Signal transduction, 85–104
    apoptosis and, 101–103
    via nuclear receptors, 90–93
Signature genes, 265
Single-stranded oligonucleotides, 189
Skeletal muscle
    anaerobic metabolism and, 107
    glucose and, 112
Small G-proteins, 95
Small nuclear ribonucleoprotein particles
    (snRNPs), 197
Smooth endoplasmic reticulum (SER), 48
Snail function, 243

SNAREs, 172
Sodium, high water concentration and, 11
Somatic cell nuclear transfer (SCNT), 227
Somatic cells, 269
Sonic hedgehog (shh), expression, 243
Son of sevenless (SOS), 101
Sound, 50, 51
Spare receptors, 88
Specification, 239–246
Specificity constant, for enzymes, 63
Spemann, Hans, 238
Spemann organizer, 238, 239, 241
Sperm, 235
Sperm flagellum, 52
S phase, of cell cycle, 210, 212
Spinal cord injury
    adult stem cells and, 228
    embryonic stem cells and, 226
Spindle fibers, 216, 218, 222
Spliceosome, 198
Splicing, 197
Spontaneous reactions, 6
Src homology 2 (SH2) domains, 100
Standard free energy, 7, 106
Standard redox potential, 120
Starch, 21, 22, 137
Statin drugs, 73
Steady-state condition, 58
Steady-state kinetics, 58–72
    assumptions and, 59
    experimental data analysis and, 63–66
    interpretation of parameters in single
        substrate/product systems, 63
Stem cells, 224, 244–246
    adult, 225, 227–229, 245
    cancer and, 231
    embryonic, 225, 244, 255
    regenerative medicine and, 224
    reserve, 244
    sources of, 225
    tissue maintenance/repair and, 222–29,
        227
    uses/points of controversy and,
        226–229
Stereocilia, 50
Stereospecific amino acids, 14
Steroid hormone receptors, 153–158
Steroid receptor coactivators (SRCs), 156
Stomach, 12
Stomata
    C_3 plants and, 136
    C_4 plants and, 140
    CAM plants and, 142

Stroke
    adult stem cells and, 228
    excitotoxicity and, 167
Structural isomers, 22
Stylopod, 249
Substrate, enzymes and, 58, 59
Substrate/velocity curve, 62
Substrate-level phosphorylation, 110
Substrate saturation curve, 61f, 64
Succinate, 116
Succinate dehydrogenase (SDH), 116, 121
Sucrose, 22, 137
Superoxide, 123
Survival pathway, 103
Synapses, 156, 159f, 173
Synaptic cleft, neurotransmitters
    and, 161
Synaptic interactions during
    development, 174
Synaptobrevin, 172
Synaptosomal-associated protein of 25 kDa
    (SNAP-25), 172
Synaptotagmin, 173
Syncytium, 277
Syntaxin, 172
Synthase, 82, 124
Systems biology, 282–285

Tamoxifen, 89
Taq polymerase, 189
Taxanes, 193
T-box (Tbx) genes, 247
TCA cycle, 116–122
Technology, self-assembly and, 41
Telomerase, 229
Telomeres, 229
Telophase, 216
Template strands, 185, 188, 193
Termination, 204
Termination codons, 204
Terminology. *See also* Glossary
    interdisciplinary research and, 283
    tissue-specific, 55
Ternary complex, 67
Tetrapod limb, 247, 250
Tetrasaccharides, 22
Thermal denaturation, 181
Thermodynamics, 2–5
    first law of, 3
    open/closed thermodynamic systems
        and, 3
    second law of, 4
Thermus aquaticus, 189

Threonine, 81
Thylakoid lumen, 135
Thylakoids, 128
Thymidylate synthase, 193
Thymine, 25, 178
Tight junctions, 276
Tissue, 278–281
    damage to, stem cells and, 227
    maintenance/repair of, 222–229
    regeneration of, 244
Tissue engineering, 279
Tissue-specific language, 55
Tissue stem cells, 224
Titin, 14
Topoisomerases, 188, 192
Totipotent stem cells, 225, 245
Transactivation functions, 92
Transaminases, 69
Transcription factors, 90–93, 266
Transdifferentiation, 229
Transfer RNAs (tRNAs), 193, 235
Transgenic plants, 141
Translational termination, 204
Translation process, 48
    Heme control of, 204–206
    initiation of, 200, 202
    interferon control of, 206
    termination of, 204
Translocation, 121
Transmembrane proteins, 41
Transmitter-gated channels, 147
Transmitting nerve cells, 157
Transpeptidation, 203
Transplantation, 251
    graft vs. host disease, 43
    regenerative medicine and, 245
Transporters, 161
Transport vesicles, 171
Trans site of Golgi apparatus, 48, 171
Tricarboxylic acid cycle (TCA), 106,
    116–118
Triglycerides, 30
Trisaccharides, 22
$tRNA_i^{met}$ initiator, 200
Truncated Bid (tBid), 103
Trypsin, 81
Tubulin, 51
Tumor necrosis factor (TNF), 102, 250
Tumors, 230
Tumor suppressor genes, loss of, 231
Turner syndrome, nondisjunction
    and, 222
Turnover number, for enzymes, 63

Tyrasine, 81
Tyrosine, 164
Tyrosine hydroxylase, 164
Tyrosine kinases (TKs), 99

Ubiquitin, 215
Uncompetitive inhibitors, 77
Unconventional transmitters, 158
Unicellularity to multicellularity,
    256–262
Unipotent stem cells, 225
Unsaturated triglycerides, 30
Uracil, 25
Uridine, 25

Vaccines, phytol tail and, 128, 130
van der Waals forces, 19
Van Gogh gene, 178
Vascular tissue, 281
Vascular tissue engineering, 279
Vegetal half, 238
Vegetal pole, 238
Velocity, 62, 72–78
Ventral side, 238
Vertebrate limb development,
    246–250
Vesicle-associated membrane protein
    (VAMP), 172
Vesicles, 33, 34
Vesicular cholinergic transporter, 161
Vesicular tubular cluster (VTC), 171
Vinca alkaloids, 192
Viral infections, 231
Vitelline envelope, 236
V_{max}, 61–66, 70, 74–79
Vocabulary. See also Glossary
    interdisciplinary research and, 283
    tissue-specific, 55
Voltage-gated channels, 146, 147
Voluntary nerve fibers, 174

Waite, Gabi N., 255–286
Water, 9–14
    biologically significant molecular
        structure of, 9
    functional role in biology, 11–14
    hydrophobic interface and, 31
    peptide bonds and, 15
    photorespiration and, 139
Watson-Crick base-pairing, 178
White fat cells, mitochondria and, 44
Wnt signaling, 243
Worrell, Michael B., 37–55

Xenopus laevis, 242
Xnr1 protein, 243
Xylem, 281

Zero-order reaction, enzymatic reactions
    and, 62
Zeugopod, 249
Zeugopodium, 252

Z-form of DNA, 181
Zinc finger region, 156
Zinc fingers, 243
Zona occludens, 276
Zona pellucida, 236
Zone of polarizing activity (ZPA), 249
Z-scheme, 131
Zygote, 156, 236

CPSIA information can be obtained at www.ICGtesting.com
Printed in the USA
LVOW071104300113

317892LV00008B/58/P